OPTIMAL AUDIO AND VIDEO REPRODUCTION AT HOME

Optimal Audio and Video Reproduction at Home is a comprehensive guide that will help every reader set up a modern audio-video system in a small room such as a home theater or studio control room.

Verdult covers everything the reader needs to know to optimize the reproduction of multichannel audio and high-resolution video. The book provides concrete advice on equipment setup, display calibration, loudspeaker positioning, room acoustics, and much more.

Detailed, easy-to-grasp explanations of the underlying principles ensure the reader will make the right choices, find alternatives, and separate the rigid from the more flexible requirements to achieve the best possible results.

Vincent Verdult is currently Senior Information Security Consultant at the Ministry of the Interior and Kingdom Relations, The Hague, the Netherlands. He was born in 1974 in the Netherlands. From 2001 to 2005, he was an assistant professor in systems and control engineering at Delft University of Technology, the Netherlands. He has always had a keen interest in audio and video engineering. In his spare time, he has used his engineering knowledge to design and build his own subwoofers, create and install his own acoustical devices, and optimize his audio-video setup in six different rooms over the last 20 years.

OPTIMAL AUDIO AND VIDEO REPRODUCTION AT HOME

Improving the Listening
and Viewing Experience

Vincent Verdult

Routledge
Taylor & Francis Group
NEW YORK AND LONDON

First published 2019
by Routledge
52 Vanderbilt Avenue, New York, NY 10017

and by Routledge
2 Park Square, Milton Park, Abingdon, Oxon, OX14 4RN

Routledge is an imprint of the Taylor & Francis Group, an informa business

© 2019 Taylor & Francis

The right of Vincent Verdult to be identified as author of this work has been asserted by him in accordance with sections 77 and 78 of the Copyright, Designs and Patents Act 1988.

Library of Congress Cataloging-in-Publication Data
Names: Verdult, Vincent, author.
Title: Optimal audio and video reproduction : improving the listening and viewing experience / Vincent Verdult.
Description: New York, NY : Routledge, 2019. | Includes bibliographical references and index.
Identifiers: LCCN 2018056873 (print) | LCCN 2018057112 (ebook) | ISBN 9780429813184 (pdf) | ISBN 9780429813177 (epub) | ISBN 9780429813160 (mobi) | ISBN 9781138335417 (hbk : alk. paper) | ISBN 9781138335387 (pbk : alk. paper) | ISBN 9780429443800 (ebk)
Subjects: LCSH: Home entertainment systems. | Home theaters. | High-fidelity sound systems. | Digital video.
Classification: LCC TK7881.3 (ebook) | LCC TK7881.3 .V47 2019 (print) | DDC 621.388—dc23
LC record available at https://lccn.loc.gov/2018056873

ISBN: 978-1-138-33541-7 (hbk)
ISBN: 978-1-138-33538-7 (pbk)
ISBN: 978-0-429-44380-0 (ebk)

Typeset in Corbel and Myriad Pro
by Apex CoVantage, LLC

Visit the companion website: www.vincentverdult.nl

CONTENTS

PREFACE

Enjoying good sound at home has been my passion for many years. Quite early I discovered that you could drastically improve the sound of your loudspeakers by simply moving them a few inches. This was good news to me, because during my student days I didn't have a lot of money to spend on audio equipment. So I had to make do with what I had. I started to look seriously into the theory and practice of loudspeaker placement. My goal was to find the best positions in my room for my loudspeakers. It was 1994, and I was studying electrical engineering at Delft University of Technology, the Netherlands. The library of the university had a good selection of books and journals on audio engineering that gave me a jump start on my quest for knowledge about loudspeaker placement. I started experimenting with different positions in the small attic in which my girlfriend and I lived. I wrote a short summary of the basic guidelines that I discovered and published it on the Internet. I even started to think about writing a book on the topic. In fact, I wrote two or three chapters and then abandoned the project, because I needed my time to focus on my studies and pass my exams.

Improving sound reproduction at home remained my goal throughout the years. I upgraded my loudspeakers and amplifier, and I designed and built my own subwoofers. Surround sound gained my interest, and I started enjoying watching movies at home. As a result, optimal sound reproduction now involved finding the optimal placement of three front loudspeakers, two surround loudspeakers, and two subwoofers. I kept reading books, articles, and blogs to keep my knowledge up to date, and I applied this knowledge to optimize the reproduction of audio and video at my own home. Over the course of a couple of years I lived in several cities and every time we moved the quest for finding the best positions in the room for my loudspeakers was renewed. I have optimized my audio-video setup in six different rooms, using basic principles, computer simulations, acoustical measurements, and critical listening. The acoustics of some of these rooms was so bad that I even created and installed my own acoustical devices.

The book that you are reading contains all the practical knowledge that I have used myself over the years to optimize my listening and viewing experience in a number of different homes. It originated out of my own curiosity and interest to optimally enjoy music and movies in my home environment. It covers both the reproduction of audio and video to do justice to the fact that modern home entertainment systems integrate multichannel digital audio and high-resolution digital video. My goal is to help you optimally set up your audio-video system to achieve the highest attainable reproduction quality given practical constraints such as limited budget, aesthetic consideration, and rooms that are not dedicated to audio-video reproduction but need to be used for multiple purposes.

Writing this book has helped me understand the topic even better than I did before. I certainly discovered some new things that helped me further optimize my own audio-video setup. I hope it helps you too in improving your listening and viewing experience.

The content of this book is based on the work of many other people in the fields of audio and video engineering, acoustics, and perceptual psychology. I would like to thank all of them for their interesting research and useful publications, as I wouldn't have much to say without it.

Special thanks to Frank Thuijs and Rieks Bauman for putting in the time and effort to review the entire manuscript. Your suggestions and comments have helped me improve the readability of the book. Of course, any remaining issues remain my own responsibility.

Thanks to Madelon van Luijk, the love of my life, for being there whenever I need you. Without your continuous support and encouragement I never would have finished this book. Thank you also for reading several versions of the manuscript, compiling the index, and allowing me to test out everything I write about in our own living room. Let's put on some music and enjoy once more the quality of our optimized sound system.

1

INTRODUCTION

Do you enjoy watching movies and listening to music at home? I do tremendously. Not only do I decide what to watch or listen to without any fixed time schedule, I can also pause the program at any moment to get a beer or go to the toilet. As a bonus, I can avoid all kinds of nuisances like commercial breaks, people munching popcorn during the show, ticket queues, and parking hassles at the local theater.

But there is more. The best part is that the sound quality at home is far better than that at most large venues. It can be a more accurate reproduction resulting in greater pleasure. At home, you can always sit in the best seat and you have more control over the acoustical environment. In addition, many domestic loudspeakers are better in their ability to reproduce sound with all its nuances and details than those used in large venues. On the video side, it is now possible to have spectacular visual experiences at home. High-definition and ultrahigh-definition digital video sources, like Blu-ray discs, provide enough detail to produce large pictures that remain sharp.

Being able to reproduce audio and video with such high quality is extremely rewarding. It brings you closer to the artist's intention. Artists, not only musicians but also filmmakers, spend hours in the recording studio to get their sound just right. Filmmakers fine-tune the soundtrack of their movie until it conveys the right mood and evokes the intended emotions for each scene. Filmmakers also fine-tune the overall visual appearance of their movie to support the story. The carefully crafted sounds and images of these artists need to be accurately reproduced in order to preserve all the nuances of the message that they try to communicate. Hence, to fully appreciate the art that they have created you need high-quality reproduction of audio and video at home.

To achieve such high-quality reproduction you need good-quality audio and video equipment. Over the years, advances in technology have been such that high-quality reproduction is possible at moderate costs. However, the reproduction quality not only depends on good equipment, but also on the room in which you watch and listen and the positions of the display, the loudspeakers, and the audience in that room. If you don't pay attention to all of these aspects you might end up, as too many people do, with an excellent audio-video system that does not live up to its full potential. The good news is that with the correct setup, your audio-video system can easily outperform a much more expensive system that was carelessly installed. But, to get the best out of your equipment, you must be willing to adjust the layout, furnishings, and decoration of the room.

In this book I provide practical advice for setting up an audio-video system at home. The goal is to help you set up your system for *optimal reproduction* of audio and video. Optimal in the sense of achieving the best reproduction quality with *your* equipment in *your* room. Optimal does not mean perfect. Perfect reproduction is only possible in the control room of the studio where the music or movie was actually created. At home, you probably use different loudspeakers and a different display than the ones used in the studio. Furthermore, your room will differ in size, layout, and furnishings from the studio. Fortunately, you can still get near perfect reproduction, due to some peculiarities of the human hearing and visual system.

Chances are that you may not be able to implement all the advice that I provide. If you set up your system in your living room or other multipurpose room, compromises will have to be made. There will always be practical constraints, aesthetic considerations, and limits on your budget. In practice, quite often, the best position for one of your loudspeakers appears to be already occupied by an immovable piece of furniture. Aesthetically, those essential acoustical treatments look rather ugly when you put them on your wall. And, without a doubt, that really awesome projector happens to be really expensive. Obviously, if you can afford to build a dedicated listening/viewing room, less compromises have to be made. Most people don't have that option, and that is why I focus on setting up the audio and video equipment in a typical living room. That said, the advice in this book is still useful for those of you who have a dedicated room.

To cope with real-life compromises, it is important that you understand the underlying principles of the advice that I give. This enables you to make the right choices, find alternatives, and separate the rigid requirements from the more flexible ones. The practical advice in this book, more often than not, is based on scientific research and good engineering practice. However, understanding the main text only requires basic high-school physics and mathematics: I use a minimum of formulas, which I often explain using simple graphs. For those who want to delve deeper, I provide additional details in a boxed text. If you wish, you can safely skip these parts.

What can you expect in the remainder of this book? To begin, in Chapter 2, I will give you an overview of the audio and video system and I will introduce some fundamental concepts like frequency and color. Then, in Chapter 3, I will delve into the different aspects of image and sound quality. The terminology introduced in that chapter allows you to accurately describe the quality of what you see and hear. This terminology is used in Chapters 4–8 to describe how reproduction quality depends on the audio and video equipment, the program material, and the design and acoustics of the room. These chapters form the main part of the book. Here you find concrete advice on setting up your system and the reasons for doing it that way. The advice is easy to recognize, as I have summarized it in numbered recommendations. Chapter 9, which concludes this book, contains an ordered list of all these recommendations for easy reference.

Although I deal with the reproduction of both audio and video, you will notice that this book contains significantly more material about audio. The main reason is that audio is a truly three-dimensional event where the sound from multiple loudspeakers surrounds the listener and propagates throughout the entire room. By contrast, video is a two-dimensional presentation that is confined to the display that is located in front of the viewer. As a result, the room has a much larger influence on the reproduction of audio than it has on the reproduction of video. For this reason, optimal reproduction of audio tends to be a bit more involved.

2
AUDIO AND VIDEO BASICS

Moving images can be boring without sound. Try it for yourself: watch one of your favorite scenes from a movie with and without sound. The experience is quite different. Sound draws you into the movie. Sound that surrounds you draws you in even further. That is why a movie theater has loudspeakers all over the place. You can have the same surround sound experience at home. The good news is that it doesn't require as many loudspeakers as in a movie theater. But you do need properly set up audio equipment. Surround sound is one of the key ingredients to a cinema-like experience at home. And yet, surround sound alone does not make your living room a home theater; the other essential ingredient is a large video display.

In this chapter I give an overview of the components that make up an audio-video system. I explain how they work together to create sound and moving images, and I explain how you perceive their creation. A little understanding of the basic principles of sound and moving images is a good investment for the chapters to come.

2.1 The Audio-Video System

An audio-video system consists of several electronic devices. Depending on the number of devices present, the system can range from simple to complex. Let's start with a simple one and add complexity later. A basic A/V (audio-video) system is depicted in figure 2.1. It consists of a video display, several loudspeakers, a subwoofer, an A/V receiver, and a Blu-ray disc player.

The basic system allows you to view high-quality digital video on the video display and to listen to multichannel digital surround sound from the loudspeakers. In this system, a Blu-ray disc player is used to play source material, such as a movie or a concert. The player sends digital audio and video information to the A/V receiver. The A/V receiver forms the heart of the system. The receiver sends the video part to the display and the audio part to the loudspeakers and the subwoofer.

Most people use a flat-panel television as their video display. You could use such a television to reproduce sound, but I do not recommend it. The loudspeakers built into a television are of low quality. But even more important, you will not get surround sound. Most televisions use only two loudspeakers, one on either side of the display.

To really create the illusion of being present you need surround sound. Surround sound envelops you with sound. You are tricked into believing you are actually sitting in the audience during that concert. It sounds as if you are in the concert hall instead of your much smaller living room. Surround sound gives you a feeling of being in the middle of the action of a movie. The action is no longer confined to the screen in front of you. An airplane is flying right over your head from the front of the room to the back, and a thunderstorm sounds as if it is right above your house.

The basic system that we are looking at provides 5.1 channels of surround sound. The notation '5.1' means that there are five main audio channels and one *low-frequency effects* (LFE)

Figure 2.1
A basic audio-video system with a subwoofer and five loudspeakers. The letters below the loudspeakers indicate their positions in the room: three loudspeakers in front of you—left *(L)*, center *(C)*, and right *(R)*; and two loudspeakers behind you—left surround *(LS)* and right surround *(RS)*.

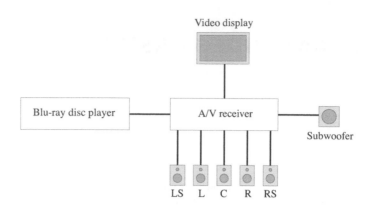

Figure 2.2
Top-down view of a room showing the loudspeaker positions in a 5.1-channel surround sound system. All the loudspeakers are approximately at the height of the listener's ears.

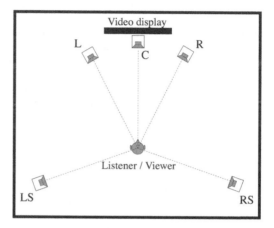

channel. The LFE channel is mainly used by filmmakers for explosions and other low rumbling sounds. Systems with 5.1 channels of surround sound are widely used and quite common (ITU-R BS.775, 2012). Plenty of source material on DVD and Blu-ray disc has 5.1 channels of sound.

The five main audio channels can be divided into three *front channels* and two *surround channels*. The loudspeakers that reproduce the front channels are in front of you close to the video display, and the loudspeakers that reproduce the two surround channels are in the back of the room and on both sides of you. This typical layout is depicted in figure 2.2. For easy reference, the front loudspeakers are labeled L (left), C (center), and R (right), and the surround loudspeakers are labeled LS (left surround) and RS (right surround).

You would expect that the subwoofer reproduces the LFE channel. It does, but that is not the complete story. A subwoofer is a loudspeaker that is specially designed for the reproduction of low frequencies. It reproduces the lowest notes of a bass guitar, the impact of a kick drum, and also the low rumbles of thunderstorms and explosions. In a domestic audio system it often reproduces the low frequencies from both the LFE channel and all of the five main channels. In fact, as I will explain in section 4.1 I strongly recommend you to use it that way.

Every A/V receiver has a *bass management* system that allows you to redirect the bass in the five main channels to the subwoofer output of the receiver. The bass management system can combine the low-frequency sound from the five main channels and add it to the LFE channel. The combined result is sent to the subwoofer output, and the main channels no longer contain any low-frequency sound. The human hearing system isn't very good at localizing low-frequency sounds, that is why you do not need a subwoofer for each of the main channels. One subwoofer placed somewhere in the room will do the job.

The subwoofer output of the A/V receiver is often called the *subwoofer channel*. It is not another name for the LFE channel. Although you can configure bass management in your A/V receiver such that the subwoofer channel equals the LFE channel, it is much more common that the two differ. Remember that the '.1' in a 5.1-channel audio-video system refers to the LFE channel and not to the subwoofer channel (AES, 2001; Holman, 2008; ITU-R BS.775, 2012).

2.1.1 Multichannel Audio

Clearly, a surround sound system is supposed to surround you with sound. To be more precise, a surround sound system has three main duties (Toole, 2018). First, to create a perception of direction of sound. Where does the sound come from? Left, right, front, or back. Second, to create a perception of distance. How far away are the things that produce this sound? In front of the loudspeakers or somewhere more remote. Third, to create a perception of spaciousness and envelopment. This important effect gives you the sense of being in a different space than your living room. You are surrounded by sound that you cannot exactly localize; it immerses you.

Quite a job for only five loudspeakers. Well, to be honest there are surround sound systems that use more than five. But five loudspeakers can surround you with sound quite effectively. Practical surround systems for domestic use rely on creating a perception, an illusion. Not every position from which a sound appears to come is occupied by a loudspeaker. Certain properties of the human hearing system are being exploited such that the surround system only needs to produce a few essential spatial cues (Davis, 2003). One trick is to add delayed versions of the sound to create a sense of distance. Another trick is to create a sound source between a pair of loudspeakers by sending the same audio signal to both loudspeakers. A sound source that appears to float between two loudspeakers is called a *phantom image*. Phantom images are used quite extensively between the front loudspeakers.

Music is often just a two-channel affair that involves only the left (L) and right (R) front loudspeakers. Traditionally, music is produced in stereo with only two channels. Phantom images are used to create a soundstage that extends from left to right between the two loudspeakers. These phantom images only appear in their intended position if you sit right in the middle at exactly the same distance from both loudspeakers. If you start to move to the left, you will reach a point where the entire soundstage collapses and sound only appears to come from the left loudspeaker.

A center loudspeaker (C) stabilizes the soundstage in front of you. Even if you do not sit exactly between the left and right loudspeakers, the sound from the center loudspeaker stays in position. The center loudspeaker always anchors the sound to the middle of the soundstage. Phantom images between the center speaker and one of the other front loudspeakers also tend to hold their position better than phantom images between the left and right loudspeakers.

When you watch a movie, the three front loudspeakers ensure that the location from which you hear the sound closely matches the location of the sound source on the screen. The center loudspeaker gets a real workout; it reproduces most of the dialog. The left and right loudspeakers are mainly used for sound effects and music.

The surround channels (LS and RS) are used in two ways. They reproduce directional sounds that need to come from the back and the sides, and they reproduce nondirectional sounds that immerse you and create a sense of spaciousness and envelopment. It is more common for the surround channels to envelop you with sound than to produce directional sounds. When you watch a movie, directional sounds from the surround channels tend to draw your attention away from the video display. Filmmakers therefore use directional sounds from the surrounds only sparingly and when they do, they use them to achieve a temporary special effect (Holman, 2008). A typical example of such an effect is an airplane flying by. The surround channels can be used to put directional sounds exactly at the positions of the loudspeakers, but phantom images between the surrounds and the front loudspeakers do not hold their position very well. They are not as stable as phantom images in the front soundstage, because of the large distance between the front and surround loudspeakers. This is another reason why filmmakers tend to use directional sounds only to convey a brief sense of movement (Toole, 2018).

In multichannel music, it is common for the front loudspeakers to create a soundstage in front of you in which you can clearly recognize the position of the different performers. The surrounds are then used to re-create the spaciousness of the venue in which the music was performed. Another approach to multichannel music has the performers all around you. A guitar at the back of the room, a piano on your left side, and a trumpet in front of you. This approach is not used very often, because most people prefer the more traditional perspective where the performers are all up front (Holman, 2008).

How many surround channels do you need? The 5.1 system uses only two (LS and RS). Localization and immersion can be improved if you use more than two surround loudspeakers, especially in large rooms. However, there will be a practical limit on the number of surround loudspeakers that people are willing to install in their homes. Equally important, filmmakers and other content providers must be willing to mix all these extra channels (Davis, 2003). As a successor to the standard 5.1-channel system, domestic surround sound systems with 7.1 and 9.1 channels are now available. The 7.1-channel system adds two surround loudspeakers at the back of the room, and the 9.1-channel system adds another two in front of you. Figure 2.3 depicts the positions of the loudspeakers in a 9.1-channel system. The additional surround loudspeakers at the back of the room are labeled LB (left back) and RB (right back); the ones in front of you are labeled LW (left wide) and RW (right wide).

A lot of source material has a limited number of surround channels. Blu-ray discs have either a 5.1 or 7.1 audio track. Almost all DVDs have 5.1 channels, but some support 6.1 channels. The 6.1-channel format is an intermediate format. It provides a single (mono) surround back channel. It is now common practice to reproduce this single channel using the two surround back loudspeakers (LB and RB) in a 7.1 system. In fact, in modern multichannel audio systems surround loudspeakers are always added in pairs, regardless of the actual number of surround channels available in the source material. In table 2.1 I present a list of the most commonly used surround formats. The labels that I use to indicate the different channels are found in table 2.2.

Figure 2.3
Top-down view showing the loudspeaker positions in a 9.1-channel surround sound system. All the loudspeakers are approximately at the height of the listener's ears.

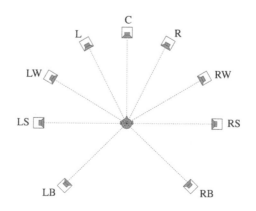

Table 2.1 Common surround sound formats and their different audio channels. The channel names are specified in table 2.2.

Format	Channels
5.1	L, C, R, LS, RS, LFE
6.1	L, C, R, LS, RS, B, LFE
7.1	L, C, R, LS, RS, LB, RB, LFE
9.1	L, C, R, LW, RW, LS, RS, LB, RB, LFE

Table 2.2 Audio channels in surround sound.

Channel	Description
L	Left front
C	Center front
R	Right front
LS	Left surround
RS	Right surround
B	Mono back
LB	Left back
RB	Right back
LW	Left wide
RW	Right wide
LFE	Low-frequency effects

You can artificially create additional surround sound channels from the channels that are available in the source material. This is called *surround upmixing*. Companies like Dolby Laboratories (www.dolby.com) and DTS (www.dts.com) have developed special signal processing methods for surround upmixing. Their upmixing methods are present in most modern A/V receivers. Upmixing even allows you to create surround sound from ordinary two-channel (stereo) material (Dressler, 2000). All the loudspeakers in your system will produce sound with surround upmixing; even if this sound is not in the original source material. Some people prefer it that way. Since they have paid good money for their loudspeakers, they expect them to be used. But there is a catch. If you use upmixing, you are not faithfully reproducing the original source material. You are taking some liberty and are interpreting the material in a different way than the artists who created it originally intended. Opinions differ. If upmixing sounds good to you, go ahead. If it doesn't, just stick to the original content. Modern A/V receivers can handle the material both ways.

Surround upmixing differs from *surround coding*. It is important to distinguish between the two. Surround coding is a method to efficiently store or transmit multiple channels of digital audio. Surround coding converts the original audio channels into a digital format that uses the least amount of storage space or transmission bandwidth. Surround coding does not create additional channels; it just encodes the existing ones. Over the years, Dolby Laboratories and DTS have developed different surround coding methods. In table 2.3 I show you the most popular ones along with some of their surround upmixing methods. In section 5.2 I explain the differences between the coding methods in more detail.

Don't get carried away with the number of surround channels. More isn't necessarily better. If you spend a fixed amount of money on loudspeakers, you can spend more on each loudspeaker if you only have to buy five speakers instead of nine. Being able to spend more on each loudspeaker enables you to buy higher quality loudspeakers. With a fixed amount of

Table 2.3 Examples of surround coding and upmixing methods.

Format	Coding examples	Upmixing examples
5.1	Dolby Digital/DTS	Dolby Prologic II
6.1	Dolby Digital EX/DTS-ES	DTS Neo:6
7.1	Dolby TrueHD/DTS-HD	Dolby Prologic IIx
9.1	–	DTS Neo:X

money, a 5.1-channel system will sound better than a 9.1-channel system. And don't forget that you have to fit all those loudspeakers in your living room and place them in their optimal positions. If a 5.1-channel system proves to be a challenge, installing a 9.1-channel system might seem nearly impossible.

The newest trend in surround sound is to create 3D sound by adding overhead and height loudspeakers. In this way, the reproduced sound not only surrounds you from all sides, but also has a height sensation. There are three major 3D surround sound formats: Dolby Atmos (www.dolby.com), DTS:X (www.dts.com), and Auro-3D (www.auro-3d.com). There are many possible loudspeaker configurations, because these formats utilize a new concept called *audio objects.* Instead of a fixed number of surround channels, the source material contains a number of audio objects that can be placed anywhere in three-dimensional space. Each object is accompanied by metadata that describe its position in space. Upon playback the surround decoder assigns the audio objects to the available loudspeakers. No matter how many loudspeakers you have, the decoder will utilize them in the best possible way. Dolby Atmos, DTS:X, and Auro-3D also include new surround upmixers that adapt any channel-based content to your particular loudspeaker layout. These upmixers ensure that every loudspeaker in your system is utilized, regardless of the number of channels available in the source material.

Dolby Atmos supports up to 24 loudspeakers at ear height and ten overhead loudspeakers. More realistically, Dolby recommends to add two or four ceiling loudspeakers to an existing 5.1, 7.1, or 9.1 system. Dolby indicates the number of overhead loudspeakers by a third digit. Thus, a 5.1.4 system is a standard 5.1 surround system augmented with four overhead loudspeakers. Figure 2.4 depicts the positions of the speakers in such a system. The Auro-3D system is based on a standard 5.1 layout and adds five height loudspeakers close to the ceiling above the L, C, R, LS, and RS loudspeakers. It also adds one overhead loudspeaker installed in the ceiling above the primary listening position. DTS:X can handle a combination of all the Dolby Atmos and Auro-3D loudspeaker positions.

Since 2014, several movies with object-based audio soundtracks have been released on Blu-ray discs. They are backward compatible and can be played on any existing 5.1, 7.1, or 9.1 system.

2.1.2 Audio and Video Equipment

Let's return to the basic 5.1-channel audio-video system from figure 2.1 and have a more detailed look at its components. The A/V receiver has to convert the digital audio data from the Blu-ray disc player into a number of electronic signals to drive the different loudspeakers. The receiver also performs bass management to direct the low-frequency sounds to the subwoofer. On the video side, the receiver sends the digital video data to the display. Figure 2.5 shows the basic 5.1-channel audio-video system that we started off with. It highlights the different signals that travel between the devices.

Digital audio and video data is usually transported using an HDMI cable. HDMI stands for High-Definition Multimedia Interface (www.hdmi.org). HDMI is supported by almost all

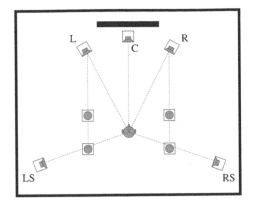

Figure 2.4
Top-down view of a room
showing the loudspeaker
positions in a 5.1.4 Dolby
Atmos system. The *L, C, R,
LS*, and *RS* loudspeakers
are approximately at the
height of the listener's
ears. The four overhead
loudspeakers are installed
near the ceiling facing
downwards.

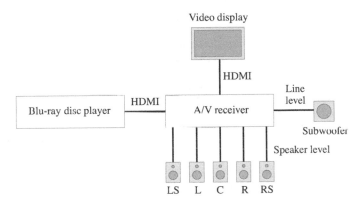

Figure 2.5
Connections in the basic
audio-video system of
figure 2.1.

modern audio, video, and computer equipment. In figure 2.5 HDMI is used to send digital audio and video data from the Blu-ray disc player to the receiver and digital video data on to the video display. Modern flat-panel televisions and front projectors are digital devices that can display digital video.

Loudspeakers and subwoofers need to be driven by an analog audio signal. There are two types of analog audio signals: a *line-level signal* and a *speaker-level signal*. A line-level signal is an electronic signal of about 1 volt. The exact level of the signal varies in time and describes the sound that a loudspeaker has to produce. However, a line-level signal is not powerful enough to drive a loudspeaker. A power amplifier that increases the signal to several tens of volts is needed. For example, a 500 watt amplifier converts the 1 volt line-level signal into a speaker-level signal of approximately 60 volts. The A/V receiver has several amplifiers built-in, one for each of the five surround sound channels. However, it does not amplify the subwoofer channel. This is a standard convention among equipment manufacturers. Almost all subwoofers have their own built-in power amplifier. That is why the A/V receiver sends a line-level signal to the subwoofer.

It is useful to think of the A/V receiver as a device that consists of two main building blocks: an *A/V controller* and a *multichannel power amplifier*. The A/V controller gets digital audio and video data from a source component, such as a Blu-ray disc player. You can connect several source components to the controller. The controller allows you to easily switch between them using the front panel controls or a remote. The A/V controller processes the digital audio data. It performs surround decoding and upmixing, and bass management, and sets the volume of each of the individual audio channels. The controller also converts the digital audio data into an analog line-level signal for each of the audio channels, including the subwoofer channel. The electronic circuits used for this conversion are called *digital-to-analog converters* (DACs). On the video side, the controller is also able to do some digital processing of the video data and sends it to one or more connected displays. The second building block of a receiver, the multichannel power amplifier, converts the analog line-level signals from the DACs to more powerful speaker-level signals that drive the loudspeakers.

Manufacturers of high-end audio equipment give you the option to replace the A/V receiver with two separate components, an A/V controller and a multichannel power amplifier, each in its own chassis. Figure 2.6 shows such a setup. Why would you want to buy two components instead of one? First, it will be more flexible. You could combine components of different brands and later on you could selectively upgrade your system replacing the components one by one. Second, separate components can offer better performance than A/V receivers can. Putting the electronics into separate chassis reduces the possibility of interference between the circuits. The power amplifier generates large powerful signals that could easily distort the delicate low-level signals that the controller is processing (Duncan, 1997). Separate power amplifiers can offer more power than receivers. A typical receiver provides between 50 and 120 watts per channel, while separate amplifiers can offer more than 200 watts per channel. You could even go further in separating components and buy a single-channel amplifier for each audio channel. Instead of one five-channel amplifier you would have to buy five monoblocks. The more amplifiers you buy, the more expensive your system becomes. Going down this route you will reach a point where you pay thousands of dollars for just a very small increase in performance (Harley, 2015; McLaughlin, 2005). A single high-quality multichannel amplifier is often a better choice.

A/V receivers are cost effective and convenient. They take up less space and reduce the number of cables in your system. If you do not want to spend thousands of dollars on just amplification, you should seriously consider buying a high-end receiver instead of a separate controller and amplifier. There are many receivers available that offer excellent sound quality at moderate costs (Harley, 2015).

You should look beyond the components of an audio-video system and consider the different functions that each component provides. The trend in lifestyle audio equipment is to integrate many functions into a single component. Many brands offer devices that combine an A/V controller, multichannel amplifier, and Blu-ray disc player into one single chassis. Often such a device is packaged together with a bunch of low-quality loudspeakers as a 'home theater in a

Figure 2.6
Audio-video system with an A/V controller and a multichannel power amplifier.

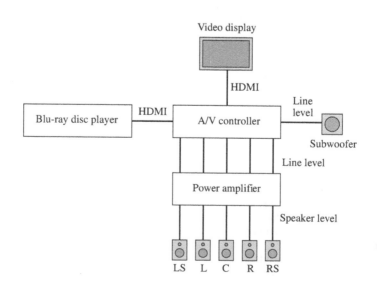

box'. If you want high-quality sound, don't buy such a package. You are much better off if you combine such an all-in-one device with separately bought high-quality loudspeakers.

The *active loudspeaker* is another example where different functions are combined into one device. An active loudspeaker has a built-in power amplifier. You can connect the active loudspeaker directly to an A/V controller. Most subwoofers are active loudspeakers. A *soundbar* is also an active loudspeaker. It integrates the left, center, and right front loudspeakers into one horizontally wide enclosure that is mounted on the wall underneath the video display. Some soundbars come with separate surround loudspeakers, others with a virtual surround system. There are no surround loudspeakers in a virtual surround system; instead an additional set of front loudspeakers is integrated into the soundbar and aimed at the sidewalls. The sound that bounces off the walls is meant to surround you, but the immersive quality is not as good as having dedicated surround loudspeakers (Briere and Hurley, 2009). Meridian Audio has taken the active loudspeaker one step further. Their high-quality loudspeakers accept a digital audio signal. Each Meridian loudspeaker performs its own digital audio processing, digital-to-analog conversion and amplification (Meridian, 2012).

The basic audio-video system in figure 2.5 only has a Blu-ray player as its source component, but, of course, you can connect all kinds of different source components to your A/V controller or A/V receiver. Source components come in many varieties. They retrieve audio and video information from a storage medium or broadcast signal. Examples are: CD player, DVD player, portable media player, personal video recorder, satellite receiver, cable set-top box, laptop, tablet, smartphone, and gaming console. Some components in your audio-video system do double duty. A television is not only a video display, it is also a source component that receives television broadcasts. Similarly, most A/V receivers include an AM/FM radio tuner. Modern televisions, A/V receivers, and Blu-ray players can also connect to your home computer network and the Internet. These smart-devices take the definition of a source component to another level. They can contain apps that you can use to access audio and video stored on other devices in your home network, for example the collection of audio files on your personal computer. They also enable you to access content from the Internet or from dedicated content providers such as Netflix. Adding network connectivity to your audio-video system also allows you to control your system using apps on your tablet or smartphone. Many equipment manufacturers are offering free apps that provide complete and intuitive control of their components. These apps can even replace the traditional infrared remote controller (AES, 2012).

In this all-digital world, is there any room left for old-fashioned analog audio sources like the turntable? Well, it depends. All A/V receivers have analog audio inputs, but that does not mean that you should use them. Most A/V receivers convert the analog input signals back to digital, because they cannot process the analog audio signal. For example, most receivers can only perform bass management on digital signals. If you, like some dedicated audiophiles, prefer the sound of vinyl records over modern-day digital audio, you are often better off with a dedicated analog audio system. In this book I focus on digital audio and video sources. For best performance, digital sources should always be connected digitally to your A/V receiver.

2.1.3 *Factors That Influence Reproduction Quality*

Audio and video signals travel through your audio-video system from component to component. At the end of the line, the video display lights up with a moving image and the loudspeakers produce the sound. But it doesn't stop there. Sound and light travel through your living room towards you. When the sound arrives at your ears and the light reaches your eyes, nerve signals are generated that stimulate your brain to create a perception.

Many factors influence the quality of what you perceive. You have to look at the entire chain to understand these factors. Figure 2.7 presents a schematic overview of the chain. Reproduction quality depends on the source material, the audio-video equipment, and the

Figure 2.7
Functional decomposition
of the audio-video
reproduction system.

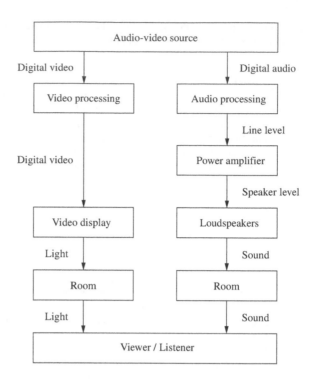

environment. The audio-video equipment is not listed in terms of components, because components often combine different functions. Instead, figure 2.7 includes the major functions that the different components provide. Video processing can be done in the source component, the A/V receiver, the A/V controller, or the video display. In practice all of them contribute to the end result. Audio processing is typically performed in the A/V receiver or A/V controller. The power amplifier can be part of the A/V receiver, built into an active loudspeaker, or exist as a stand-alone device.

High-quality reproduction starts with good source material. The technical quality of the source is of paramount importance. If you start out with low-quality material, you will never hear or see the full capabilities of your system. Source material is offered in different delivery formats. Choose the one with the highest quality. For example, a Blu-ray disc has a more detailed video image than a DVD. Audio files also differ in quality. They are often compressed to save storage space and network bandwidth. Heavily compressed audio files, like MP3 files, do not sound as good as the original uncompressed audio.

High-quality video requires a high-quality video display. The display has the largest influence on the perceived quality of video. The video processing in the source components and the A/V receiver influence the overall quality to a lesser degree, but high-quality processing is still important to get that last bit of performance out of your system. The viewing environment also influences the perceived quality. Video displays perform best in a dim or dark environment. Therefore, it is important that you can control the amount of ambient light in your room. For optimal quality, you should also pay attention to the position of the display with respect to the position of the audience and the positions of the loudspeakers. The audience should be able to view the display from a comfortable angle and from a distance that allows them to see every detail in the picture. Sound and picture form a coherent whole only if the display is centered between the left and right front loudspeakers.

High-quality audio requires good loudspeakers. Technology has matured such that most audio equipment does not degrade the audio signal in any perceptual way (Rumsey, 2002;

Winer, 2018). However, loudspeakers, even those of the highest quality, always distort and degrade the signal. Therefore, you need to choose your loudspeakers wisely. They have the largest influence on the overall sound quality (Moore, 2013; Toole, 2018; Winer, 2018). This does not mean that you do not have to pay attention to the other audio components. Carefully matching the amplifier to your loudspeakers is also important. Since digital signals are almost immune to distortions, the sound quality mainly depends on the quality of the audio components that process analog audio signals. Besides the loudspeakers and the amplifiers, the digital-to-analog converters (DACs) in the A/V receiver or A/V controller influence sound quality (Harley, 2015; Pohlmann, 2011).

The cables that you use to connect the audio-video equipment have the least effect on sound quality, at least if you compare their influence with that of the equipment itself. Dedicated audio and video cables can be ridiculously expensive while having only a marginal effect. The high prices bear little relationship to the cost of designing and manufacturing the product. In fact, any decently constructed cable will do just fine. Many myths surround special audio cables, and many audiophiles really believe they can hear an improvement when using expensive cables even when science and logic suggests that there should be no audible difference (Russell, 2012; Waldrep, 2017; Winer, 2018). You have been warned!

The environment in which you reproduce sound has a huge impact on the perceived sound quality. Everything seems to matter: the size, shape, and furnishing of the room; the construction of the walls, floor, and ceiling; and the positions of the loudspeakers and the listeners in the room. The interaction between the room and the loudspeakers can be so influential that it can dominate the overall perceived sound quality (Bech, 1994; Everest and Pohlmann, 2014; Toole, 2018). Moving the loudspeakers around the room can drastically change the sound. So much, that an excellent loudspeaker in the wrong location can sound significantly worse than an optimally positioned lower quality loudspeaker (Colloms, 2018; Newell and Holland, 2007; Olive et al., 1994). Hence, controlling the environment in which you reproduce sound should be your priority. The environment forms a significant part of the reproduction chain.

You are also part of the reproduction chain. You are both the listener and the viewer of the end result. How you perceive the reproduced sound and images is important. The human visual and hearing systems are the final arbiters in judging reproduction quality. They work in a complex way. Not every difference in reproduced sound and images can be perceived, and some differences are far more important than others. Understanding this final part of the reproduction chain requires some background knowledge on how the illusions of moving images and sound are created.

2.2 Moving Images

Moving images on a video display are not moving at all. What you are looking at is actually a series of still images presented to you in rapid succession. If the images are displayed at a sufficiently high rate then an illusion of movement is perceived by your brain. Your brain integrates the flashing sequence of images into one continuous moving image. This is called *flicker fusion* (Anderson and Anderson, 1993). A video display typically shows 50 or 60 still images per second. Every still image consists of a large number of very small picture elements, called *pixels* (Poynton, 2012). A high-definition (HD) video display has 1080 horizontal lines of 1920 pixels, and an ultrahigh-definition (UHD) display has 2160 lines of 3840 pixels. If you get up very close to your display you can see that the image is nothing more than a rectangular grid of small pixels. If you move farther away from the display, you will reach a certain distance where your eye can no longer resolve the individual pixels and they blend into one smooth image. Each pixel is a small light source that can change its brightness and its color. The video display creates moving images by emitting light from its pixels. If you use a direct-view display, like a flat-panel television, you are directly looking at the pixels. If you use a video projector, you are looking at a projection screen that reflects the light from the pixels in the projector.

Figure 2.8
Reflection of light. *On the left*, a specular reflection off a shiny surface. The angle of incidence matches the angle of reflection. *On the right*, a diffuse reflection off a matt surface. Light is scattered into multiple directions.

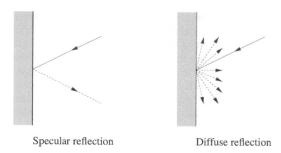

Specular reflection   Diffuse reflection

Figure 2.9
Parts of the human eye.

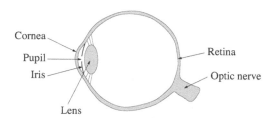

Without light you see nothing. All visual experiences require light, not only the ones from a video display. Light from a light source, such as the sun, illuminates your surroundings. The light reflects off the objects around you and travels to the eyes where it is converted into nerve impulses that create a visual experience in your brain. Light travels in straight lines, like a beam of light from a flashlight. In air it moves at approximately 300 000 km/s. Too fast for us to see it moving. Light that falls on a shiny object, such as a mirror, is reflected in one direction. A mirror reflects all the light back towards you such that you can see your own mirror image. Light that reaches a surface at a certain angle is reflected such that the angle of incidence matches the angle of reflection. This is illustrated in figure 2.8. Matt surfaces reflect light in all directions. They scatter the light and are much less directional. Therefore, you cannot see your mirror image in a matt surface. The amount of light that is reflected determines the appearance of an object. The less light reflected, the darker the object appears.

Light rays from the video display travel in straight lines towards your eyes. The rays enter the eye through the cornea, pass through the lens, and illuminate the retina. Figure 2.9 shows the different parts of the human eye. The cornea together with the lens bend the light rays such that a focused image is projected onto the retina. The lens can change shape in such a way that you can focus on objects at different distances. The retina converts light into neural signals that are relayed to the brain by the optic nerve. For this purpose the retina contains a large number of photoreceptors (Purves and Lotto, 2011).

2.2.1 Luminance and Brightness

The photoreceptors in your eye can handle a large range of light intensities. The intensity of light that a surface returns to the eye is called *luminance*. Luminance is expressed in candela per square meter (cd/m²). The luminance scale is designed to match the sensitivity of the human eye (Poynton, 2012; Purves and Lotto, 2011). The maximum luminance produced by a cinema screen is typically about 50 cd/m²; for a flat-panel display it is about 200 cd/m²; and for the newer high dynamic range (HDR) flat-panel displays it can be 1000 cd/m² or more. Figure 2.10 puts these figures into perspective. The cinema screen has a low maximum luminance, because it is meant to be viewed in a darkened room and a very bright screen in a dark room would induce unwanted eye strain. An HDR display has an extended luminance range that is mainly used to produce more realistic highlights in small areas of the screen. The extended luminance range allows emissive light sources, such as the sun, and specular reflections from bright shining objects to be much brighter than other parts of the screen. This adds depth and dimension to a two-dimensional picture. While on an HDR display the highlights are much brighter than on a standard display, they almost always occupy only a small portion of the screen. As a result, the overall image will not appear much brighter, since the average brightness stays about the same (ITU, 2017).

The human eye contains two different types of receptors: *rods* and *cones*. Rods operate predominantly at low luminance levels, and cones operate best at higher levels. The rods are meant for night vision and produce a colorless visual image that lacks detail. Cones, on the other hand, are capable of producing highly detailed colorful visual images. Rods are present throughout the retina of the eye, but cones are mainly concentrated in a small central region of the retina called the *fovea*. Objects that you want to see in detail need to be projected onto this small region. Therefore, your eyes are frequently moving around to inspect your surroundings (Purves and Lotto, 2011). Such eye movements also happen frequently when you are watching a big video display, but you are mostly unaware of it.

At a certain average luminance level of your surroundings, your eyes can handle a luminance range of about 1000:1 (Poynton, 2012). Thus, the highest luminance level that you can perceive is about 1000 times stronger than the lowest luminance level that you can perceive at the same time. This 1000:1 range can be shifted up and down along the luminance scale shown in figure 2.10 by adaptation. The eye adapts to the average luminance level, such that at any average level the range of discernible luminance levels remains approximately 1000:1. Adaptation involves a change in pupil diameter, a photochemical process in the rods and cones, and some neural mechanisms. The darker the surroundings, the larger the diameter of your pupils. Adaptation from dark to bright is faster than from bright to dark. It can take several minutes for your eyes to adapt to darker surroundings. Movie theaters often slowly dim the lights before a movie starts, to allow you some time to adapt.

The luminance levels that can appear in real life are much larger than the maximum luminance that a video display can produce. In digital video the objective is not to produce the real-life luminance of the scene that was captured, but to produce a luminance that is proportional to the real-life luminance up to a certain maximum. Hence, a video display reproduces

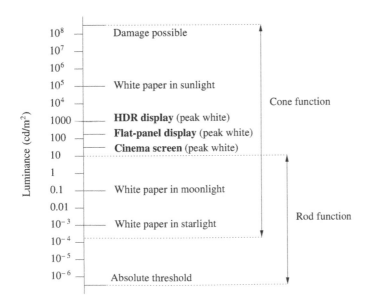

Figure 2.10
The range of luminance levels relevant for human vision. Maximum luminance of a flat-panel display and a cinema screen compared to the luminance of white paper in different lighting conditions. *On the right side* the operating ranges of cones and rods are indicated.

Adapted from Purves and Lotto (2011), figure A.4.

relative luminance. Relative luminance, denoted by *Y*, is the real-life luminance divided by the maximum luminance of the display. It is always a number between 0 and 1 (Poynton, 2012).

The perception of luminance is called *brightness.* It is important to distinguish between the two. Luminance can be measured and expressed as a number; brightness is a visual sensation that only exists in your brain. Brightness is a subjective quantity that cannot be measured. In general, the higher the luminance of a surface, the brighter it appears. But, the human visual system has some idiosyncrasies. A classic example is called *simultaneous brightness contrast.* It shows that the brightness of a surface not only depends on its own luminance but also on the luminance of its surroundings (Poynton, 2012; Purves and Lotto, 2011). Figure 2.11 shows two gray rectangles with exactly the same luminance. Nevertheless, the rectangle on the left looks brighter than the one on the right. The difference in brightness is due to the difference in background luminance. A gray rectangle on a dark background appears brighter than the same rectangle on a lighter background.

2.2.2 *Color*

Color is a perception; it is not a property of a physical object. Color only exists in your brain. The physical quantity that gives rise to color vision is the *wavelength* of light (Purves and Lotto, 2011). Light is a periodically oscillating electromagnetic wave. The distance that such a wave travels during one period of its oscillation is called the wavelength. Visible light has wavelengths between approximately 400 and 700 nm (nanometer). As you can see in figure 2.12, different wavelengths give rise to different color sensations. Visible light only takes up a small portion of the electromagnetic spectrum; there exist many other types of electromagnetic radiation.

Figure 2.11
Simultaneous brightness contrast. The two gray rectangles have the same luminance. The rectangle on the *left* appears to be brighter because of the darker background.

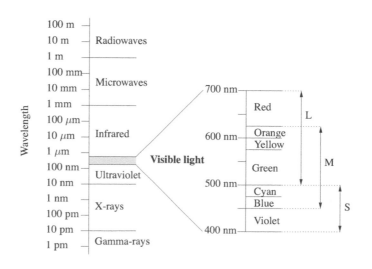

Figure 2.12
Visible light and the electromagnetic spectrum. Different color sensations correspond to different wavelengths of visible light. *On the right side* the operating ranges of the three different types of cones *(L, M, and S)* are indicated.

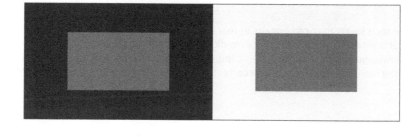

Natural light from the sun consists of a relatively even mixture of multiple wavelengths between 400 and 700 nm. You perceive this light as colorless or white light. When white light shines on a surface, certain wavelengths are absorbed and others are reflected. The wavelengths in the reflected light give the surface its color appearance. For example, if only the larger wavelengths are reflected the surface appears red. Compared to white light, colored light contains only a small fraction of the wavelengths between 400 and 700 nm. If colored light shines on a surface, the color appearance of the surface not only depends on the wavelengths it can reflect, but also on the wavelengths that are present in the light that illuminates it. For example, a blue surface that is illuminated by red light appears to be black.

Color perception depends on three distinct types of cones in the retina of the eye. Each type of cone responds to a different range of wavelengths. These ranges are illustrated in figure 2.12. The L-type cones respond to long wavelengths that result in red color sensations; the M-type cones respond to medium wavelengths that result in green; and the S-type cones respond to short wavelengths that result in blue. The relative activation of the three cone types determines the colors that you see. This explanation of color vision is called *trichromacy theory*. Although it cannot explain the perception of all color sensations it is successful in explaining many aspects of color vision (Purves and Lotto, 2011).

A video display produces colors by mixing red, green, and blue light in certain proportions. Because of the trichromacy theory of color vision it is possible to produce a large range of colors from such a mixture (Poynton, 2012). As long as the three different types of cones are activated in the same way as with the original color, the mixture of red, green, and blue is indistinguishable from the original color. In a flat-panel display, every pixel is divided into three subpixels, a red, a green, and a blue one. Because of the limited acuity of your eyes this mosaic of colored subpixels blends into one color when you are at a certain distance from the display (Lee et al., 2008). Projectors also produce colors by mixing red, green, and blue light, but in a different way as I will show you in section 6.1.3. Note that there exist displays that use more than three colors, but this is not very common at the moment (Bodrogi and Khanh, 2012).

The three primary components red, green, and blue are called the *primary colors*, often denoted simply as *RGB*. By varying the intensity of these components different colors are produced on the video display. I show some important combinations in table 2.4. The table uses the relative intensities of the three primary components. Each component can range from 0 to 1. Black corresponds to all three components having zero intensity. White corresponds to all three components having maximum intensity. Between black and white, you can obtain every shade of gray, as long as you keep the intensities of the three components equal. For example, a mid gray is obtained when all three components have a relative intensity of 0.5. Different shades of red, green, and blue can be obtained by varying one component and keeping the other two components at zero intensity. Cyan, magenta, and yellow

Table 2.4 Producing colors from the red (*R*), green (*G*), and blue (*B*) primaries.

Color	R	G	B
Black	0	0	0
Red	1	0	0
Green	0	1	0
Blue	0	0	1
Cyan	0	1	1
Magenta	1	0	1
Yellow	1	1	0
White	1	1	1

are obtained from a mixture of two primary colors. Cyan, magenta, and yellow are called the *secondary colors*. Cyan is obtained from green and blue. If you add the primary color red to the secondary color cyan, you get white light. Therefore, red and cyan are said to be complementary colors. Similarly, green is the complement of magenta, and blue is the complement of yellow.

The range of colors that a video display can reproduce depends on the particular choice of the red, green, and blue primary colors. A standard choice is needed to ensure that color reproduction stays constant among different displays. The current standard for HD video sources can be found in recommendation ITU-R BT.709 (2015), commonly abbreviated as Rec. 709. The colors for standard-definition (SD) and ultrahigh-definition (UHD) video are different and can be found in recommendations ITU-R BT.601 (2011) and ITU-R BT.2020 (2015), respectively. These recommendation documents specify the primary colors in terms of the 1931 CIE *chromaticity diagram* (CIE S 014–1, 2006). Let me explain how such a diagram works. Take a look at figure 2.13. The CIE chromaticity diagram is the inverted U-shape that is closed at the bottom with a straight line. Every point inside this closed shape corresponds to a visible color. The inverted U-shape is called the *spectral locus.* The points on this locus correspond to light that consists of only one wavelength. The locus starts at the bottom left at 400 nm, moves to the top, and ends at the bottom right at 700 nm. Mixing light of different wavelengths produces the colors that are represented by the points inside the closed shape. Points outside the closed U-shape are not associated with any color. The line at the bottom that closes the inverted U-shape corresponds to the purple and magenta hues that can only be obtained by mixing light with short and long wavelengths. The cross in the middle represents the *white point:* an even mixture of light of different wavelengths. Every color in the chromaticity diagram can be represented by two chromaticity coordinates x and y.

The primary colors red, green, and blue form a triangle in the chromaticity diagram. Figure 2.13 shows the triangle that results from the specification of the primary colors in Rec. 709 (ITU-R BT.709, 2015). The video display can only produce the colors that are inside this triangle. The triangle is called the *color gamut* of the display. Typically, the range of colors that a display can produce is only a small fraction of all the colors within the inverted U-shape.

Figure 2.13
CIE 1931 chromaticity diagram with coordinates x and y. The *triangle* indicates the Rec. 709 video gamut. The *cross* in the middle represents the white point.

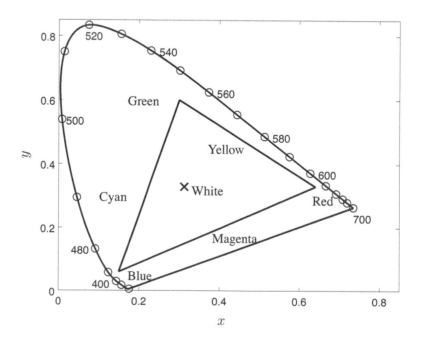

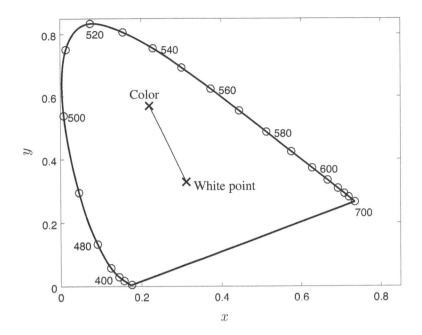

Figure 2.14
Hue and saturation in the
chromaticity diagram.
Rotating the point labeled
Color around the white
point will change the hue.
Moving it along the line
towards the white point
will decrease its saturation.
Moving it farther away
from the white point
increases its saturation.

The color produced by a pixel of the video display can be described using the two chromaticity coordinates x and y. Besides color, the pixel also produces a particular luminance. Therefore, a complete description of the light from the pixel not only involves the chromaticity coordinates x and y, but also the relative luminance Y. Alternatively, the light from a pixel can be described as a combination of the three primary colors *(RGB)* of the display. However, such a description is only unique if in addition the relative luminance and the chromaticity coordinates x and y of each of the three primary colors and the white point are specified.

Another, somewhat more intuitive, way to describe the light from a pixel is to specify it in terms of its *relative luminance, hue,* and *saturation.* The hue of the pixel determines *which* color. It is the attribute according to which the pixel appears to be similar to red, green, or blue; or a combination of two of them. Saturation is the colorfulness of the color. You can think of it as the amount of a particular color (Bodrogi and Khanh, 2012; Poynton, 2012). With the help of figure 2.14, I show you how hue and saturation can be related to the chromaticity diagram. The point labeled *Color* can be specified in relation to the white point in terms of the distance to the white point and the rotation around the white point. The distance to the white point equals the saturation of the color. The farther away from the white point, the more saturated the color becomes. The rotation around the white point determines the hue of the color. If you move the point labeled *Color* in a clockwise fashion around the white point, it first goes from green to yellow; then it becomes red followed by magenta, blue, and cyan; and finally it returns back to green.

2.3 Sound

Sound is both a physical event and a perception of your brain. As a physical event, sound is a vibration that propagates through the air or through another medium, such as water or wood. Sound is produced by vibrating or moving objects, like the leaves of a tree rustling in the wind or the vibrating string of a violin. There are many more examples of sound. Animals, humans, and all kinds of man-made and natural objects produce sound. In an audio-video system the loudspeakers are the devices that produce sound. A loudspeaker has one or more diaphragms, usually cone- or dome-shaped, that can move backward and forward in response to the electrical signal from the amplifier. These diaphragms set the air molecules in the direct

Figure 2.15
Parts of the human ear.

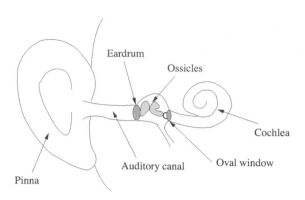

vicinity of the loudspeaker in motion. The air molecules start to vibrate. They are squeezed together or pulled apart depending on the direction of movement of the loudspeaker's diaphragms. Vibrating air molecules pass their vibration onto neighboring air molecules and as a result sound propagates through the air. The air molecules themselves stay approximately at the same position; it is only the vibration that propagates through the air. This moving vibration is called a *sound wave.*

When a sound wave reaches your ears, the vibration of the air molecules is converted into nerve impulses that your brain translates into a perception. Figure 2.15 shows the basic parts of the human ear. The external ear, called pinna, collects sound and directs it down the auditory canal. At the end of this canal, a thin membrane starts to vibrate due to the arriving sound. This membrane is the eardrum. Three tiny bones, the ossicles, transfer the vibration of the eardrum to a second membrane: the oval window of the cochlea. The cochlea is a bony spiral-shaped construction that is filled with fluid. Running along the length of this spiral is the basilar membrane that contains hair cells that send nerve impulses to the brain. The nerve impulses coming from the two ears are cross-compared in the brain and result in a perception of three-dimensional sound (Moore, 2013; Zhang, 2015).

The human hearing system has incredible powers of perception, but at the same time it has its limitations. It can be astonishingly acute in its ability to detect a particular difference in sound, while at the same time it is surprisingly casual towards some other differences. The interaction between the ear and the brain is rather complex. Several parts of the brain are involved. The brain imposes structure and order on a sequence of sounds using both logical prediction systems and emotional reward systems. The nerve impulses from both ears are analyzed and categorized, based in part on the memory of previous sound experiences. The nerve impulses are translated in a perceptual illusion and also in an emotional response (Levitin, 2006).

2.3.1 Frequency, Pitch, and Timbre

Sounds vary widely in their complexity. There are simple sounds, like the ringing of a bell or a clap of your hands; more complex sounds, like a person speaking or a certain melody played on the piano; and there are complex mixes of many different sounds, such as when all instruments in a symphony orchestra play together. All these sounds, from simple to complex, are just vibrations carried through the air. Sound can be analyzed by looking at how the magnitude of these vibrations changes in time. Figure 2.16 shows a typical example of a complex sound; it is a 2 second fragment from a piece of pop music. The horizontal axis corresponds to time and the vertical axis to magnitude. The figure clearly shows that the magnitude increases and decreases in time. It goes from positive to negative. The positive parts correspond to squeezing the air molecules together, while the negative parts correspond to pulling them apart. The magnitude is a function of time. In engineering, such a function is called a *signal*, and if it relates to sound it is common to refer to it as an *audio signal*. More details of the audio signal can be seen in figure 2.17, which zooms in on the part between 0.5 and 0.6 seconds. I have expressed the magnitude of the audio signal in both figures as a normalized level between 1 and −1, because in this section I am more interested in the shape of the signal than its absolute magnitude. The shape of the signal determines which kind of sound we are hearing. Is it a violin or a piano? The absolute magnitude influences the loudness of the sound, which is the topic of section 2.3.2.

Certain properties of sound are difficult to understand by just looking at the shape of an audio signal in time. Therefore, it is common engineering practice to look at signals in the so-called *frequency domain* (Brown, 2015). At first, you might find this notion a bit abstract, but I assure you that it is a convenient way to explain the reproduction of sound in your listening room. Let's start at the beginning. The concept of frequency is related to a mathematical function called the *sine*. The sine is an audio signal with a simple shape. This shape is shown in figure 2.18. The sine is a periodic signal; it keeps repeating its shape in time. The figure shows only one period.

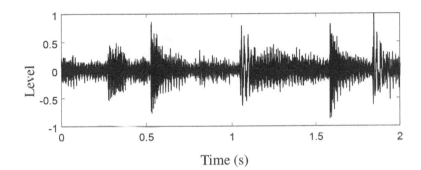

Figure 2.16
Example of a complex audio signal.

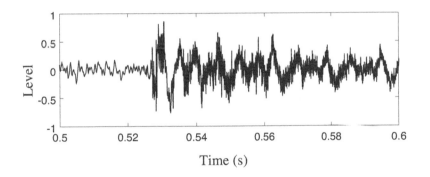

Figure 2.17
The same audio signal as in figure 2.16, but with a different time scale to show more details in the part of the signal between 0.5 and 0.6 seconds.

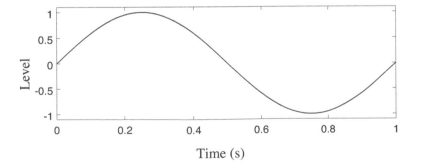

Figure 2.18
One period of a sine.

A sine can be uniquely described by just three parameters: *frequency, amplitude,* and *phase.* The frequency of a sine is simply the number of times its shape repeats itself in 1 second. In other words, it is the number of periods per second. Frequency is expressed in Hertz (Hz). The sine in figure 2.18 has a frequency of 1 Hz. Figure 2.19 shows some sines with different frequencies. The amplitude of a sine is the maximum value that it reaches. The sines in figures 2.18 and 2.19 all have amplitude 1. The phase describes the point at which the sine starts at time instant zero. It is equivalent to a shift in time. It is common to express this time shift in degrees (like an angle) instead of seconds. One reason is that in this way the shift becomes independent of the frequency of the sine. One period of a sine corresponds to 360°. Therefore, a phase of 360° corresponds to a time shift of a whole period. Similarly, a phase of 180° corresponds to a time shift of half a period, and so on. Figure 2.20 shows some sines with different phases. Because the sine keeps repeating its period, there is no need for phases larger than 360°. The starting point of the sine can always be specified as a number between 0° and 360°. The sines in figures 2.18 and 2.19 all have phases of 0°.

No matter how complex the shape of an audio signal, it can always be decomposed into a sum of sines with each sine having a particular frequency, amplitude, and phase. Such a

Figure 2.19
Sines with different frequencies. *From top to bottom*: 2 Hz, 3 Hz, and 4 Hz. The higher the frequency, the more periods of the sine fit within a time interval of 1 second.

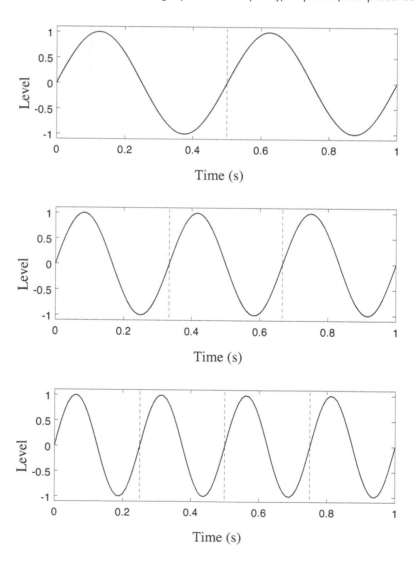

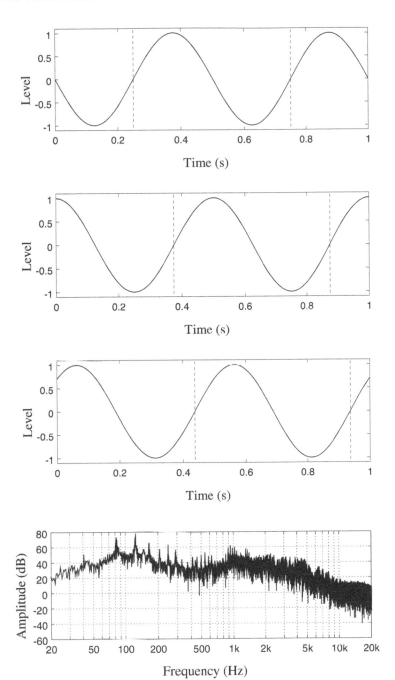

Figure 2.20
Sines with different phases. *From top to bottom*: 180°, 90°, and 45°. All these sines have frequency 2 Hz and amplitude 1. Since a phase of 180° corresponds to a shift of half a period, a sine with a phase of 180° is exactly the opposite from a sine with phase 0° (the *top panel* in figure 2.19).

Figure 2.21
Amplitude spectrum of the example audio signal in figure 2.16. The audio signal can be decomposed into a sum of constituting sines. The amplitude spectrum shows the amplitude of these sines at each frequency: it shows the frequency content of the audio signal.

decomposition is called *Fourier analysis*. Each sine in the decomposition can be thought of as a frequency component of the original audio signal. The amplitude of the sine indicates its relative strength. Looking at all the amplitudes together you get an idea of the frequency content of the audio signal. A plot showing all the amplitudes is called the *amplitude spectrum*. Figure 2.21 shows the amplitude spectrum of the example audio signal from figure 2.16. The two peaks around 100 Hz dominate the spectrum. They account for the large oscillations in the time signal. Note that in the figure the amplitudes are expressed in decibels (dB), a concept that I explain in more detail in section 2.3.2.

A complete frequency domain view of an audio signal consists of two graphs: the amplitude spectrum and the *phase spectrum*. As you might have guessed, the phase spectrum is a plot that shows the phases of each of the frequency components. If you add all frequency components and use the amplitudes from the amplitude spectrum and the phases from the phase spectrum, you get exactly the original audio signal back. Together, the amplitude and phase spectrum completely describe the original audio signal. In audio engineering, the phase spectrum is often omitted, because it is a bit more difficult to interpret than the amplitude spectrum. As a result the amplitude spectrum is often simply referred to as the *frequency spectrum*. However, it is important to remember that for a complete description, you need both the amplitude and the phase spectrum. In fact, there are many different audio signals that have exactly the same amplitude spectrum; you need the phase spectrum to be able to distinguish them in the frequency domain.

The human hearing system analyzes sound by splitting it into frequency components. It acts as a Fourier analyzer. So, Fourier analysis is not only a convenient engineering tool, it is also an important mechanism in our perception of sound. The frequency analysis is carried out by the basilar membrane in the cochlea. This membrane contains hair cells that are frequency selective, such that each part of the membrane is sensitive to a certain range of frequencies. It is as if the basilar membrane contains a map of frequencies ranging from low to high, stretched out along its length. You could compare it to a piano keyboard where in a similar way the notes that you can play are stretched out along the length of the keyboard (Levitin, 2006; Moore, 2013).

The range of frequencies relevant for the reproduction of sound starts at 20 Hz and ends at 20 000 Hz. The upper limit is commonly referred to as 20 kHz, which is short for 20 kilo Hertz. The range from 20 Hz to 20 kHz roughly corresponds to the frequencies that we, humans, can hear. The lower limit for hearing a pure sine is about 20 Hz. Sounds below this frequency can sometimes be felt as a vibration if they are very intense (Moore, 2013). The upper frequency limit is a subject of some controversy. For most people the upper limit is lower than 20 kHz, because the ability to detect sounds above 15 kHz quickly deteriorates with age (Moore, 2013). Under controlled conditions only a minority of people can hear a very subtle difference between music with and without frequencies above 20 kHz (Hamasaki et al., 2004). Nevertheless, audio equipment has been introduced that can reproduce frequencies up to 48 kHz or even 96 kHz. If the reproduction of frequencies above 20 kHz makes any difference at all, it will be a subtle difference that not everyone can hear. Therefore, it is safe to take 20 kHz as the practical upper limit of hearing.

Changes in frequency are perceived in a *logarithmic* way. This means for example that the interval between 100 Hz and 200 Hz is perceived to have the same length as the interval between 1000 Hz and 2000 Hz, despite the fact that the calculated length of the second interval is ten times larger than that of the first. The human hearing system favors frequency ratios above frequency differences. A frequency ratio of 2:1 is called an *octave*. Each octave is perceived to have the same length. Octaves play a special role in music: when the frequency of a musical note is shifted up by one octave it sounds remarkably similar to the unshifted note (Levitin, 2006). The audible frequency range can be divided into 10 octaves. Because of the logarithmic perception of frequency, the frequency axis in a graph is usually in a logarithmic scale, like the one in figure 2.21. Such a scale gives equal weight to the low frequencies between 20 Hz and 200 Hz, the mid frequencies between 200 Hz and 2 kHz, and the high frequencies between 2 kHz and 20 kHz. Another example of a logarithmic frequency scale is given in figure 2.22 along with the division in low, mid, and high frequencies. To make the concept of frequency less abstract, I have included the frequency ranges of some common musical instruments in this figure.

Pitch is a perception of sound that allows you to order sounds on a musical scale. Pitch depends mainly on the frequency content of the sound. It is a subjective quantity that only exists in the brain. A single note played on a musical instrument results in a perception of pitch. Several

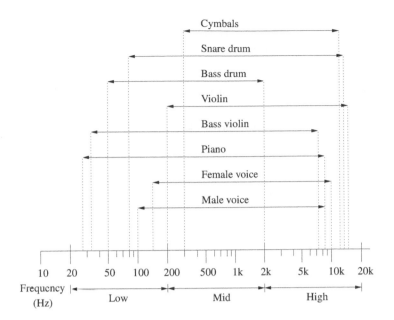

Figure 2.22
Frequency ranges of some musical instruments.

Adapted from Ballou (2015), figure 48.9.

notes played in a sequence result in a perception of melody. Pitch is one of the primary means by which musical emotion is conveyed. Musical notes with frequencies in the range from 55 Hz to 2000 Hz produce the strongest sensations of pitch. Sounds above about 5 kHz are not perceived as having pitch and do not produce melodies (Levitin, 2006; Moore, 2013).

A note played on the piano and the same note played on the violin are perceived to have identical pitch, but they do sound quite different from each other. The sound of the two notes differs in *timbre*. Like pitch, timbre is a subjective quantity. It allows you to distinguish different musical instruments and to recognize different voices. Timbre is the perceived difference between two sounds that are similarly presented and have the same pitch and loudness (Moore, 2013). Timbre depends on the entire frequency spectrum of the note, while pitch is determined only by the *fundamental frequency*. The fundamental frequency is the lowest and also the loudest frequency component of the note. It is accompanied by several *overtones* that are higher in frequency. Each overtone has a particular frequency, amplitude, and phase. The pattern of overtones that accompanies the fundamental frequency gives each musical instrument its own characteristic sound. The frequencies of overtones are often simply multiples of the fundamental frequency, but this is not always the case.

2.3.2 Sound Pressure and Loudness

The loudness of a sound depends on its frequency content and on its intensity. *Sound intensity* is the sound power that passes through an area of one square meter. The term 'power' is used in physics to denote energy per second. Thus, *sound power* equals sound energy per second (Everest and Pohlmann, 2014; Moore, 2013). Sound power and sound intensity are often used interchangeably with the implicit assumption that the area through which the sound passes equals one square meter. Sound intensity or power is measured in decibels (dB) and expressed as a *sound pressure level* (SPL). As the name suggests, SPL depends on the magnitude of the pressure changes in the air (see box 2.1). The human ear can deal with a huge range of pressure changes from 0.00002 Pa (Pascal) near the threshold of audibility, to 200 Pa where the sound intensity is so large that it can instantly damage your ears. The decibel is used to cope with this enormous range in a more convenient way. It compresses the range of pressure changes into the equivalent SPL range that starts at 0 dB and ends at 140 dB. Take a look at figure 2.23 to get an idea of what these numbers mean. For example, the average SPL in the cinema is 85 dB with occasional excursions to the maximum of 105 dB (DCI, 2008).

Figure 2.23
The range of sound
pressure levels relevant for
human hearing. The safe
exposure levels on the *right*
***side* are recommended by**
the WHO (1999) and the
EPA (1978).

Adapted from Toole (2018),
figure 4.3; safe exposure levels
after Blomberg and Lewis (2008);
cinema sound levels after DCI
(2008).

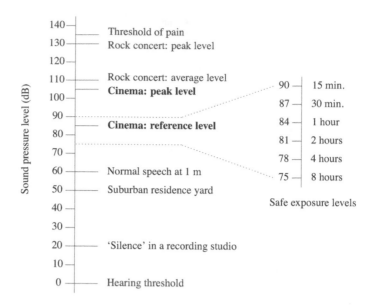

BOX 2.1 SOUND PRESSURE LEVEL

Sound intensity is proportional to the square of the magnitude of the pressure variations. Sound intensity is commonly expressed in decibels as a sound pressure level (SPL).
The SPL for a certain pressure p can be computed by comparing this pressure to a reference pressure p_0 as follows:

$$L = 10 \log_{10}\left[\frac{p^2}{p_0^2}\right] = 20 \log_{10}\left[\frac{p}{p_0}\right]$$

The reference sound pressure p_0 equals 20 μPa (0.00002 Pa), which corresponds to the
threshold of human hearing for a sound that consists of a sine with a frequency of 1 kHz
(Everest and Pohlmann, 2014; Moore, 2013).

The decibel converts a power ratio into a difference. You find some typical examples listed
in table 2.5. For example, if two sounds have a 3 dB SPL difference, this means that one of
them has twice the sound power of the other. Thus, a 3 dB increase in SPL means that the
sound power has doubled. Similarly, a 3 dB decrease in SPL means that sound power has
halved. A decrease is often denoted with a negative number. Hence, a change of –3 dB means
a decrease of 3 dB. The definition of SPL is such that the lowest sound power that humans can
hear corresponds to 0 dB SPL. This lowest sound pressure level is called the *threshold of audibility*. When you use the decibel to express an absolute sound level, the threshold of audibility
is used as a reference. For example, the sound power that corresponds to 20 dB SPL is 100
times larger than the sound power at the threshold of audibility (0 dB SPL).

Loudness is a subjective quantity that allows you to order sounds on a scale extending from
quiet to loud (Moore, 2013). In general, the higher the SPL, the louder the sound. However,
the frequency content of the sound also influences loudness. Our ears are most sensitive to
frequencies between 1 kHz and 5 kHz. Sounds outside this range of frequencies, either lower
or higher, do not appear as loud when they have the same SPL. Figure 2.24 shows how loudness depends on frequency. The curves in this figure are the so-called *equal loudness contours*

Table 2.5 Examples of decibel changes and the corresponding power ratios.

dB change	Power ratio
0 dB	1:1
1 dB	1.26:1
3 dB	2:1
6 dB	4:1
10 dB	10:1
20 dB	100:1
30 dB	1000:1
40 dB	10 000:1

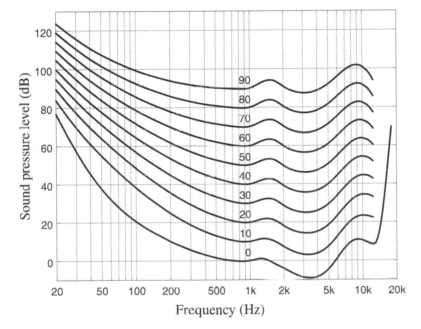

Figure 2.24
Equal-loudness contours.
The different curves
correspond to different
loudness levels from 0
phon to 90 phon. The
curve indicated by 0
phon corresponds to the
threshold of audibility.
Sounds below this curve
cannot be heard. Note that
this threshold rises quickly
above 15 kHz.

Based on ISO 226 (2003).
Permission to reproduce extracts
from ISO 226 is granted by
BSI. British Standards can be
obtained in PDF or hard copy
formats from the BSI online
shop: www.bsigroup.com/Shop
or by contacting BSI Customer
Services for hardcopies only:
Tel: +44 (0)20 8996 9001, Email:
cservices@bsigroup.com.

(ISO 226, 2003; Suzuki and Takeshima, 2004). Each curve shows how much SPL is needed for a sine of a specific frequency to sound as loud as a sine of 1 kHz. By definition the loudness level of a 1 kHz sound equals its SPL. At other frequencies the loudness level typically differs from the SPL. Loudness level is measured in phon.

The equal-loudness contours show that our ears are less sensitive to low-frequency sounds than to mid and high frequencies. Low-frequency sounds require more SPL to sound as loud as mid and high frequencies. For example, the curve of 40 phon shows that at 20 Hz you need 100 dB SPL to equal the loudness of a 40 dB SPL sound at 1 kHz. A difference of 60 dB. Thus, to achieve effective reproduction of low frequencies, your loudspeakers should be capable of producing sufficient SPL. Therefore, low-frequency loudspeakers often have a large diaphragm that allows them to move plenty of air. Similarly, the bass instruments in an orchestra are often the largest ones (Colloms, 2018). The equal-loudness contours are not only higher in level at low frequencies, they also converge. This means that low frequencies decrease faster in perceived loudness than mid and high frequencies as the overall sound level drops. In other words, if you lower the volume control of your audio system, you will not only reduce the overall loudness, but you will also reduce the amount of low frequencies with respect to the mid and high frequencies. You hear less bass. The tonal balance between the low frequencies and the mid/high frequencies is altered. To avoid altering the tonal balance, you should reproduce sound such that it is as close as possible to the original sound level. At the mid and high frequencies the equal-loudness contours are almost parallel. In this region, a change of 10 dB in SPL corresponds roughly with a 10 phon change in loudness level. Such a change is perceived as a doubling or halving of the loudness (Toole, 2018).

Loud sound can permanently damage your ears. It can result in partial or complete deafness or in a persistent ringing in the ears called *tinnitus*. Hearing loss from loud sound mostly occurs at frequencies between 3 kHz and 6 kHz (Blomberg and Lewis, 2008). Despite an increasing amount of scientific research on hearing loss, it is currently still irreversible. You should therefore protect your ears from loud sounds. The World Health Organization (WHO, 1999) and the U.S. Environmental Protection Agency (EPA, 1978) recommend a maximum exposure of 75 dBA SPL for a period of 8 hours. The SPL level in their recommendation is expressed in dBA, which means that A-weighting has been applied. Such a weighting takes into account the fact that the ear is less sensitive to low frequencies than to mid and high frequencies. The higher the SPL the shorter the safe exposure period. For every 3 dB rise in SPL level, the safe exposure period should be halved. This is illustrated in figure 2.23 on the right. To avoid permanent damage, you should avoid sound levels above 110 dB as much as possible. It is a good idea to wear earmuffs or earplugs during leisure and occupational activities that involve high sound levels; for example when using power tools (chain saw, lawnmower, drill), participating in motorsports (motorcycle, motorboat), or being at a shooting range. Conventional hearing protection is less suitable for activities where sound quality is important, because it attenuates high frequencies more than low frequencies. You are better off using so-called *musician's earplugs* that reduce sound level in all frequencies equally (Niquette, 2006). Many professional musicians use such earplugs during rehearsals.

2.3.3 Sound Propagation

Sound travels in straight lines from its source, like a ray. At an average room temperature of 20°C (68°F) sound propagates through the air with a speed of 344 m/s (1129 ft/s). Since every sound wave can be decomposed into a sum of sine waves, the propagation of a complex waveform can be studied by looking at the propagation of each of its constituting sines. A sine wave with a frequency of f Hz repeats itself every $1/f$ seconds. In other words, one period of the sine wave lasts $1/f$ seconds. The distance that a sine wave travels in one period is called the *wavelength*. It is denoted by λ (lambda) and can be computed as

$$\lambda = \frac{c}{f}$$

with the speed of sound $c = 344$ m/s. The wavelength determines how sound propagates through rooms and around obstacles. What happens to the sound depends on the relative size of the room and the obstacles with respect to the wavelength. Things can be large or small relative to the wavelengths under consideration. This means that the propagation of low-frequency sounds can be quite different from the propagation of high-frequency sounds. Figure 2.25 shows the wavelengths for the range of audio frequencies from 20 Hz to 20 kHz.

Figure 2.25
Frequency and wavelength.

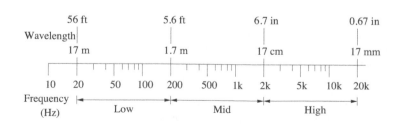

When a sound wave hits a large surface, part of it gets reflected, part of it gets absorbed, and some part of it gets transmitted. The amounts of reflection, absorption, and transmission depend on the acoustical properties of the surface (Cox and D'Antonio, 2016; Everest and Pohlmann, 2014). A hard surface reflects most of the sound, and a porous surface absorbs most of the sound. Sound transmission mainly depends on the thickness and the mass of the surface. The sound waves are reflected off a smooth surface in a similar way as light is reflected off a shiny surface; as you can see on the left side of figure 2.8. Just as with light, the angle of reflection equals the angle of incidence. A nonsmooth surface can scatter sound in multiple directions in a complicated fashion; similar to the diffuse reflection of light on the right side of figure 2.8. Scattering only happens when the surface irregularities are large compared to the wavelength of the sound.

An object that is much larger than the wavelength of sound reflects the sound. Directly behind a large object an acoustic shadow is formed. This is illustrated on the left of figure 2.26. Around the edges of the object the sound changes direction: it is bent around the object. This bending of sound is called *diffraction*. Because of diffraction sound can travel around corners and obstacles. By contrast, an object that is much smaller than the wavelength of sound allows sound to pass essentially undisturbed. The sound wave wraps around the edges and continues as if nothing has happened. This is illustrated on the right of figure 2.26 (Brown, 2015; Everest and Pohlmann, 2014). Low-frequency sounds (large wavelengths) will travel through your living room without being affected by most of the objects in the room. Only the larger furniture, such as couches, have sufficiently large dimensions to match the corresponding wavelengths. Hence, your subwoofers can be hidden behind all sorts of objects without their sound being obstructed. On the other hand, high-frequency sounds (small wavelengths) will not tolerate any object in their path. It is therefore important that you have an unimpeded view of the high-frequency drivers of your loudspeakers.

What happens when sound passes through an opening also depends on the relative size of the opening with respect to the wavelength of the sound. An opening that is large compared to the wavelength passes the sound, but on all sides of the opening an acoustic shadow is formed. This is illustrated on the left side of figure 2.27. Diffraction around the edges results in some part of the sound being radiated into the shadow area, but overall the sound continues its way as a beam that is about as wide as the opening. The opening is restricting the directions in which the sound continues its way. An opening that is small compared to the wavelength of sound acts as a point source that radiates sound in all forward directions. This is illustrated on the right side of figure 2.27.

Figure 2.26
Propagation of sound around objects. *On the left,* **the object is larger than the wavelength; it obstructs the propagation of sound.** *On the right,* **the object is smaller than the wavelength; it allows the sound to pass undisturbed.**

Figure 2.27
Propagation of sound through openings. *On the left,* **the opening is larger than the wavelength; it restricts the directions in which the sound propagates.** *On the right,* **the opening is smaller than the wavelength; it allows the sound to propagate in all forward directions.**

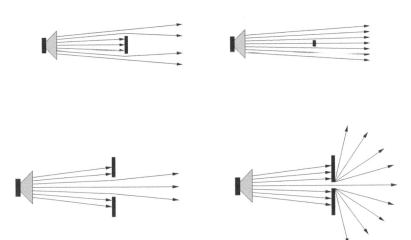

Figure 2.28
Loudspeaker directivity. As the frequency increases the loudspeaker becomes more directional.

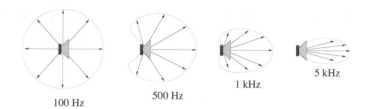

100 Hz   500 Hz   1 kHz   5 kHz

Figure 2.29
The inverse-square law. Doubling the distance from the sound source results in the sound spreading through four times the area.

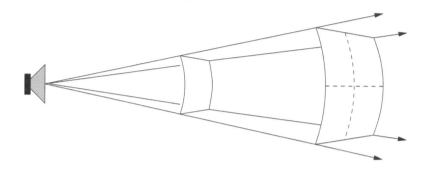

The directions into which sound is radiated by a loudspeaker diaphragm depend on the relative size of the diaphragm with respect to the wavelength of sound. The effect is similar to sound passing through an opening (Toole, 1988). At a low frequency where the loudspeaker diaphragm is small with respect to the wavelength, the sound is radiated in all directions. As the frequency increases the wavelength becomes smaller, and the diaphragm becomes large with respect to the wavelength. As a result the sound is beginning to beam ahead. Figure 2.28 shows how the *directivity* of a loudspeaker changes with frequency (Colloms, 2018; Newell and Holland, 2007). Most loudspeaker systems use a number of differently sized loudspeaker diaphragms. The larger ones only reproduce the low frequencies while the smaller ones are used for the mid and high frequencies. The size of the diaphragms is matched as much as possible to the wavelengths that they radiate in order to maintain wide dispersion across a large range of frequencies. Nevertheless, even loudspeakers with multiple diaphragms become more directional when the frequency increases. Musical instruments exhibit similar behavior. They radiate low-frequency sound in all directions, but the mid and high frequencies are only radiated in narrow beams. The direction and width of these beams depend on the size and orientation of the major sound radiating elements of the instrument (Toole, 1988).

As sound propagates over a certain distance, its intensity decreases. The inverse-square law dictates that sound intensity decreases with the square of the distance (Brown, 2015; Everest and Pohlmann, 2014). In other words, if the sound travels twice as far, its intensity is reduced four times. The reason is that the area over which the sound spreads has increased four times, as shown in figure 2.29. As a result of the inverse-square law, each doubling of the distance reduces the SPL of the sound by 6 dB (see table 2.5). Most conventional loudspeakers roughly follow the inverse-square law. In a room sound can sustain slightly higher intensity than this law predicts because of the multitude of reflections in the room (Toole, 1988).

2.3.4 Room Acoustics

The room in which you listen has a profound effect on the sound that you hear. The acoustics of a room determine how sound behaves in the room. The acoustics depend on the qualities of the room such as its size, shape, and furnishings. At low frequencies, the wavelengths of the sound are comparable to the dimensions of the room. At low frequencies, the room and the positions of the loudspeakers in the room determine the quality of the sound that you hear. By contrast, at high frequencies, the quality of the sound depends primarily on the quality of

BOX 2.2 THE SCHROEDER FREQUENCY

In large rooms with a strongly diffuse sound field, the transition from the region of wave acoustics to the region of ray acoustics is given by the Schroeder frequency (Schroeder, 1996):

$$f = 2000\sqrt{\frac{T}{V}}$$

where T is the *reverberation time* in seconds (see section 8.2), and V is the volume of the room in m³. However, in small rooms, such as a typical domestic living room, diffuse sound fields do not exist and therefore the calculated Schroeder frequency may be too low (Baskind and Polack, 2000). Typical rooms with a size of 30–70 m³ (ITU-R BS.1116, 2015) and a maximum reverberation time of 0.4 s have a Schroeder frequency between 200 and 300 Hz (Toole, 2018).

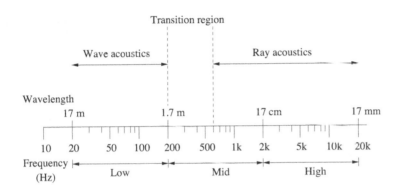

Figure 2.30
Frequency regions relevant for the behavior of sound in domestic rooms. At low frequencies sound behaves like waves, while at high frequencies sound behaves like rays. In the transition region both types of behavior occur.

the loudspeakers. In between these two extremes, there is a transition region where all these aspects matter (Toole, 2006, 2018). The transition region starts at the so-called *Schroeder frequency* (see box 2.2). In most domestic living rooms it lies somewhere between 200 and 300 Hz. As rooms get larger, the Schroeder frequency drops. Typically, the transition region ends somewhere between 400 and 600 Hz.

At mid and high frequencies above the transition region, sound behaves as a collection of rays. Each loudspeaker produces a beam of sound that reflects off the surfaces of the room and off all the objects in the room that are larger than the wavelength of sound. You should visualize sound at these frequencies as rays that bounce around the room changing direction each time they hit a large surface. At low frequencies below the transition region, sound behaves more like a wave that propagates between the surfaces of the room. These waves develop particular resonance patterns in the room that result in an uneven reproduction of low frequencies. At frequencies within the transition region, both ray and wave behavior can be observed. The three different frequency regions are shown in figure 2.30.

Mid and High Frequencies

Mid and high frequencies travel like rays through your room. These rays start at each loudspeaker as a diverging bundle of sound. Each time a ray encounters an object, a wall, the ceiling, or the floor, it is reflected and its direction is changed. At each reflection some of the sound energy is lost. In addition, sound energy is lost because of the inverse-square law. As the ray bounces around in the room its intensity decreases and it will eventually cease to exist (Everest and Pohlmann, 2014).

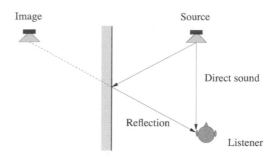

A reflecting surface acts as an acoustical mirror. The incoming ray is reflected such that the angle of incidence equals the angle of reflection. Another way to look at it is to imagine that the reflected ray originates from a mirror image of the sound source, as shown in figure 2.31. This mirror image lies at the same distance from the reflecting surface as the real sound source, but on the opposite side.

Reflected sound differs in a number of ways from the sound that reaches you directly. First, the reflected sound arrives from a different direction. Second, the reflected sound arrives somewhat later in time because it has to travel a larger distance than the direct sound. Third, the intensity of the reflected sound is lower, because it has traveled a larger distance (inverse-square law) and has lost some energy at the reflecting surface. How much energy depends on the acoustical properties of this surface. Finally, the frequency content of the reflected sound differs from the direct sound, because of two reasons. First, the reflecting surface might absorb some frequencies more than others. Second, the reflected sound has lost some high-frequency content because of the directivity of the loudspeaker (remember figure 2.28). The sound ray that causes the reflection leaves the loudspeaker at a much larger angle than the direct sound (see figure 2.31). At this angle high frequencies are attenuated because the loudspeaker becomes more directional as the frequency increases.

The total sound that you hear in a room is a complex mix of the direct sound from the loud-speakers and a succession of multiple reflections arriving from different directions. To make sense of this complicated sound field, it is useful to divide it into three components (EBU-Tech 3276, 1998; Toole, 2018):

1. *The direct sound:* sound that travels directly from the loudspeaker to the listener.
2. *The early reflections:* sounds that have been reflected only once on their way to the listener. Typically, these sounds consist of reflections from the floor, ceiling, and the sidewalls of the room. In most domestic rooms they arrive at the ears of the listener within 20 ms of the direct sound. To get an idea of how they propagate through the room, I show you the early reflections for only the left loudspeaker in figure 2.32. In practice of course all loudspeakers in your audio system generate such reflections.
3. *The late reflections:* sounds that have been reflected multiple times on their way to the listener. They arrive from all kinds of directions. Because they have been reflected multiple times and have traveled quite a distance, they have a much lower sound intensity than the direct sound and the early reflections. These low-level reflections are referred to as the *reverberation* of the room.

It is amazing how the human hearing system deals with the complicated sound field caused by all these reflections. In the majority of domestic rooms, we seem to be able to adapt to the sonic signature of the room and hear 'through the room'. Despite the fact that the sound field consists of many reflections, we can identify the sound source itself and even localize sounds appropriately (Toole, 2006, 2018). To a certain extent, we like reflections. They create a sense of space, and they enrich the timbre of the sound. The right amount of reflected sound adds a pleasant richness to musical sounds. It prevents the sound from appearing dull. It also improves speech intelligibility. However, too much reflected sound is a bad thing as it impairs the clarity and definition of the sound (Benade, 1985; Toole and Olive, 1988).

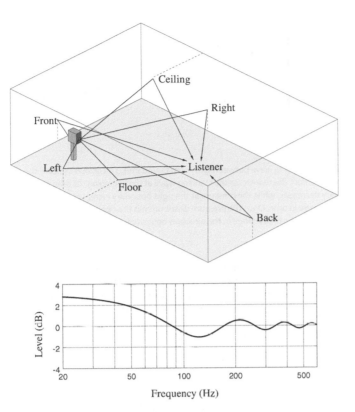

Figure 2.32
Early reflections from the left loudspeaker on their way to the listener in a rectangular room. The sound reflects on the front and back walls, the left and right sidewalls, the floor, and the ceiling.

Figure 2.33
The change in the low-frequency response of a loudspeaker when it is placed at a distance of 1 m (3.3 ft) from a room boundary.

The *precedence effect* (Haas, 1972; Wallach et al., 1973) explains why we hear a single sound instead of a sequence of separate sounds from the reflections in the room. Due to this effect, the sound that arrives first at our ears dominates the impression. All sounds that arrive within a certain short time interval are fused into one perception. For complex sounds, this fusion interval may be as long as 40 ms (Moore, 2013). All sound appears to come from the direction of first arrival. The reflections do not alter the localization of the sound. They only increase the loudness and add a pleasant sense of broadening to the sound source. Reflections that arrive after the fusion interval may be perceived as separate sounds if they are intense enough. If they arrive very late with respect to the first sound they are perceived as echoes.

Low Frequencies

Low frequencies travel through your room like waves. At these frequencies, the loudspeakers are small compared to the wavelengths and therefore they radiate sound in all directions. The floor, ceiling, and walls of the room reflect the sound such that the sound travels as a wave from one room boundary to the other. It is like dropping a small stone in a pool of water. The stone generates a circular wave that travels in all directions towards the edges of the pool where it is reflected.

The low-frequency sound emitted by a loudspeaker is modified by the reflection of the sound from nearby room boundaries. This change in low-frequency behavior is called the *adjacent-boundary effect* (Allison, 1974; Waterhouse, 1958). It occurs when the distance between the loudspeaker and the boundary is small compared to the wavelength of the sound emitted. The reflection from the nearby boundary interferes with the direct sound from the loudspeaker such that the power output of the loudspeaker changes at certain frequencies. The frequencies that are affected depend on the distance between the wall and the loudspeaker. Figure 2.33 shows the change in frequency response for a loudspeaker that is 1 m (3.3 ft) away from a wall. The frequencies below about 80 Hz are amplified by the room, and above 80 Hz the frequency response becomes irregular with peaks and dips at different frequencies. In a typical room, things are more complicated since the loudspeaker is close to three different room boundaries:

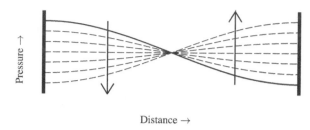

Figure 2.34
Pressure changes for a standing wave between two room boundaries. The *arrows* indicate the change in pressure with time. The *thick line* represents the standing wave at the beginning. As time progresses the pressure at the left boundary decreases, while the pressure at the right boundary increases. The *broken lines* are snapshots of the pressure in time. Once the pressure distribution is the opposite of the thick line, the process reverses and the pressure starts to change in the direction opposite to the directions indicated by the arrows.

the floor, the wall behind the loudspeaker, and the nearest sidewall. All these boundaries interfere with the low-frequency response of the loudspeaker. For an optimal reproduction of low-frequency sound it is important to carefully position the loudspeaker with respect to these nearby boundaries. I give you some practical advice on this topic in section 4.2.5. For now it suffices to remember that the low-frequency response is altered by the position of the loudspeakers with respect to the nearest room boundaries.

Certain low frequencies can resonate in the room (Everest and Pohlmann, 2014; Gilford, 1959; Jones, 2015b). This works as follows: the sound from the loudspeaker travels to a room boundary where it is reflected and travels back in the opposite direction. At the other side of the room it reaches the opposite room boundary and gets reflected again. The sound travels back and forth between the two opposing room boundaries. When the distance between these boundaries equals half a wavelength, each subsequent reflection adds energy to the original wave and a *resonance* develops. The sound reflects back on itself creating a *standing wave* as shown in figure 2.34. At the room boundaries the maximum variation in air pressure occurs, while at a point exactly between the two walls the pressure is always zero. This point is called a *null*. At a null the round trip to the boundary and back equals half a wavelength. As a result the reflection always arrives with a 180° phase shift and the sound at the null is canceled. The positions of the null and the points of maximum pressure variations do not change in time, they stay at the same position between the boundaries. Their positions define the 'standing' wave. What does change in time is the pressure at the points other than the nulls. This pressure goes up and down, and follows the familiar sine function.

Between two opposing room boundaries, multiple resonances can develop. The first resonance occurs at a frequency $f_o = c/2l$, where l equals the distance between the two boundaries and c is the speed of sound (344 m/s). Subsequent resonances are multiples of this fundamental frequency f_o, that is, resonances occur at $2f_o$, $3f_o$, $4f_o$, and so forth. Figure 2.35 shows the pressure distribution of the first four resonances between two walls that are 4 m (13 ft) apart. For each resonance frequency (43 Hz, 86 Hz, 129 Hz, and 172 Hz) the pressure maximums and nulls occur at quarter-wavelength distances (2 m, 1 m, 0.67 m, and 0.5 m, respectively). Note that at each null the polarity of the pressure variation changes. When the pressure goes up on one side of the null, the pressure goes down on the other side. All resonances have maximum pressure variation close to the room boundaries. This means that at the room boundaries the low-frequency resonances will be at their loudest. You can experience this yourself. Play a piece of music with a lot of low frequencies. Listen how the amount of low-frequency content changes while you move closer to one of the walls and away from it.

Every room has multiple resonance modes. In a rectangular shaped room, they occur between the ceiling and the floor, between the front and the back wall, and between the two sidewalls.

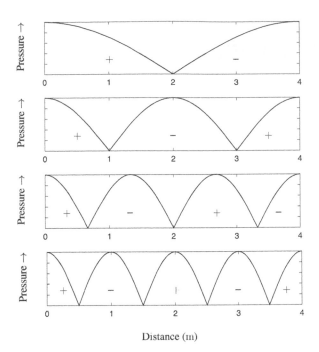

Distance (m)

Figure 2.35
Pressure distribution for standing waves between two walls that are 4 m (13 ft) apart. *From top to bottom*: 43 Hz, 86 Hz, 129 Hz, and 172 Hz. The + and − indicate that at each null the pressure variation changes polarity.

Each pair of room boundaries has a particular sequence of room resonances. To complicate matters further, room resonances can also result from sound waves retracing their own path between four or even all six boundaries. Not only rectangular rooms have resonance modes; rooms of all shapes have them.

Room resonances distort the reproduction of low frequencies. Some low frequencies are reproduced much louder than others. The effect strongly depends on the position of the listener in the room. The resonance frequencies are at their loudest at the pressure maximums, while at a null they might almost disappear. In practice, complete cancellation at a null never happens, because the walls are not perfect reflectors. Room resonances can create large peaks and dips in the low-frequency response. Fluctuations up to 30 dB are not uncommon (Everest and Pohlmann, 2014). Resonances also store energy in the room, such that when the sound source is turned off, the room may slowly release this energy. As a result the bass may lack specific pitch definition and sound as a low rumble. At low frequencies the room significantly changes the sound from the loudspeakers. It is not easy to achieve high-quality sound reproduction at these frequencies. As you will discover in Chapters 4 and 8 it requires the use of multiple subwoofers and careful positioning of all loudspeakers and all listeners in the room.

3

OPTIMAL REPRODUCTION

What is optimal reproduction of audio and video? In the introduction to this book I stated that optimal reproduction means that you achieve the best possible quality with your equipment in your living room. Perfect reproduction is not the aim here. Perfect reproduction would mean that the audio and video is reproduced in exactly the same way as in the studio control room. As you know from the previous chapter, the room in which you listen has an enormous influence on the sound quality. Your living room will differ from the studio control room and thus sound different. As a result, perfect reproduction is not possible. The good news is that the rooms do not need to sound identical, as long as certain quality aspects of the reproduction stay the same. These quality aspects are the topic of this chapter.

Sound quality is what we hear, not what we measure. It is a multidimensional phenomenon: it is made up of several aspects (Gabrielsson and Lindström, 1985; Letowski, 1989; Rumsey, 2002). Measurements alone cannot completely describe all relevant aspects of sound quality. Although measurements are helpful, and frequently used, they are limited in their description. Therefore, sound quality is often described in words. A whole family of descriptive terms exists. I encourage you to get acquainted with these terms, because they allow you to recognize and describe the different aspects of sound quality more easily. Learning the appropriate terms sharpens your ability to hear subtle differences. At first, learning to recognize the different nuances might seem daunting, but I ensure you that it is well worth the effort. Once you know what to listen for, your appreciation for sound that is well reproduced increases. It is quite similar to learning how to taste wine and describe the different nuances in flavor.

Image quality also consists of several aspects. A big difference with sound quality is that most aspects of image quality can be easily measured. Therefore, describing image quality is less complicated than describing sound quality. In addition, the viewing environment has less influence on the reproduction quality and is much easier to control.

When you reproduce source material that contains both audio and video, the image quality influences the perceived sound quality and vice versa. There is no dominant factor. Audio quality and video quality are equally important to the overall audiovisual quality (Pinson et al., 2011). Therefore, to have an optimal audiovisual experience, you need to optimally reproduce both audio and video.

3.1 Film and Video Production

It is of no use to talk about the quality of video reproduction without knowing exactly what you are trying to reproduce. Therefore, let's take a quick look at how video material is being created. The production of film or video is an art form. The artist creates a sequence of visual images that are meant to stimulate your brain, tell a story, and evoke emotions. Often the images are meant to represent a scene that you can encounter in real life, but not necessarily so. Many feature films contain special effects that are created using computer-generated imagery (CGI). Computer graphics animation can even be used to create an entire movie that looks like a comic book. These examples illustrate that a sequence of video images does not need to have a basis in reality. Therefore, the quality of video reproduction is not judged by how well it represents real life. What matters is how well it represents the artist's vision.

Most video material is being recorded using a camera that is pointed at a real-life scene. The camera projects the three-dimensional world onto a two-dimensional canvas. The artist decides which part of this three-dimensional scene is being recorded. The position and movement of the camera determines how the scene is framed. The type of lens on the camera can make the angle of view wide or narrow. The camera can be focused on a small part of the scene, or the whole scene can be rendered sharp. Part of the scene can be lit, and part of it can be dark. These are all examples of artistic decisions made by the filmmakers.

Today, video material is either shot using a digital camera or a traditional film camera. Digital is more common, but some filmmakers stick to film because they like its distinctive look. Anyway, when film is used, it is almost always converted into digital video for postproduction (EDCF, 2005). The main reason is that digital files are much easier when it comes to editing scenes, adding visual effects, and applying color correction. The days of cutting film with scissors and sticking scenes together with tape are over (Piper, 2008). Digital processing is cheap and offers many possibilities. Digital video can be processed and copied without any loss of quality. Quite a contrast with the old-fashioned methods of optical copying where each copy of the film loses some color and sharpness.

A digital camera operates very similar to the human eye. A lens projects an image onto a sensor that responds to light. The sensor consists of a grid of millions of tiny photoreceptors. Each photoreceptor generates an electrical signal that is proportional to the amount of light that it receives. To capture color, the sensor is covered with a color filter array that consists of alternating red, green, and blue filters. The filters are chosen such that they correspond to the three primary colors that human color vision is based on (recall the trichromacy property described in section 2.2.2). The most commonly used color filter array is the *Bayer filter* (Bayer, 1976). As shown in figure 3.1, this array has twice as many green filters as there are red and blue filters. The reason is that the human eye is most sensitive to green light, because both the *L*- and *M*-type cones respond to the corresponding wavelengths (see figure 2.12).

Shooting is only a small part of making a movie or video program. A lot of work is done after the actual shooting in postproduction. The pieces of the movie are put together. Scenes are edited. Computer-generated imagery and sound are added. The postproduction imaging workflow is completely digital. Today, a lot of background interiors and exteriors are generated using computers and merged with separate shots of the actors. In addition, every frame is digitally manipulated to fine-tune the lighting, contrast, and color. The filmmaker uses such manipulations as an important tool to affect the emotional response of the audience and to communicate a sense of drama (Piper, 2008).

The filmmaker's attention to detail in lighting, contrast, and color would be completely lost if video displays and viewing conditions vary too much. The filmmaker would be chasing a moving target during postproduction, because the movie would look different in different theaters. The audience would never see the subtle nuances. To ensure the same look everywhere, video standards have been introduced. The video material will look the same in every studio and theater as long as these standards are being adhered to. The most important standards for digital cinema and digital video come from organizations, such as the Digital Cinema Initiatives (dcimovies.com), the Society of Motion Picture and Television Engineers (smpte.org), the International Telecommunications Union (itu.int), and the European Broadcasting Union

R	G	R	G
G	B	G	B
R	G	R	G
G	B	G	B

Figure 3.1
Alternating red *(R)*, green *(G)*, and blue *(B)* filters in a color filter array of a digital imaging sensor. This particular pattern is called a Bayer filter.

(ebu.ch). These standards play a major role in the remainder of this book, because they are essential if you want to achieve optimal reproduction of video at home.

3.2 Image Quality

Image quality is the term used to describe how good or how bad the image from your video system looks. Image quality has several aspects, which are shown in table 3.1. I have arranged the aspects into three different groups: *tonal quality, detail,* and *spatial quality.* Think of all the aspects as the language that you use to describe video quality. In the chapters that follow I use this language to describe how changes in equipment and setup will influence reproduction quality. But first, let's look at what these image-quality aspects actually mean.

One warning before we start: image quality strongly depends on the source material. Obviously, if you are watching a black-and-white movie, your video system is not going to show you any color. It would not make sense to conclude that reproduction quality has suffered because of this. It was the intention of the filmmaker not to use any color in the movie. Film-makers can deteriorate image quality on purpose to achieve a particular artistic goal. For example, they can introduce a blue color cast to achieve a cool setting or intentionally blur the image to achieve a dreamlike quality. Do not confuse such artistic manipulation with reproduction quality. Remember that optimal reproduction is all about optimally communicating the artist's vision. There is no need to correct for artistic use of image quality. However, in all other cases your video system must be able to achieve the best possible image quality. Not every scene from the movie should look blue and blurry, only the ones that the filmmaker intentionally created that way.

3.2.1 Tonal Quality

The tonal quality of an image is about luminance and color. In photography and video, the word 'tone' is used to describe a particular shade of gray or a particular color. Tonal quality is one of the most important descriptors of image quality (Brennesholtz and Stupp, 2008).

Brightness

Brightness is the subjective impression of luminance. The apparent brightness of the video display depends on the amount of light it produces. How bright is bright enough? How much luminance do you need? Well, it depends on the viewing conditions. A display that produces too much light will look dazzling in a dark room. You will be squinting all the time, which is not very comfortable. Projection displays are designed to be viewed in a dark environment, while flat-panel displays are designed to be viewed in an environment that has some remaining ambient light. Consequently, comfortable viewing requires a projection display to be less bright than a flat-panel display. However, if the display is not bright enough, it will look washed out, lacking punch and definition. So clearly, there must be an optimal choice.

Table 3.1 Overview of image-quality aspects.

Tonal Quality	Brightness
	Contrast
	Tonal definition
	Grayscale accuracy
	Color accuracy
Detail	Spatial definition
	Temporal definition
Spatial Quality	Uniformity
	Geometry

Table 3.2 **Optimal luminance and contrast values for projection and flat-panel displays. Based on specifications for Digital Cinema (DCI, 2008; SMPTE RP 431–2, 2011) and HDTV (EBU-Tech 3320, 2017; ITU, 2009b; VESA, 2017). For SDR video the peak luminance is measured with a 100% white screen. For HDR video the peak luminance is measured with a 100% white patch that covers only 10% of the entire screen area (on a black background).**

	Target	Tolerance
Peak luminance projection (full-screen)	48 cd/m²	38–58 cd/m²
Peak luminance SDR flat panel (full-screen)	100 cd/m²	70–200 cd/m²
Peak luminance HDR flat panel (10% patch)	1000 cd/m²	600–10 000 cd/m²
Minimal sequential contrast (full-screen)	2000:1	500:1
Minimal simultaneous contrast SDR	200:1	100:1
Minimal simultaneous contrast HDR	10 000:1	6000:1

BOX 3.1 LUMINANCE

Luminance L_v is measured in cd/m² in the SI measurement system. In North America the foot-Lambert (ftL) is often used instead. They are related as follows:

$$1\,\text{ftL} = \frac{1}{\pi}\,\text{cd}/\text{ft}^2 = \frac{1}{\pi(0.3048)^2}\,\text{cd}/\text{m}^2 \approx 3.42626\,\text{cd}/\text{m}^2$$

Table 3.2 shows the recommended peak luminance values for both projection and flat-panel displays. The listed target value is the one you should aim for, but everything will still be fine as long as you stay within the tolerance specified. In section 4.3.1 I describe how you can measure the peak luminance of your display. Luminance is measured in candela per square meter (cd/m²), also called *nits*. In North America the foot-Lambert (ftL) is often used instead of cd/m². The relation between the two is explained in box 3.1.

High dynamic range (HDR) flat-panel displays are capable of creating a much higher peak luminance than *standard dynamic range* (SDR) displays. The maximum luminance capability of an HDR display is only used for small areas of highlights in the picture, the so-called specular highlights (see also section 2.2.1). The recommended luminance for these highlights is 1000 cd/m² or more (ITU-R BT.2100, 2017). The average luminance of a scene on an HDR display will be about the same as on an SDR display, only the highlights will be much brighter. Convincing HDR images are not really possible on a projection system. Currently, projectors intended for home theater use lack the increased luminance and contrast ranges.

Contrast

Contrast ratio is the ratio between the highest and lowest luminance that a display delivers. Contrast has an enormous influence on the perceived image quality. A high-contrast image evokes a sense of realism, and the image may even appear to be sharper (Poynton, 2012). Contrast ratio can be computed by dividing the luminance of a white area (L_{white}) by the luminance of a black area (L_{black}):

$$C = \frac{L_{white}}{L_{black}}$$

The luminance of a white area corresponds to the peak luminance of the display. Since you should fix the peak luminance at the recommended value from table 3.2, contrast mainly depends on the luminance of black.

Good contrast is achieved if the black tones in the image are really black and not gray. Deep and full blacks result in an overall image that looks crisp and punchy. Poor blacks, on the other hand, result in a washed-out picture that looks flat and bland. In practice, many factors conspire to increase the luminance of black, thereby impairing contrast. The black tones will never be completely black. Light from the lightest areas of the screen will spill into the darker areas, due to the scattering of the light in the display and the reflection of light back to the screen from objects and surfaces in the room. In addition, the ambient light in the room lightens up the areas on the screen that should remain black (Poynton, 2012).

Contrast can be specified in different ways. It is important to distinguish between *sequential contrast* and *simultaneous contrast*. To determine sequential contrast (also called *on/off contrast*) L_{white} is measured by displaying a completely white image and L_{black} is measured by displaying a completely black image. Simultaneous contrast (also called *checkerboard* or *intraframe contrast*) is measured using a checkerboard pattern of white and black patches. The most commonly used pattern is the ANSI pattern shown in figure 3.2. Simultaneous contrast is obtained by dividing the sum of the luminances of the eight white patches by the sum of the luminances of the eight black patches. Simultaneous contrast is typically much lower than sequential contrast, because the light emitted by the white patches increases the luminance of the black patches.

Both sequential contrast and simultaneous contrast should be as high as possible. The minimum values that you should aim for are given in table 3.2. The target values indicate excellent performance. But I should warn you that they can be difficult to achieve. Achievable contrast very much depends on the viewing environment. You should at least try to achieve the values given as tolerance. In section 4.3.1 I provide some practical tips to increase contrast in your viewing environment.

Tonal Definition

Tonal definition is used to describe the number of different tones you can distinguish in a video image. Between the extremes of black and white lie several shades of gray. In real life an almost infinite number of variations exists. But in digital video the number of shades that can be represented is limited. A sufficient number of shades is needed to accurately represent areas in the picture with gradual changes in luminance. The difference between adjacent shades should be small enough, such that the area of gradual change is perceived as a smooth transition from dark to light. When an insufficient number of shades is used, such areas will not be smooth, but will contain stripes and bands of different luminance steps. This effect is called *banding* or *contouring*. Figure 3.3 shows how a gradual change from black to white will look if only 11 shades of gray (including black and white) are used. A similar effect can occur in areas of gradual color changes; this effect is called *posterization*. Tonal definition in relation to gradual color changes is also referred to as *color resolution* (Bodrogi and Khanh, 2012).

Figure 3.2
ANSI checkerboard pattern
for measuring simultaneous
contrast.

Figure 3.3
Example of banding when
only 11 shades of gray are
used to represent a gradual
transition from black to
white.

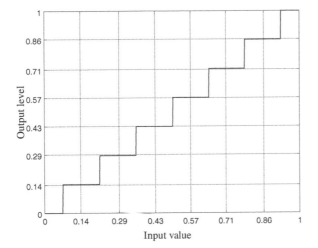

Figure 3.4
Quantization with eight
levels. The output of the
quantization process is
obtained by rounding the
input values to the nearest
quantization level.

Tonal definition depends on the number of different levels that the red, green, and blue components of a pixel can obtain. Recall that the luminance and color of each pixel in a digital video image arise from a mixture of red, green, and blue (see table 2.4). Gray is obtained when the intensity of the red, green, and blue components are equal; different shades of gray are obtained by varying this intensity; colors are obtained by mixing unequal amounts of red, green, and blue.

The number of different levels in the red, green, and blue components of digital video are the result of a process called *quantization*. This process works as follows: the sensor in a digital camera converts the light that falls onto the red, green, and blue subpixels (see figure 3.1) into an electrical voltage that takes a range of continuous values. Let's say any value between 0 and 1 volt. Quantization converts this range of values into a finite number of fixed levels. Figure 3.4 shows an example with eight quantization levels. Every input value is rounded to the nearest quantization level. The output always consists of one of the eight quantization levels; the values in between are not used. The difference between the input and the output of the quantization process is called the *quantization error*. The maximum error occurs when the input value lies exactly in the middle of two quantization levels. The quantization error can be made smaller by increasing the number of quantization levels (Pohlmann, 2011; Poynton, 2012). Summing up, tonal definition depends on the number of quantization levels used for the red, green, and blue components.

In digital video the different quantization levels are coded using *bits*. A bit is a binary digit that can have either the value of 0 or 1. It is the basic unit of information in the digital world. Bits can be stored and processed as a voltage that is either 'on' or 'off'. The precise value of this voltage doesn't matter as long as the distinction between 'on' and 'off' can still be made. Consequently, storing information as bits instead of precise voltages yields a certain degree of immunity against disturbances and distortions. Table 3.3 illustrates how eight quantization levels can be represented using a binary representation of 3 bits. In general, the number of levels that you can represent with n bits equals 2^n. The number of bits used to represent the

Table 3.3 Example of a 3-bit binary representation. Note that it is customary to indicate the first quantization level as 'level 0' instead of 'level 1'.

Level	Binary
0	000
1	001
2	010
3	011
4	100
5	101
6	110
7	111

Figure 3.5
Quantization with
3 bits. The output of the
quantization process is
obtained by rounding the
input values to the nearest
quantization level and
encoding this level with
3 bits.

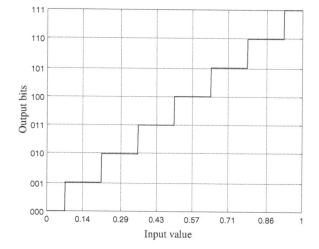

quantization levels is called the *word length* or *bit depth*. Figure 3.5 shows how quantization can be used to convert a continuous value between 0 and 1 into a digital representation with 3 bits. Note that the only difference between figure 3.5 and figure 3.4 are the labels used on the output axis.

How many bits are needed to encode digital video? For high-quality video a minimum of 8 bits per color component is needed (Poynton, 2012). DVD and HD video such as Blu-ray disc and HDTV uses 8 bits (ITU-R BT.709, 2015). UHD video uses 10 or 12 bits (ITU-R BT.2020, 2015). More bits means better tonal definition. Table 3.4 shows the number of quantization levels per color component and the total number of colors. With 8 bits, the number of levels in each component is $2^8 = 256$; so the total number of colors that can be represented becomes $256 \times 256 \times 256 \approx 16.8 \times 10^6$.

Grayscale Accuracy

Grayscale accuracy deals with the distribution of the different shades of gray between black and white. In SDR content grayscale accuracy mainly depends on a display control called *gamma*. This control determines how 'fast' the shades change when you move across the entire scale from black to white. Figure 3.6 contains an example grayscale that consists of 11 shades of gray. The figure shows what happens to this grayscale for a high and low setting of gamma. A gamma that is too low results in a washed-out picture, because the darker gray

Table 3.4 The number of bits, quantization levels, and colors in digital video.

Number of Bits	Number of Levels	Total Number of Colors
8	256	16.8×10^6
10	1024	1073.7×10^6
12	4096	68719.4×10^6

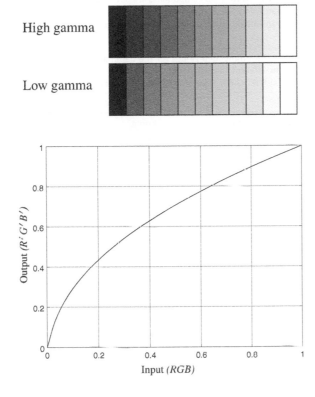

High gamma

Low gamma

Figure 3.6
Comparison of a high and low gamma setting. The choice of gamma determines the distribution of the different shades of gray between black and white. The largest effect occurs in the darkest levels.

Figure 3.7
The SDR opto-electronic transfer function. The components R, G, and B are transformed into R', G', and B', respectively.

tones are not dark enough. A gamma that is too high results in loss of detail in the darker parts of the image, because the darker gray tones are too dark to be clearly distinguishable. This is called *black crush* or *loss of shadow detail*. At the correct setting of gamma the darker gray tones are as dark as possible without impairing shadow detail. It is a delicate balance.

Gamma is related to the way the red, green, and blue components are represented in digital video. Before quantization, the *RGB* components are transformed with a nonlinear function that is related to the way human vision distinguishes different luminance levels. This function is known as the *opto-electronic transfer function*. This transfer function ensures that the quantization error is equally perceptible across the entire grayscale. As a result the quantization error has minimum perceptual impact on the video quality. Without this transfer function, the quantization error would be more perceptible in the darker regions compared to the lighter regions. The transfer function for SDR video (ITU-R BT.709, 2015) is shown in figure 3.7. Although the opto-electronic transfer function has a positive effect on the perception of the quantization error, it also distorts the distribution of shades across the entire grayscale. Therefore, to obtain an accurate reproduction of the grayscale, the effect of the opto-electronic transfer function needs to be reversed before the video is displayed. Gamma correction takes care of this. The entire processing chain is shown in figure 3.8.

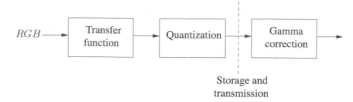

Storage and
transmission

Figure 3.8
SDR video processing chain. First, the *RGB* components are put through the opto-electronic transfer function that mimics human vision. Next, they are quantized and become digital. The resulting digital video can be stored and transmitted, but it needs to be gamma corrected before being displayed.

Table 3.5 Recommended values for the gamma setting (γ) of a video display in different ambient light conditions (DCI, 2008; EBU-Tech 3320, 2017; ITU-R BT.1886, 2011; SMPTE RP 431–2, 2011; Poynton, 2012).

Application	Ambient Light	γ
Digital photography	bright	2.2
HDTV	dim	2.4
Digital cinema	dark	2.6

Figure 3.9
Gamma curves that map the input video signal to relative luminance values. The uppermost curve represents the inverse of the SDR opto-electronic transfer function. The other three curves, from top to bottom, represent $\gamma = 2.2$, $\gamma = 2.4$, and $\gamma = 2.6$.

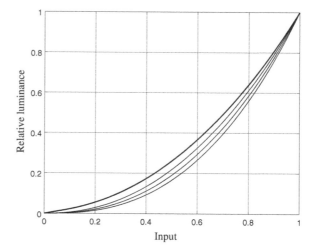

Contrary to what you might expect, gamma correction is not the exact inverse of the opto-electronic transfer function. The optimal gamma correction depends on the ambient light level. If the video would be reproduced in a dark or dim environment using the exact inverse of the transfer function, the image would appear to lack contrast. This is the perceptual effect of simultaneous brightness contrast in action (see figure 2.11): the luminance level of the environment influences the appearance of the picture (Poynton, 2012). Therefore, the correct setting of the gamma control depends on the viewing conditions. The darker the environment the higher the gamma setting. The recommended settings are given in table 3.5. The gamma symbol γ represents a numerical parameter that describes the nonlinearity of gamma correction. In figure 3.9 the different settings of γ are compared to the inverse of the opto-electronic transfer function.

The rendering of luminance in HDR content material is not based on gamma curves. HDR content has a much larger luminance range (1000 cd/m² or more) than SDR content has (less than 100 cd/m²). This expanded range of luminance values calls for a different mapping than

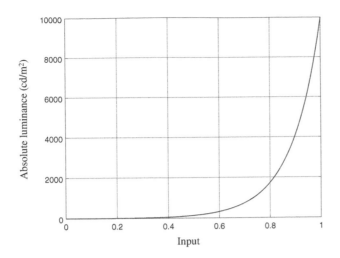

Figure 3.10
SMPTE ST 2084 transfer function for HDR video that maps the input video signal to absolute luminance values (ITU-R BT.2100, 2017; SMPTE ST 2084, 2014). The maximum luminance value is 10 000 cd/m², which is much more than current display technology is capable of, but it makes the HDR transfer function future proof.

the SDR opto-electronic transfer function, because the SDR transfer function is not a good approximation for human vision over an extended range of luminance values (ITU-R BT.2100, 2017).

There are currently a number of different standards for rendering HDR content; the two most common ones are *HDR10* (cta.tech) and *Dolby Vision* (dolby.com). HDR10 support is mandatory for UHD Blu-ray players, while Dolby Vision is optional (BDA, 2015b). Both HDR10 and Dolby Vision use the SMPTE ST 2084 (2014) transfer function, also known as the PQ (Perceptual Quantization) transfer function (ITU-R BT.2100, 2017), instead of the gamma-based curves from figure 3.9. Figure 3.10 shows this HDR transfer function. An important difference is that the gamma curve maps *relative luminance* while the SMPTE ST 2084 transfer function maps *absolute luminance*. Let me explain. The maximum output value of the video signal after gamma correction represents relative luminance. This allows each display to map this maximum value to its own peak luminance value. By contrast, the SMPTE ST 2084 transfer function defines how each value of the video signal should be mapped to absolute luminance, regardless of the peak luminance capabilities of the display. Since not every display has the same peak luminance capabilities, both HDR10 and Dolby Vision use static metadata embedded in the digital video signal that describes the absolute luminance range of the content. This metadata is used by the display to perceptually map the content to its own, possibly limited, luminance range. In addition to the static metadata, Dolby Vision uses dynamic metadata that is content dependent and is allowed to vary from frame to frame.

Another standard for rendering HDR is *HLG* or *Hybrid-Log-Gamma* (Borer and Cotton, 2015; ITU-R BT.2100, 2017). It has been jointly developed by BBC and NHK for broadcasting purposes. Its transfer function is shown in figure 3.11 and consists of a hybrid curve that applies a standard gamma curve to the lower luminance range and a logarithmic curve for the higher luminance values. Similar to the gamma curve, HLG maps relative luminance; thus the maximum output value of the video signal is mapped to the peak luminance output of the display. HLG doesn't use metadata and retains playback compatibility with regular SDR displays.

Color Accuracy

Colors should be accurately reproduced. They should be displayed in the way that the filmmaker intended. Vibrant colors should stay vibrant; muted colors should stay muted. The colors need to have the right amount of brightness, tint, and saturation. They should look natural. Problems with color reproduction often show up in skin tones. We are very sensitive

Figure 3.11
HLG transfer function
for HDR video that maps
the input video signal to
relative luminance values
(Borer and Cotton, 2015;
ITU-R BT.2100, 2017).

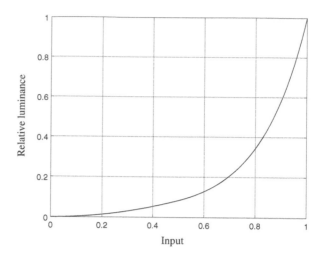

Table 3.6 Chromaticity coordinates x and y according to ITU-R BT.709 (2015), DCI (2008), and ITU-R BT.2020 (2015).

	Rec. 709		DCI-P3		Rec. 2020	
	x	y	x	y	x	y
Red	0.640	0.330	0.680	0.320	0.708	0.292
Green	0.300	0.600	0.265	0.690	0.170	0.797
Blue	0.150	0.060	0.150	0.060	0.131	0.046

to the accurate reproduction of skin tones. Too much color saturation will make people look flush. Problems with tint will make people look either too pink or too green.

Accurate color reproduction requires that the primary colors R, G, and B are consistent among displays. The standard primary colors used for SD video are specified in Rec. 601 (ITU-R BT.601, 2011), those for HD video in Rec. 709 (ITU-R BT.709, 2015), and those for UHD video in Rec. 2020 (ITU-R BT.2020, 2015). You do not need to worry about this difference, because your video display will automatically adapt its primary colors depending on the source material. Table 3.6 and figure 3.12 compare the HD and UHD color specifications (see section 2.2.2). They also show the DCI-P3 color gamut, which is the standard color gamut used in digital cinema (DCI, 2008; SMPTE RP 431–2, 2011). This gamut is considered to be an intermediate step between the Rec. 709 and Rec. 2020 gamuts and is currently the largest color gamut that modern video displays are able to reproduce. For the time being, we lack the technology to reproduce all the colors of the Rec. 2020 specification. Note that the triangle for UHD video is much larger than the other two triangles. Consequently, UHD video can reproduce many more colors. Be warned though that the perceptual difference is much smaller than figure 3.12 suggests. The main reason is that the CIE diagram is not perceptually uniform: the distance between two points that can just be distinguished in color is much larger in the top half of the diagram than it is in the bottom half.

Accurate color reproduction not only depends on the accuracy of the primary colors, but also on the *white point* that is used. The white point is the color of gray. It is the color that you get when the three components R, G, and B contribute equally (Poynton, 2012). Having the correct white point is perceptually important, because your visual system compares all other colors with this reference. The recommended white point is the same for all video formats.

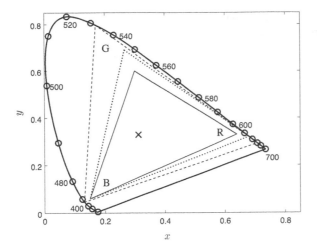

Figure 3.12
CIE 1931 chromaticity
diagram with Rec. 709
gamut *(smallest triangle,
solid lines)*, DCI-P3 gamut
*(middle triangle, dotted
lines)*, and Rec. 2020 gamut
*(largest triangle, broken
lines)*. The *cross* in the
middle corresponds to the
D65 white point.

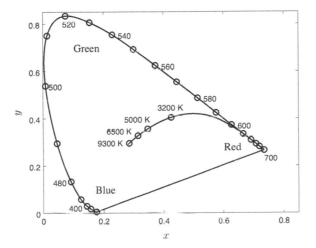

Figure 3.13
CIE 1931 chromaticity
diagram with color
temperature.

Its chromaticity coordinates are $x = 0.3127$ and $y = 0.3290$. It is indicated by the cross in figure 3.12.

An alternative way to specify the white point is by *color temperature*. Displays often have a configuration setting that you can use to change color temperature. The color temperature relates to the temperature of a hot object that emits light. It is expressed in kelvin (K). Figure 3.13 shows the chromaticity coordinates for different values of the color temperature. As the temperature increases the white point shifts from red to blue as indicated by the curve. This might be a bit counterintuitive, because in artistic terms we associate red with higher temperatures and blue with lower temperatures. Color temperature works the other way around. The recommended white point ($x = 0.3127$ and $y = 0.3290$) corresponds to a color temperature of 6500 K. This color temperature approximates daylight at noon. It is also referred to as CIE standard illuminant D65 (CIE S 014–2, 2006).

3.2.2 Detail

Another image-quality aspect is *detail*. High-quality video shows a lot of detail. The picture appears sharp. It has distinct lines, crisp textures, and intricate details. It is neither soft nor blurry. It has excellent resolution, clarity, and definition. Detailed pictures impart a real-life quality to the reproduced video. Plenty of detail should be visible in static objects and scenery, but also in moving objects and during camera movements. To distinguish between these two situations, I use the term *spatial definition* to describe the amount of detail in a static picture and the term *temporal definition* to describe apparent detail in pictures that change quickly over time.

Spatial Definition

The more pixels, the more detail you can see in an image. Therefore, spatial definition mainly depends on the number of pixels. The number of pixels in a video image is called the *resolution*. To obtain excellent spatial definition you need a high-resolution video display and high-resolution source material. The space between adjacent pixels in a video display should be as small as possible to avoid the *screen-door effect*. If the space between the pixels is not small enough, the individual pixels become visible and the picture looks as if you are viewing it through a mesh screen (Brennesholtz and Stupp, 2008).

Spatial definition suffers if there are local faults in the picture. Such faults or artifacts come in different varieties and have different causes (Bodrogi and Khanh, 2012; Brennesholtz and Stupp, 2008; Taylor et al., 2009):

- *Pixel faults:* Individual pixels of a video display can fail to operate, causing a pixel to remain black or constantly lit.
- *Moiré pattern:* A conspicuous pattern can result on the display if the resolution of the display is too low to render a certain pattern of closely spaced lines: the grid of pixels interferes with the pattern in the source material and a new pattern emerges.
- *Blocking:* Small square patches of color that obscure detail in the picture can result from poor video compression or errors during the transmission of digital video.
- *Ringing:* Ripples and halos around sharp edges occur when too much edge enhancement or sharpening is applied during video processing.
- *Mosquito noise:* Small dots that are buzzing around high-contrast edges can result from poor video compression. They look somewhat like a swarm of insects; hence the name. They are often only visible in solid areas of the picture.
- *Film grain:* A layer of low-contrast grain can occur if the source material has been shot on photographic film. It is often only visible in the darker parts of the picture. Some people appreciate a small amount of film grain, because it gives the picture a certain cinematic appeal.

Temporal Definition

Digital video is made up of a sequence of still images. The number of images per second determines the temporal definition of the video. Too few images per second result in the perception of *flicker* and in the breakdown of the illusion of viewing a continuous moving image. A sufficiently fast rate of subsequent images is needed to give you the impression of smooth movement and correct timing.

The perception of flicker depends on the frame rate, the size of the screen, and the luminance of the screen. The *frame rate* is the number of images per second. The larger and the brighter the screen, the higher the frame rate must be to avoid visible flicker (Bodrogi and Khanh, 2012; Poynton, 2012). When television systems were first introduced, they derived their frame rate from the AC power line frequency: 50 Hz in Europe and 60 Hz in North America. These frame rates are sufficiently fast to avoid visible flicker (Salmon et al., 2011). Computer screens are usually much brighter than a television screen, because they are used in a bright office environment. A frame rate of more than 70 Hz might be needed to avoid flicker for such a bright screen. On the other hand, in the darkness of the cinema were the screen luminance is limited, a frame rate of 48 Hz is sufficient to avoid flicker. In the early days of film the 48 Hz rate was considered too expensive: it used too much film material. Therefore, films were shot with a 24 Hz frame rate and upon projection every frame was shown twice to achieve a display rate of 48 Hz (Brennesholtz and Stupp, 2008; Poynton, 2012).

Showing a single frame multiple times increases the frame rate but can result in a motion artifact called *judder*. It introduces a certain jerkiness to the image: the movements of objects and the camera appear to be discontinuous instead of smooth (Brennesholtz and Stupp, 2008). More advanced techniques called *frame rate interpolation* can be used to increase the frame rate and at the same time reduce judder effects (see section 6.3.5). To avoid judder in the cinema, experienced filmmakers avoid rapid camera movements and track moving objects with the camera on a dolly (Roberts, 2002).

If the frame rate is too low, moving objects appear blurred. *Motion blur* results in loss of detail in moving images. The visible resolution of a moving object appears to be less than the resolution of this object when it doesn't move. The reason is that modern displays do not flash an image but keep it displayed until it is time to display the next frame. Your eyes typically track the average position of a moving object. Thus, during the time between two frames the object is stationary, while your eyes are still trying to move with it. As a result the image of the object is moving on the eye's retina and appears to be blurred. The larger the screen and the faster the object moves, the more severe the blurring becomes (Brennesholtz and Stupp, 2008).

3.2.3 Spatial Quality

The picture presented on a video display should have a uniform appearance over its entire area and it should be rectangular without any geometrical distortions.

Uniformity

The luminance should be completely uniform across the entire display area. In practice it is not: it decreases when you move from the center to the edges. Typically, the change is larger in projection systems than in flat-panel displays (Bodrogi and Khanh, 2012; Brennesholtz and Stupp, 2008). For a flat-panel display the luminance at the edges should be at least 90–95% of the luminance at the center (EBU-Tech 3320, 2017; ITU, 2009b). For a projection system the requirements are less restrictive: the luminance difference is allowed to be 70–90% (DCI, 2008).

The color appearance should also be uniform across the entire display area. In section 6.2.5 I explain in detail how color differences can be measured. There I introduce the measure ΔE_{uv}^* to express a color difference in a single number that is perceptually relevant. Regarding color uniformity, the difference between the edges and the center of the screen should be such that $\Delta E_{uv}^* \leq 4$ (EBU-Tech 3320, 2017).

Geometry

A video image is rectangular with a certain aspect ratio between its horizontal and vertical dimension (see section 5.1.1). It should be displayed without any *geometrical distortion*, even when the aspect ratio of the source material does not match the aspect ratio of the display. Some displays stretch an image in the horizontal or vertical direction to fill the entire screen. However, this modification can severely distort the picture. Circles will become ovals. People will look extraordinarily tall or fat depending on which direction is stretched.

All the pixels of the video material should be displayed. You might think that this is obvious, but it isn't. Many displays cut off the outer parts of the picture. This behavior stems from the early days of television when the outer part of the image often contained artifacts and distortions. Today, with high-quality digital video the picture is free of such artifacts and there is no longer any excuse for not showing the entire picture.

Projection systems can have several other geometric distortions. In these systems the projector should be carefully aligned with the screen (see section 4.4), and the optical parts should be of high quality.

3.3 Sound Production

Creating music is an established art form. Designing sound for movies is also considered to be an art. Not everyone realizes that the actual recording of sound and music is an art as well. It is not only a technical affair; it requires several artistic choices to be made. Sound can be recorded using a microphone. The nature of the recorded sound depends on the specific type of microphone used. Since sound travels through three-dimensional space, the recorded sound also very much depends on the position of this microphone with respect to the sound source, the directivity of the sound source, and the acoustics of the environment. Clearly, the recorded sound very much depends on the choices made by the recording engineer. Stereo and multichannel sound requires the use of multiple microphones to capture sound for each of the individual channels. Again, there are choices to be made. The engineer could use one microphone for each channel and position the microphones such that a sense of the acoustical environment is captured. Or the engineer could use many more microphones than channels and record each microphone on a separate audio track. These tracks are then mixed together to create the different channels such that the combined result sounds pleasing. This multitrack recording method offers an enormous amount of flexibility. It allows the engineer to build up a sound mix gradually. For example, different instruments could be recorded at different times and in different places, their relative levels can be fine-tuned, special effects can be added one track at a time, and so forth (Milner, 2009).

Not all sound is recorded using microphones. It is also possible to create sound electronically on a computer, a synthesizer, electric guitar, or some other electronic instrument. The sound can be recorded directly from the audio output of these devices without the use of a microphone. This electronically generated sound is only converted into acoustical sound waves when it is being played through loudspeakers. However, the direct recording of electronically generated sound doesn't always produce the result that the artist is after. Therefore, quite often electronically generated sound is first converted into acoustical sound using a loudspeaker, and subsequently recorded by placing a microphone in front of the loudspeaker. For this reason, electrical guitar, electrical bass, and certain electrical organs are often recorded using a dedicated amplifier and loudspeaker where the instrument, amplifier, and loudspeaker have been designed together to create a unique sound (Ballou and Ciaudelli, 2015).

Today, most audio recording and mixing is done digitally. In the studio, the analog audio signal from a microphone or electronic instrument is converted to a digital audio signal for further processing. Digital audio signals can be replayed and mixed without any loss of sound quality. Special effects can be added and removed. Mixing and effect settings can be memorized and altered with just a click of a button. Digital recording is not only convenient, it is also far more accurate than analog recording. It does not add any coloration to the sound. Some people claim that analog recordings sound 'better' than digital recordings. They like analog better because it sounds fuller compared to a digital recording. Indeed this difference exists, but it is the analog recording method that introduces coloration, not the digital one. People who favor analog tape and vinyl records like this coloration. It has nothing to do with high fidelity or accurate reproduction (Waldrep, 2017; Winer, 2018).

3.3.1 Recording and Mixing

How do you capture the essence of acoustical sound? It requires the careful positioning of one or more microphones and a certain amount of control over the acoustical environment in which the recording is being made. Placing microphones is more art than science. There is not just one way to do it right. It is subjective and somehow depends on the preference of the recording engineer (Ballou and Ciaudelli, 2015).

Proper placement of a microphone starts by taking into account the directivity of the sound source. Most sound sources radiate sound unevenly in different directions. Different frequency ranges with different strengths are radiated in different directions. Low frequencies are often radiated in all directions, while the higher frequencies only radiate in certain directions (comparable to the directivity of loudspeakers as depicted in figure 2.28): for example, from the top plate of a violin, from the bell of a trumpet, or from the tone holes of a woodwind. A microphone can only sample a tiny portion of the complex three-dimensional sound field that surrounds a sound source. Great skill is needed to place the microphone such that what we hear in the recording is representative of what we hear in real life (Ballou and Ciaudelli, 2015; Benade, 1985; Toole, 2018).

Sound recordings are influenced by the acoustics of the space in which they are made. Every room imprints its own acoustic signature. The sound from the source is reflected by the room boundaries and other objects in the room. The microphone records a mixture of direct and reflected sound. The distance between the microphone and the source affects their relative strengths. The distance also affects how much sound is picked up from other sound sources in the room. To isolate a particular instrument, the microphone should be placed very close to it. To record the ambience of the room, a larger distance is needed such that multiple reflections are picked up. At a larger distance two spaced microphones can be used to directly generate a stereo recording. Often a combination of techniques is used. For example, horns and woodwinds are usually recorded using one microphone nearby. Piano and strings are typically recorded with two spaced microphones at a distance. Drums are recorded using multiple microphones: one for each drum combined with two microphones suspended over the entire set to capture the cymbals (Ballou and Ciaudelli, 2015).

The audio signals from the different microphones are mixed together to create a combined sound that expresses the intentions of the musicians and engineers. They listen to the mix through a set of loudspeakers in the studio control room, react to what they hear, and adjust the mix. They put the different instruments in different audio channels, such that the instruments are positioned in space. Typically, they also add multiple delayed versions of the sound to simulate multiple reflections of an acoustical environment. These artificial reflections provide the glue to tie all the instruments together in a pleasant-sounding end result (Holman, 2008; Winer, 2018). How does this compare to a live performance? Classical music is performed without any amplification in a concert hall. The acoustical properties of this hall are an integral part of the sound that you experience. The recording is only an imitation of the sound that you would hear during the concert. The recording engineer tries to create the illusion that you are actually there. By contrast, popular music is often performed live using loudspeakers. The environment in which the performance takes place is less important than it is in classical music. The recording engineer uses spatial cues to create an acoustic fiction that does not need to be related to an existing natural environment. It should just sound right (Benade, 1985; Rumsey, 2002).

Sound for movies consists of dialog, music, and effects. Dialog is often recorded on set during filming. A microphone on a boom is held above the actors. During postproduction mistakes on the part of the actor and bad recordings are replaced in a process called *automated dialog replacement* (ADR). Music is added to create a mood and atmosphere that fits the story. It is recorded in a special studio where the musicians and the conductor watch the movie on a big screen during their performance. *Sound effects* are all the sounds in a movie that are not dialog or music. Typical examples are footsteps, chirping birds, wind, traffic, and explosions. Sound effects are being recorded by foley artists. These artists use an extensive collection of props to create sounds in the studio that match the action on the screen. They also record and collect sounds outdoors in many different environments. During postproduction the final sound mix of the movie is created by integrating dialog, music, and sound effects. Creating sound for movies is definitely an art form. It involves a whole team of creative people working together. There are Academy Awards (Oscars) for the best achievements in sound editing, sound mixing, and film music (Masters, 1999; Piper, 2008).

Mastering is the final creative stage for every recording, be it music or a movie soundtrack. A mastering engineer prepares the final sound mix for its release on a delivery medium (for example CD, DVD, or Blu-ray disc). The mastering engineer listens critically to the sound and makes the final adjustments to the dynamic range and the frequency balance. For this purpose, the mastering room is equipped with high-quality loudspeakers. The acoustical properties of the room are somewhere between that of a studio control room and a typical domestic room. A skilled mastering engineer has a clear understanding of how the final mix will sound in a wide range of other circumstances compared to how it sounds in the mastering room. Based on this knowledge the engineer improves the sound of the mix both technically and artistically (Katz, 2015; Newell and Holland, 2007).

3.3.2 Sound Reproduction

Exact reproduction of the sound field of a live performance is not an attainable goal. The sound field produced with a handful of loudspeakers will bear little resemblance to the sound field generated by the original instruments. The differences are just too many. First, the positions of the loudspeakers in the room do not necessarily match those of the instruments. Second, loudspeakers radiate sound in a very well-defined way, while acoustical musical instruments radiate sound in a complex way with different parts of the instrument radiating different frequency ranges in different directions. Finally, the room in which you listen to the loudspeakers is often different from the one in which the instruments were played. Note that even when the loudspeakers and instruments are in the same room, the sound field that they generate will be very different. Because they are placed in different positions and radiate sound in different ways, the reflections in the room will be different, as will be the effect of the room boundaries and the room resonances (see section 2.3.4).

The goal is to reproduce the sound from the loudspeakers in the studio control room as close as possible. Reproduction of sound using loudspeakers should be considered a performance in its own right. In the control room, the sound engineers and the other artists involved have created a specific listening experience. It can be an imitation of a live performance, but it often is an artistic creation in itself (Moulton, 1990; Newell and Holland, 2007; Toole, 2018).

How do you recreate the sound from the studio control room using your own loudspeakers in your own room? It would require a similarity of loudspeakers and room acoustics. But, this is difficult to achieve in practice. The acoustics of a typical living room do not match the acoustics of the typical studio. In addition, loudspeakers for domestic use often differ from those used in the studio. To make matters even worse the recording industry lacks meaningful standards for playback circumstances. Many recording engineers choose loudspeakers that instead of being neutral add a certain quality to the sound. This means that only those who listen to the same loudspeakers will hear the intended sound quality (Newell and Holland, 2007; Toole, 2018). In practice, there are significant differences in sound between different studio control rooms (Mäkivirta and Anet, 2001). Therefore, recordings can vary widely in their sound quality, and what you hear at home will almost always be different from what was heard in the studio.

The movie industry has taken steps to improve on the lack of standards that prevails in the music industry. Sound for movies is being mixed in a standardized environment that resembles a typical cinema. Mastering for DVD and Blu-ray disc releases is being done in a small room with loudspeakers intended for domestic use. While not all factors that influence sound reproduction are rigorously controlled, it still is an important step forward (Toole, 2018).

These days many people listen to music through headphones. While this may be pleasant and convenient, it differs a lot from listening through loudspeakers. The spatial quality of the sound is completely different, because there are no reflections from the listening room. In addition, the pinnae of the ears are being bypassed. The pinnae normally generate important cues for localization of sounds. When these important spatial cues are missing, the instruments and especially the singers are not positioned in front of you, but mostly inside the head between the two ears. Therefore, many listeners prefer to listen to music reproduced by loudspeakers (Zhang, 2015).

3.4 Sound Quality

Sound quality is the term used to describe how good or how bad the sound from your sound system is. Obviously, the quality of reproduced sound strongly depends on the quality of the source material. Unfortunately, the quality of source material varies widely due to the lack of meaningful standards for the reproduction of sound in studio control rooms. Since you cannot know how the material sounded in the studio, you cannot decide whether your reproduction at home is correct. It is not possible to reproduce an exact copy of the sound at home (Toole, 2018). However, not all is lost.

Recognizing deficiencies in reproduced sound is a universal skill. We have a common preference of what sounds right and what sounds wrong. There exists a common body of listening preferences. Without exactly knowing how the source material sounded in the studio, we have the remarkable ability to distinguish between the quality of the source material and the quality of the reproduction. Thus, the absence of deficiencies in the reproduction can be used as a measure for sound quality. Our musical taste does not influence this ability. We can even detect flaws in the reproduction of music that we do not like (Toole, 2018).

Sound quality can be described in many different ways. The overall impression of how closely the reproduced sound matches the 'original' sound is called *fidelity*. The overall impression that concentrates on the lack of aggravations and annoyances in the reproduced sound is called *pleasantness* (Gabrielsson and Lindström, 1985; Toole, 1985). Besides these overall qualifications, sound quality can be described in more detail by looking at different aspects of sound. Table 3.7 presents a hierarchical overview of the relevant aspects. The main division is between *timbral quality* and *spatial quality* (Letowski, 1989). Timbral quality aspects allow you to judge that two sounds that have the same pitch, loudness, and spatial quality are dissimilar. Timbral quality is also called *sound color* (EBU-Tech 3286-s1, 2000). Spatial quality aspects allow you to describe the distribution of sound sources and the size of acoustical space. The timbral quality aspects contribute 67% to the overall sound quality; the spatial quality aspects of the sound coming from the front contribute another 25%, and the remaining 8% comes from the spatial quality of the surrounding sound (Rumsey et al., 2005).

In addition to the different quality aspects listed in table 3.7, the overall sound quality depends on the presence or absence of extraneous sounds (EBU-Tech 3286-s1, 2000; Gabrielsson and

Table 3.7 Overview of sound-quality aspects.

Timbral Quality	Spectral balance	Fullness
		Brightness
		Coloration
	Clarity	Source definition
		Time definition
	Dynamics	Transient impact
		Compression
Spatial Quality	Localization	Definition of sound images
		Soundstage width
		Soundstage distance
		Soundstage depth
	Immersion	Presence
		Envelopment

Lindström, 1985; Toole, 1985). The reproduced sound should be free of unwanted sounds that originate from the reproduction equipment, the room, or outside. You do not want to be disturbed by air-conditioning hum, traffic noise, rattling objects, and so forth.

3.4.1 Spectral Balance

Spectral balance is the balance between the different frequencies in the audible frequency spectrum (AES, 2008). A sound reproduction system should reproduce all audible frequencies in an equal amount. It should not emphasize certain frequencies or frequency ranges. On a large scale, spectral balance may refer to the balance between the low-frequency (20–200 Hz), mid-frequency (200–2000 Hz), and high-frequency (2–20 kHz) ranges. On a much finer scale, it describes the presence or absence of certain peaks and dips in the frequency response. Such irregularities in the frequency response can audibly distort the sound. This distortion is called *coloration*.

Fullness

The amount of low-frequency sounds and their balance with respect to the mid- and high-frequency sounds is called *fullness* (Gabrielsson and Lindström, 1985; Toole, 1985). Good sound is neither too full nor too thin. Excessive bass in reproduced sound is a constant reminder that you are listening to a reproduction. It sounds too heavy and weighty. It distracts you from other higher frequency sounds. On the other hand, too little bass sounds thin, lean, and lightweight. Too little bass robs music of its rhythm and drive, and sound effects like explosions lose their impact (Harley, 2015).

The amount of bass is a matter of great importance. About 30% of the overall sound quality depends on the quality of the low-frequency reproduction (Toole, 2018). The total amount of bass that you hear strongly depends on the interaction between the loudspeakers and the room. Room resonances and reflections from nearby room boundaries play their part (see section 2.3.4). Loudspeakers that are capable of producing the most extended bass response achieve high-quality ratings in controlled listening tests (Toole, 1986b).

How low should you go? Ideally, your sound system should go all the way down to 20 Hz. However, most musical instruments, including bass guitar, bass violin, and bass drum produce negligible output below 30 Hz. Audible bass below 30 Hz is relatively rare. Only pipe organs, synthesizers, and special effects go that low (Fielder and Benjamin, 1988). Reproduction of the 30–60 Hz frequency region is very important. Despite the fact that it is only 30 Hz in size, it does represent a whole octave containing many musical notes. You do not want to miss them. Good low-frequency extension preserves a sense of musical scale and avoids colorations in the upper-bass region known as 'boomy bass' (Colloms, 2018).

Brightness

The balance of the high-frequency sounds with respect to the mid- and low-frequency sounds is called *brightness* (Gabrielsson and Lindström, 1985; Toole, 1985). Good sound is neither too bright nor too dull. The high frequencies, also called *treble*, should sound as a natural extension of the mid frequencies. They should not sound as if they are 'riding on top of them'. An excessive amount of high frequencies can add a certain shrillness to the sound. By contrast, not having enough energy at the high-frequency range makes the sound dull, dark, or rumbling (Harley, 2015).

Coloration

Coloration is the distortion of timbre. Coloration introduces an identifiable tonal quality to the sound that doesn't sound natural. It arises when the frequency response of the sound production system has substantial peaks and dips in certain frequency ranges (AES, 2008).

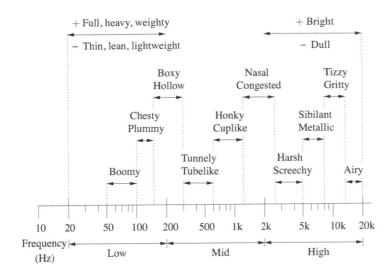

Figure 3.14
Approximate frequency ranges for different types of coloration.

Based on AES (2008), Colloms (2018), Duncan (1997), Harley (2015), and Katz (2015).

The frequency response describes how your sound system reproduces the different frequencies. Your system should not add any tonal quality of its own; it should reproduce all the different frequencies in the amount dictated by the source material. A graph of the frequency response of your system should be a flat line without any peaks or dips. A peak in the frequency response means that your sound system amplifies a certain range of frequencies. A dip means that your system suppresses a certain range of frequencies. Unfortunately, in practice every sound system has peaks and dips in its frequency response. Your loudspeakers and their acoustical interactions with the room are the major offenders. Figure 3.14 shows a number of terms that can be used to describe coloration in certain frequency ranges.

Coloration in the mid frequencies is very annoying. The mid frequencies include many of the frequencies of the human voice and the most important musical notes of a wide range of instruments (see figure 2.22). The sound quality in this frequency range has a large effect on the correct reproduction of timbre. Good midrange sound is smooth, liquid, and free of any coloration. Unwanted midrange coloration in your sound system is easy to spot using a high-quality recording of a choir, piano, or male speaking voice (Harley, 2015).

The amount of *softness* of the high frequencies is an important quality aspect of sound reproduction. The human ear is very sensitive to the frequencies between 2 kHz and 5 kHz (see figure 2.24). Good high-frequency sound is neither overly subdued and mild nor excessively hard and shrill (Gabrielsson and Lindström, 1985; Toole, 1985). The high frequencies can sound a bit aggressive, but they should never be grainy. Grainy high frequencies have an unpleasant coarseness or roughness overlapping the instrumental and vocal timbres. Grainy coloration of high frequencies in your sound system is easy to spot using a high-quality recording of a flute, violin, or saxophone (Harley, 2015).

The amount to which coloration is audible depends on three factors: 1) the frequency of the peak or dip, 2) its frequency bandwidth (broad or narrow), and 3) its amplitude (strength). Broad peaks are more easily heard than narrow peaks. The smaller the bandwidth of the peak, the higher the amplitude needs to be in order to be audible. Peaks at high frequencies are more easily heard than peaks at low frequencies. The lower the frequency of the peaks, the higher the amplitude needs to be in order to be audible (Toole and Olive, 1988). Figure 3.15 shows some examples of the detection thresholds for peaks of different amplitudes and bandwidths. Interesting enough, peaks and dips are not equally audible. Peaks in the frequency response are far more audible than dips having the same amplitude and bandwidth (Bücklein, 1981). The human hearing system seems to be able to tolerate narrow dips even if

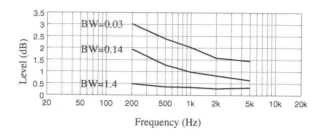

Figure 3.15
Detection thresholds for amplitude deviations in the frequency response. The three curves correspond to peaks with a different bandwidth (BW) in octaves. A peak with a certain bandwidth cannot be heard if its amplitude lies below the indicated line.

Based on data from Toole and Olive (1988), fig. 30.

they are very deep. It is as if we are able to somehow reconstruct this missing information. By contrast, large narrow peaks always distract us and result in coloration of the sound.

3.4.2 Clarity

Clarity refers to the ability to hear and distinguish different instruments and voices within complex orchestrations and the ability to distinguish individual notes and words. The sound reproduction should be clear, distinct, and pure. It should not be diffuse, muddy, mushy, or thick (Gabrielsson and Lindström, 1985; Toole, 1985). Every detail and nuance in the reproduced sound should be clearly audible. Other words for clarity are *definition* and *transparency* (EBU-Tech 3286-s1, 2000). The ability to distinguish words in speech and song is called *intelligibility*.

Clarity is often a trade-off between smoothness and resolution of detail. Too much detail in the reproduced sound makes it aggressive, analytical, and fatiguing. Too little detail and it becomes dull and uninvolving. Good reproduction of detail is sometimes called *vivid* (Harley, 2015).

Source Definition

The subjective impression that different instruments or voices sounding simultaneously can be identified and distinguished is called *sound source definition* (EBU-Tech 3286-s1, 2000). In everyday life, the sound reaching your ears is typically a mix of multiple sound sources. The human hearing system is usually quite capable of identifying and distinguishing the individual sound sources that make up this mix. Your hearing system uses many different cues to achieve the perceptual separation, such as pitch, timbre, time structure, loudness, location, and changes therein (Moore, 2013). Good sound reproduction enables you to clearly distinguish the individual sound sources based on such cues.

Time Definition

The subjective impression that individual short sounds in rapid succession can be identified and differentiated is called *time definition* (EBU-Tech 3286-s1, 2000). Individual sounds should be distinguishable in time and you should be able to hear how each of these sounds develops over time. Sounds constantly change. The way a sound starts is called the *attack*. The way it evolves after that is called the *flux,* and the way it ends is called the *decay*. Figure 3.16 shows this particular time structure. The attack can be sudden or gradual, and it usually creates energy at many different frequencies that are not simple multiples of each other. The flux is a more stable phase that often consists of a fundamental frequency that defines a certain pitch and a series of overtones that are multiples of this fundamental frequency. The timbre of a sound not only depends on the flux, but is also influenced by the time structure of the attack and decay (Levitin, 2006). Accurate reproduction therefore requires that your sound system reproduces well-defined attacks and decays, and does not diffuse or muddle them.

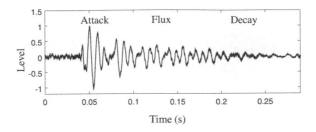

Figure 3.16
Example of an audio signal
that illustrates the attack,
flux, and decay.

Good sound reproduction enables you to clearly perceive attacks, pitches, and other details in the sound (Gabrielsson and Lindström, 1985; Harley, 2015; Toole, 1985).

Time definition is extremely important in the reproduction of low frequencies. Good bass is tight, fast, well-defined, and punchy. You should be able to hear all the individual bass notes, and the attack, pitch, and decay of each of these notes. Drums and percussion should sound tight having a sudden impact and punch. Good reproduction of low frequencies requires fast decays. If the bass doesn't decay fast enough, it will sound muddy, fat, thick, and slow. Bass that continues too long robs the music of its rhythm and sounds like a dull roar that blurs the underlying details. Room resonances (see section 2.3.4) are the most common cause of slowly decaying bass (Harley, 2015; Toole, 2018).

3.4.3 Dynamics

Dynamics refers to the rendition of loudness changes. It encompasses the ability of the sound system to play loud without audible distortion or compression. A sound system having good dynamics reproduces loud transients with a sense of dynamic power and a feeling of liveliness in the reproduced sound. Good dynamics also means that during the softer passages, the transients retain their clarity and a certain sense of attack (AES, 2008; Colloms, 2018).

Transient Impact

Good rendition of loud, quickly changing sounds is called *transient impact* (AES, 2008). Quickly changing sounds that have a sudden attack and a rapid decay are called *transient sounds*. A typical example is the whack of a snare drum. Good reproduction of transients provides an over-all sense of slam, impact, and power (Harley, 2015). If your sound system is up to the job, you should feel the dynamic impact of the sound in your body. This is especially satisfying with bass-heavy popular music or movies with lots of explosions and other low-frequency special effects.

Compression

Compression can lead to the undesirable reduction of the dynamic range. Compression reduces the range between the strongest and weakest sound levels. As a result, the overall presentation may become flat and uninvolving. The clarity of the sound suffers and during loud peaks an overall sense of strain prevails. One way that compression occurs is when the sound system reaches its maximum output level. The sound level of the loudest sounds becomes limited, thus reducing the overall dynamic range. Loud and compressed sound often results in listener fatigue and a strong desire to turn down the volume. Compression becomes especially annoying when it induces an audible modulation effect where midrange frequencies seem to drop in loudness when bass notes are present (AES, 2008).

Unfortunately, many commercial music recordings are heavily compressed during the mastering process. Some music producers find that compression makes their product more competitive when played on the radio. The compressed audio signal gives the illusion of increased loudness but at the expense of the overall sound quality. All the sound is crammed closer to the maximum level (Deruty and Tardieu, 2014; Katz, 2015; Milner, 2009; Vickers, 2010). Figure 3.17 compares a compressed audio signal with an audio signal that has a good dynamic

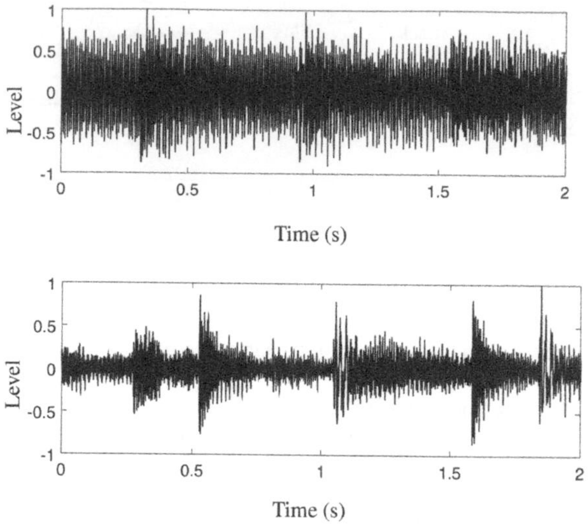

Figure 3.17
Example audio signals.
The signal at the *top* has a
compressed dynamic range
compared to the signal at
the *bottom*. The top signal
will therefore sound louder,
while the bottom signal
has a better overall sound
quality.

Figure 3.18
Specification of sound
source direction. *On the
left,* a top-down view
where the azimuth specifies
the position of the sound
source in the horizontal
plane. *On the right,* a side
view where the elevation
specifies the position in the
vertical plane.

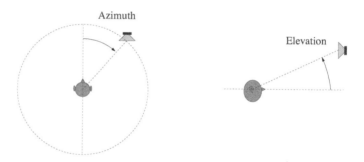

range. High-quality recordings that have sufficient dynamic range and little compression often sound quieter. Movie soundtracks also sound a lot softer. The sound for movies always contains very loud and very soft passages. To make everything fit within the level range of the audio signal, the overall sound level must be on the low side. For this reason, Blu-ray discs and DVDs have a much lower overall loudness level than most CDs or downloaded audio files (Taylor et al., 2009). Remember to adjust the volume setting on your amplifier when you switch from watching a movie to listening to music.

3.4.4 Localization

Localization refers to the ability to judge the direction and distance of a sound source (AES, 2008; Moore, 2013). The direction of a sound source can be specified by *azimuth* and *elevation,* as depicted in figure 3.18. The azimuth specifies the position of the sound source in the horizontal plane. It is the angle between the sound source and a line that runs from your nose to the back of your head. It works like a compass: 0° is directly in front of your face, 90° is at the side of your head opposite your right ear, and 180° is behind your head. The elevation is used to describe the vertical position and is the angle between the sound source and the horizontal plane: 0° is directly in front of you; 90° is directly above your head.

Your ears localize sound sources based on three different cues. Figure 3.19 illustrates them. The first cue is the *interaural time difference* (ITD). The sound arrives a bit later in time at the ear that is farthest away from the sound source. The second cue is the *interaural level difference* (ILD). The sound level is lower at the ear that is farthest away from the sound source. The third cue is the frequency-dependent interaction of the sound field with the torso, head, and

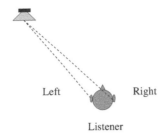

Figure 3.19
Localization of a sound source on the left side of the head. The distance from the sound source to the right ear is larger than the distance to the left ear. Therefore, the sound at the right ear arrives later in time and is lower in level. In addition, the sound at the right ear has a different frequency spectrum, because the higher frequencies are obstructed by the head and torso. Finally, the sound at the right ear hits the pinna at a different angle compared to the angle at which the sound hits the left ear.

pinna. These three cues together provide a certain redundancy, so that even under difficult conditions you are still able to accurately locate sound sources. The dominant localization cue depends on the frequency content of the sound. Frequencies well below 1 kHz diffract around the head, because their wavelengths are long compared to the size of the head (see section 2.3.3). For these frequencies interaural time difference is an important cue. Higher frequencies do not diffract around the head and are therefore partly obstructed by the head. At these frequencies the interaural level difference becomes more useful (Moore, 2013).

Front-back discrimination and localization in the vertical plane (elevation) is possible because of the interaction of the sound field with the torso, head, and pinnae. Together they modify the frequency spectrum of the sound arriving at the ears. This spectral modification varies with the angle at which the sound reaches the ears. Some frequencies are emphasized, others are attenuated. It results in a unique frequency response for each direction in space, which is called the *head-related transfer function* (HRTF). The torso and the head influence the frequency response above about 500 Hz where the wavelengths are small compared to the dimensions of the human body. The pinnae only interact strongly with sound waves that have wavelengths smaller than a few millimeters. As a result, the influence of the pinnae is limited to frequencies above 6 kHz (Moore, 2013; Shaw, 1974).

It is easier to localize sound sources in a room than outdoors in free space. The room boundaries and the objects in the room reflect the sound from the source. As a result, your ears are supplied with multiple repetitions of the original sound that arrive at different times and different angles (see section 2.3.4). This particular pattern of reflections greatly helps in determining the right position of the source in the room (Benade, 1985).

Most music recordings and video-based content create the illusion of a soundstage in front of you. This soundstage originates mainly from the left, center, and right front loudspeakers. The surround loudspeakers are mainly used to enhance the feeling of envelopment. Directional sound from the surround loudspeakers is limited and often only used to create temporary special effects (see also section 2.1.1). Typically, the frontal soundstage is made up of several individual sound sources as depicted in figure 3.20. Each source has a particular position within this soundstage. The soundstage itself is perceived to have a certain width, depth, and distance from the listener (Rumsey, 2002).

Definition of Sound Images

The extent to which different sources of sound are spatially separated and positionally defined is called *definition of sound images* (Toole, 1985). Good sound reproduction gives you the impression that sound sources have well-defined positions around you. All sound sources

Figure 3.20
Soundstage width, depth,
and distance.

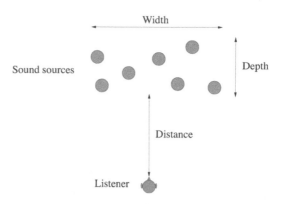

should be appropriately placed for the type of source material, and they should have an apparent size that fits. For example, a single singer may have a sharply defined position, while a whole section of violins is expected to result in a much broader sound image (AES, 2008; EBU-Tech 3286-s1, 2000). Strong reflections in the listening room or in the recording have the ability to make a sound source appear larger. This is called *apparent source width* (ASW). It explains why in sound reproduction a section of violins can sound larger than the actual width of the loudspeaker reproducing them. This broadening of sound sources is governed by early reflections that arrive within 80 ms of the direct sound, come from the sidewalls, and have significant frequency content above 500 Hz (Toole, 2018).

Our ability to determine the position of a sound source is more precise in the horizontal plane (azimuth) than in the vertical plane (elevation). In the horizontal plane we are able to determine the position in front of us within 1° to 3°. Moving away to the side and the rear it becomes more difficult and on average we can only achieve an accuracy of 10° to 20°. In the vertical plane, our accuracy ranges from 5° to 20° for sources in front of us and becomes gradually worse above and behind us where it ranges from 30° to 40° (Toole, 1988).

Good sound reproduction allows you to hear the sound images as distinct objects that appear to hang in three-dimensional space. The images can be localized, and they are clearly separated. The sound should be clear, open, and transparent, neither veiled nor opaque. Good image separation results in focused, tight, and sharp images. Bad image separation, on the other hand, is described as blurred, congested, and confused (Harley, 2015).

Sound images should be directionally stable. They should not move from their intended positions with changes in pitch, loudness, or timbre (AES, 2008; EBU-Tech 3286-s1, 2000). Furthermore, the reproduction should be free from abnormal spatial effects that do not occur in common experience. For example sounds should not be heard inside your head, stretched between you and the loudspeaker, split-up in multiple positions, or have no location at all (Toole, 1985).

Soundstage Width

Soundstage width refers to the perceived width of the ensemble of sound images in front of you from the extreme left to the extreme right. Good sound gives you the impression of a frontal soundstage that is neither too wide, nor too narrow. The soundstage should be wider than 30° but not go beyond 90° (AES, 2008; EBU-Tech 3286-s1, 2000). The soundstage should be continuous across its width. The display of sound images should be perceived as an integrated whole, without any illogical groupings of images and without any large gaps between them (Toole, 1985).

Soundstage Distance

Soundstage distance is the perceived range between you and the ensemble of sound sources in front of you (Rumsey, 2002). It is sometimes referred to as *nearness* (Gabrielsson and

Lindström, 1985). Recordings that contain real or simulated reflections can create the illusion of sound sources at different distances than the distances to the loudspeakers. Our hearing system uses the relative strengths of the direct sound and its reflections as an important cue for distance. The farther away the source, the lower the overall sound level and the stronger the relative level of the reflections (Toole, 1988).

Soundstage distance is largely determined by the recording itself. A forward presentation means that the sound appears in front of the loudspeakers. The sound is immediate, vivid, and lively. A laid-back presentation means that the sounds appear to come from slightly behind the loud-speakers. The sound is lush, gentle, and easygoing. The sound should neither be too forward nor too laid-back. When the sound is too laid-back it becomes bland and uninvolving. But when it is too forward it becomes aggressive and it will make you turn the volume down (Harley, 2015).

Soundstage Depth

Soundstage depth is the subjective impression that the soundstage has an appropriate front-to-back depth. You are able to distinguish sound sources at various distances (Rumsey, 2002; Toole, 1985). A soundstage that lacks depth sounds flat, because it has all the sound sources at the same distance.

Depth is a subtle aspect of sound reproduction that is difficult to attain. Our ability to accurately judge the distance of sound sources is rather limited. Errors of 20% or more are not uncommon. Important cues for estimating sound source distance in a room are the differences in intensity, time delay, and frequency content of the direct sound and the reflections (Moore, 2013). Consequently, soundstage depth strongly depends on the presence of reflected sound in the recording. These reflections decay over time and become less and less strong. To achieve the correct sensation of depth in reproduced sound, your sound system must be able to accurately reproduce these low-level decaying reflections. Only truly excellent loudspeakers are able to create soundstage depth. Lower quality loudspeakers have significant cabinet and diaphragm resonances that easily overpower the low-level depth cues in the recording (Colloms, 2018).

3.4.5 Immersion

Immersion refers to the subjective impression of being immersed in the sound field. It is a global description of the sound field, and it is harder to conceive than the direction, size, and distance of sound sources. Immersion has two different attributes: *presence* and *envelopment*. Presence refers to the extent that you and the reproduced sound sources are present in an acoustical space. You are able to sense the boundaries of this acoustical space. Envelopment refers to the sensation of being completely surrounded by sound. You are right in the middle of an acoustical space that extends all around you. Presence and envelopment are closely related but not the same (Rumsey, 2002).

You don't need to exactly replicate the sound field of a real acoustical space in your listening room. Sound field immersion is already a fact when you only reproduce the key spatial cues from a limited number of directions. A high-quality two-channel recording can give you a strong sense of presence. This sensation is mainly controlled by the ratio between the direct sound and the reflections in the recording. Envelopment requires multiple loudspeakers around you. A strong sense of envelopment occurs when these loudspeakers reproduce recorded sounds with appropriate delays from appropriate directions such that these sounds complement the frontal soundstage (Holman, 2008; Toole, 2018).

Some people believe that immersion requires a diffuse sound field. It doesn't. As said before it suffices to reproduce the right amount of delayed sounds from the appropriate directions. This is good news because a diffuse sound field can only occur in a large hall, not in a typical domestic room that is way too small. In a diffuse sound field, sound is expected to arrive from all directions with equal probability. Because of this property, a diffuse sound field does sound spacious and enveloping (Toole, 2018).

Presence

Presence is the subjective impression that you are in an acoustical space together with the sound sources. Other words that are used to indicate presence are *perspective* (Toole, 1985) and *spaciousness* (AES, 2008; Gabrielsson and Lindström, 1985). With a good rendition of presence the sound is open, has palpable dimensions, and fills the room. You have the feeling that you are at the performance with the musicians close to you. A less perfect rendition sounds closed, shut up, and narrow. It gives you the impression that you are close, but still looking on; there is a certain distance between you and the sound sources. In the worst rendition of presence, the sound is constrained to the locations of the loudspeakers and you seem to be outside of the acoustical environment 'looking in' through a narrow opening (Gabrielsson and Lindström, 1985; Toole, 1985).

A good rendition of presence allows you to judge the size of the acoustical environment in which the recording was made. Are you present in an intimate jazz club or in a large stadium? A good rendition also allows you to judge whether the acoustical environment is highly reflective or absorbent. This aspect is called *reverberation*. A highly reflective room has many late reflections that are relatively strong with respect to the direct sound (see section 2.3.4). Strong late reflections cause the room to sound reverberate. A lack of strong late reflections makes the sound dry. A small room may be very reflective, but it should still sound small. Size and reverberation are two different aspects of the rendition of an acoustical environment in sound reproduction. Together they are referred to as *ambience* reproduction (AES, 2008; EBU-Tech 3286-s1, 2000).

Envelopment

Envelopment is the sensation of being surrounded by sound. This aspect can be subdivided into source-related envelopment and environmental envelopment. Source-related envelopment means that you can hear several sound sources around you. Environmental envelopment means that you are immersed in reverberant sound or in diffuse sound that cannot be localized (Rumsey, 2002). A typical example of the latter is applause from an audience all around you. Environmental envelopment is often referred to as *listener envelopment* (LEV). It is governed by late reflections that arrive more than 80 ms later than the direct sound, come at you from the side and the back, and have significant frequency content between 100 Hz and 1000 Hz. The reproduction of these spatial cues for envelopment requires the use of surround loudspeakers that are placed to the sides and slightly behind you (Toole, 2018). A good rendition of envelopment wraps you inside a continuous sound field and gives you the impression of an integrated whole. There are no holes in ambiguously located sounds around you, and localizable sounds move smoothly from one place to the other (Harley, 2002).

3.5 Listening Tests

Judging sound quality by simply listening to the sound remains important, because not all aspects of sound quality can be measured in a meaningful way. Despite a lot of research over the years, not everything that we hear can be correlated to an objective measurement. The present state of our knowledge is rather limited (Colloms, 2018; Harley, 2015; Lipshitz and Vanderkooy, 1981). For example, how would you measure soundstaging or envelopment? Even today, a listening test is the ultimate arbiter of sound quality. Often, it is the only way to discover how audio equipment, a room, or a particular placement of the loudspeakers really sounds.

Listening tests must be highly controlled to achieve meaningful results. We humans are easily fooled and without strict controls the results of a listening test will not be objective, consistent nor repeatable (Lipshitz and Vanderkooy, 1981). Expectations and preferences can largely distort the conclusions drawn from a listening test. You might like the reputation of one brand of equipment better than the other. From an aesthetic point of view you might prefer a certain position in the room for your loudspeakers. You have spent a lot of money on your new amplifier, so it must be better than the old one. Under such circumstances, it becomes difficult to establish whether subtle differences in sound quality are real or imagined (Winer, 2018). Be careful with reviews in the audio press that describe sound quality of audio equipment. These reviews are often not much more than opinions, because the reviewers didn't use the appropriate controls to avoid expectations and preferences distorting the outcome of their listening tests.

Despite the pitfalls, I recommend that you use the listening test as a standard tool to evaluate sound quality. Always listen to loudspeakers and amplifiers before you buy them. Use a listening test to evaluate the changes that you make to loudspeaker positions or room acoustics. Determine whether the changes really improve your sound.

Evaluating sound quality by careful and critical listening is different from listening for pleasure. You focus on particular quality aspects of the sound, like those described in section 3.4. Everybody can distinguish good sound from bad sound, but recognizing subtle differences is a learned skill that improves with practice (Harley, 2015). It helps if you give a grade to each of the aspects that you are listening for. You could use a five-point grading scale for this purpose: the higher the score the better the sound quality (ITU-R BS.1284, 2003). Adhere to the following guidelines to get meaningful results from your listening test:

- *Use revealing source material.* The source material has an enormous influence on perceived sound quality (Harley, 2015; Lipshitz and Vanderkooy, 1981; Toole, 1985). Problems with certain sound-quality aspects are only revealed when you use appropriate source material. For example, a recording of a single trumpet will not reveal problems in the low-frequency reproduction. Neither will it reveal much about the ability of the system to play loud without compression. However, it can tell you a lot about coloration in the mid and high frequencies. Use source material that reveals specific sound-quality aspects: orchestra music for soundstaging and dynamics, solo piano and acoustic guitar for clarity and coloration, acoustic bass for time definition, organ music for bass extension, rock and blues for transient impact, and so on (Harley, 2015).
- *Change only one thing at a time.* You must keep all other factors the same, otherwise you do not know what causes the perceived change in sound quality. For example, make sure you sit in the same spot when comparing different loudspeakers. Also place them in the exact same position in the room and use the same amplifier.
- *Match your listening levels.* Sound quality varies with listening level. As the volume is turned up the sound becomes fuller, more present, and enveloping. Changing only one thing at a time must therefore also be applied to your listening level. A close match is crucial if you want to get meaningful results when comparing two different situations (Harley, 2015; Toole, 2018). The equal-loudness contours (figure 2.24) show that the spectral balance is different at different playback levels. The louder sound always has relatively more bass. In addition, the louder sound permits more of the low-level spatial cues to be heard. They become loud enough to raise above the threshold of audibility, and hence the louder sound has a different spatial quality. The preferred listening level at home is 78 dB on average with peaks of 96 dB (see box 3.2). You can measure your listening level using either a dedicated SPL meter or an app on your smartphone.
- *Listen to A/B/A.* Listen not only to situation A followed by situation B, but also the other way around. It is important to reverse the comparison. Your hearing system adapts itself to situation A and partly compensates for its flaws. Just after you switch from A to B, your hearing system is still compensating for the flaws of situation A. This compensation results in the wrong impression of situation B. A second or third reversal between A and B is needed to be able to judge their relative merits (Colloms, 2018).
- *Make sure you are fit and relaxed.* Perceived sound quality depends on your mood, state of relaxation, and fitness. Make sure you are up to the job. It is often better to judge sound quality at the beginning of the day when you are still fresh. Note that a recent cold or other illness can also influence your perceptions (Colloms, 2018; Harley, 2015; Winer, 2018).

BOX 3.2 REFERENCE LISTENING LEVEL

Since sound quality changes with playback level, a reference listening level is needed to ensure consistent sound reproduction. The standard for the cinema is 85 dB on average with peaks of 105 dB (DCI, 2008; SMPTE RP 200, 2012). ITU and EBU use a different standard, which boils down to 78 dB on average with peaks of 96 dB for each channel (EBU-Tech 3276, 1998; ITU-R BS.1116, 2015). This is much lower than the cinema reference, but most people find the cinema reference way too loud for enjoyable listening at home. People prefer a sound level at home that is 8–10 dB lower than the cinema reference level of 85 dB (Holman, 2008; Toole, 2018). Therefore, an average level in the 75–78 dB range seems to be more reasonable for home listening.

One final warning: only shift into critical listening mode when you are evaluating sound quality (Harley, 2015). When you are listening or watching for pleasure, focus on the music or the movie, *not* on the different aspects of sound quality. In the end it is not only sound quality that matters, but the overall experience that the source material provides. Make sure you enjoy it. If you are unable to make the distinction between listening for sound quality and listening for pleasure, you miss the whole point.

4
ROOM DESIGN

The room in which you listen and watch forms an essential part of your audio and video experience. Good equipment alone doesn't get the job done. Paying attention to the design of your room is the single most important step towards optimal reproduction of audio and video.

As we already saw in section 2.3.4, your room influences the reproduction quality to a large extent. Your audio experience depends on the acoustical properties of your room, on the position of the loudspeakers, and on where you are seated in the room, whereas your video experience depends on the darkness of the room and on where you sit with respect to the video screen.

In this chapter, I will show you how to optimally design your room. I explain how to determine the optimal positions for your loudspeakers, where to place your video screen, and how to seat a small audience. It is easy to optimize the experience for a single seat, but most likely you want to enjoy music and movies in the company of others. Optimizing the experience for all seats in the room is a bit more complicated. The approach that I recommend is to first optimize for the prime seat and then add the other seats around it.

Flexibility in furnishings and decoration is essential if you want to achieve the best results. But even then, practical constraints might prevent you from placing the loudspeakers or seats in their optimal positions. For example, their sonically optimal position might clash with the

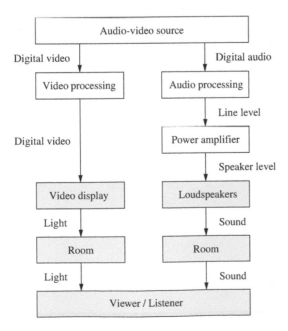

position of a door or a window. Often, the room that you selected to enjoy audio and video will be used for other purposes as well. It might be your living room or spare bedroom. This imposes its own constraints on furnishings and decorations. Thus, in a typical room, compromises have to be made. Fortunately, there are plenty of options to choose from. I will show you different options and discuss their impact on the quality of the reproduced audio and video. Surrendering a room solely for the purpose of audio and video reproduction seems to be the ideal choice, but it is costly and impractical for many. And it is not a panacea: even in a dedicated room, a few compromises need to be made.

Given all the different possibilities, it is hard to get your room design right the first time. Certain decisions that you have made before may need to be adjusted along the way. Give priority to the placement of your loudspeakers, your video screen, and your audience. Do not try to hide the equipment, but integrate it aesthetically into your room's interior design (Rushing, 2004) and make sure that you create a comfortable environment.

4.1 The Need for Subwoofers

By now it should be clear that the most important step towards good sound is to determine the optimal listening position and the positions of your loudspeakers in the room. Putting your loudspeakers in the right places should be your top priority in designing the listening room, because loudspeaker placement affects almost all sound-quality aspects, especially spectral balance, clarity, dynamics, and localization.

Proper localization of sound and other spatial effects requires that your front loudspeakers are placed in front of you and the surround loudspeakers are placed behind your back surrounding you. Figure 4.1 shows their typical positions with respect to the prime listening position. The loudspeakers don't need to be placed in exactly the same position as in this figure. There is some flexibility in placement, but the overall geometric relations should be preserved: front speakers in front of you, surround loudspeakers to the side and the back, left and right at the appropriate sides, and the center loudspeaker centered between the left and right front loudspeakers.

Figure 4.1
Typical loudspeaker positions in a 7.1-channel surround sound system. The prime listening position is the position where the lines drawn from each loudspeaker converge.

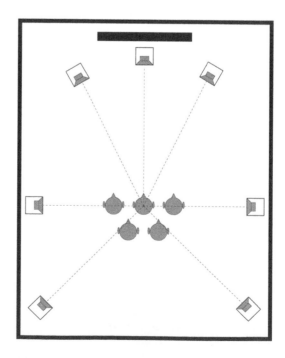

Unfortunately, the loudspeaker positions that are optimal for directional and spatial effects are far from optimal for the reproduction of low frequencies (Toole, 2006, 2018). Putting the loudspeakers at the positions for optimal spatial reproduction (shown in figure 4.1) results in an uneven low-frequency reproduction. Since each loudspeaker is placed at different distances with respect to the walls of the room, each one interacts differently with the room. Therefore, room resonances and adjacent-boundary effects (see section 2.3.4) influence each loudspeaker differently in the low-frequency range below about 200 Hz. Some frequencies will be attenuated while others will be amplified by the interaction of the room and the loudspeakers. As a result, you will hear a different quality and quantity of bass from each of the loudspeakers. Bass reproduction is at best mediocre and often much worse.

The best way to obtain better bass is to use one or several subwoofers to reproduce the lowest frequencies (Toole, 2018). Your surround processor or receiver has a feature called *bass management* that you can use to remove low bass from the front and surround loudspeakers and redirect it to the subwoofers. Typically, the crossover frequency between the subwoofers and other loudspeakers is fixed at 80 Hz, but sometimes it is better to use a slightly higher crossover frequency between 80 and 120 Hz (more on this in section 7.5.2).

Recommendation 4.1 *Keep the frequencies below 80 Hz out of your main loudspeakers, and use one or more well-positioned subwoofers to reproduce these low frequencies.*

By keeping the bass below 80 Hz out of your front and surround loudspeakers, you gain a great amount of flexibility in loudspeaker placement: you can position the subwoofers independently of the front and surround loudspeakers. Thus, you get the best of both worlds: you position the front and surround loudspeakers for optimal spatial reproduction, and you position the subwoofers for optimal bass reproduction. This strategy works, because spatial effects are mostly nonexistent at the low frequencies that the subwoofer reproduces (Martens et al., 2004; Toole, 2018; Welti and Devantier, 2006).

Most people wrongly think that subwoofers are only used to increase bass output. Well, they are often used for this purpose and it has given them a bad reputation. Many low-quality subwoofers exist that harm the overall sound quality by producing a thick boomy sound. However, bass quantity is not the primary goal here, we want to use subwoofers to improve the accuracy and pureness of the low-frequency reproduction. The good news is that high-quality pure-sounding subwoofers do exist. They are essential pieces of equipment if you want to get the best possible low-frequency reproduction.

The use of subwoofers and bass management has become commonplace (EBU-Tech 3276, 1998). It has its origin in low-cost, low-performance systems, but it has fundamental advantages for all sound systems. In fact, the traditional approach with multiple full-range floor-standing loudspeakers is not the best solution. It results in unpredictable and different bass quality and quantity from each loudspeaker. At some bass frequencies peaks and dips of 40 dB or more can occur (Welti and Devantier, 2006). By contrast, bass management with multiple subwoofers placed in the right locations results in a more uniform bass response. Not only at the prime seat, but also at the seats of the other members of the audience. However, it is not a magical solution; it does require careful placement of your subwoofers. I explain the basic guidelines for subwoofer placement in the upcoming section 4.2.6 and leave the more advanced optimization techniques to section 8.1.3.

Bass management keeps the low frequencies out of your main front and surround loudspeakers. This has a number of additional advantages that go beyond the increased flexibility in loudspeaker placement. Since our ears have a reduced sensitivity at low frequencies (remember the equal-loudness contours from figure 2.24), the reproduction of these frequencies requires large displacements of large loudspeaker diaphragms. Without the need to reproduce the lowest frequencies, the main loudspeakers can be smaller in size and are thus

Figure 4.2
Comparison of loudspeaker diaphragm displacement with and without the presence of a low-frequency signal. *At the top,* a 1 kHz audio signal is accompanied by a low-frequency signal of 60 Hz. This low-frequency signal significantly increases the diaphragm displacement compared to the situation shown at the *bottom* where only the 1 kHz audio signal is present.

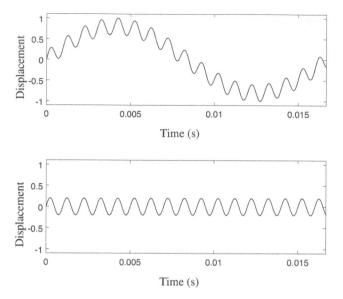

easier to integrate in your room. Furthermore, the quality of the sound reproduced by your main loudspeakers improves, because they no longer have to make the large displacements required for low bass. They can play louder without strain and their clarity in the mid frequencies greatly improves (Colloms, 2018; Harley, 2015). This is due to the fact that sound reproduction is the most accurate for small displacements of the loudspeaker diaphragm. Large displacements can introduce all kinds of nonlinear distortions (more details on such distortions follow in section 7.1.4). The example audio signals in figure 4.2 illustrate the difference in diaphragm displacement with and without the presence of a low-frequency signal. Summing up, the use of subwoofers not only improves the quality of low-frequency reproduction, but the sound quality at all the other frequencies as well. It also increases the dynamic range of your entire sound system: it can play louder without audible distortion or compression.

4.2 Floor Plan

The best way to start with the design of your room is to make a floor plan. The floor plan is a simple map that you can use to generate ideas and try out different loudspeaker and screen positions. Carefully measure your room's dimensions and draw your floor plan to scale. Make sure you include the correct position and size of the windows, doors, built-in closets, alcoves, and other structural objects; also indicate the positions of your electrical outlets, cable television connectors, and computer network sockets.

The floor plan helps you assess the suitability of the room. Does everything fit? Where can you put your video display? Can furniture and loudspeakers be located without disrupting traffic patterns? Be sure to take into account the multiple uses of the room, your lifestyle, and priorities. Most of you have to use an existing space that serves double duty as a living room. Reality often sets a limit on the number of surround loudspeakers, the maximum screen size, and the number of seats that your room can accommodate (Rushing, 2004). Even for a basic setup with five loudspeakers and one row of seats you need at least a 30 m² (320 ft²) floor area (EBU-Tech 3276, 1998; ITU-R BS.1116, 2015).

4.2.1 *Placement Options*

Every room has its own peculiarities, and therefore you need to carefully explore the different options for placing your equipment in your room. Figure 4.1 shows a typical floor plan of a rectangular listening room with five seats for the audience and seven loudspeakers surrounding them. This floor plan is a good starting point, but it is not achievable in every room. Your room may have a different size or shape. Your audience may be larger or smaller. You may have a different number of surround loudspeakers. Or, for some reason, the indicated positions for the loudspeakers conflict with traffic patterns, furniture, or structural objects in the room.

Use your own floor plan to explore your achievable placement options. You need to find suitable positions for the following five parts:

1. One video display.
2. Three front loudspeakers (left, center, and right).
3. Multiple surround loudspeakers.
4. One or several subwoofers.
5. Multiple seats for the audience.

You cannot just place these parts anywhere in the room and expect to get a good result. There are some constraints. The overall layout should look like the example in figure 4.1. The video display (part 1) and the front loudspeakers (part 2) should be in front of the audience (part 5). The surround loudspeakers (part 3) should be placed around the audience (part 5) at the back of the room. These basic relations between parts 1, 2, 3, and 5 ensure proper localization of sound; the sound appears to come from the right directions (left, right, front, and back) and the sound matches the action on the video screen. The subwoofers (part 4) do not reproduce any localizable sound. Therefore, their placement is less constrained and independent of the position of the video display (part 1) and the other loudspeakers (parts 2 and 3). The subwoofers can be placed almost anywhere in the room. The only consideration is to find a position that optimizes their sound quality. What matters is the position of the subwoofers with respect to the room boundaries, the standing wave patterns in the room, and the position of the audience (part 5) in the room.

Proper localization of sound requires a symmetrical loudspeaker layout. The optimal situation in figure 4.1 shows that the layout of the loudspeakers on the left of the audience is the mirror image of the layout on the right. Not only are the loudspeakers on both sides placed similarly with respect to the audience, they are also placed symmetrically with respect to the room boundaries. Thus, the ideal loudspeaker layout is symmetrical in two ways, with respect to the audience and with respect to the room boundaries. However, not every room allows this ideal setup. Sometimes you are forced to break up the symmetry in one way or the other. An example of an asymmetrical placement with respect to the room boundaries is shown in figure 4.3. While

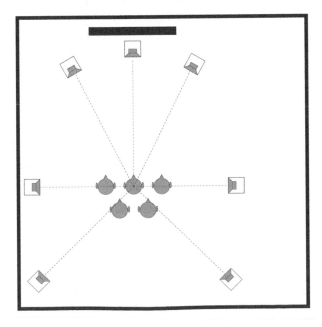

Figure 4.3
Symmetrical loudspeaker arrangement placed asymmetrically in the room.

Figure 4.4
Symmetrical front
loudspeaker arrangement
with an asymmetrical
surround loudspeaker
arrangement.

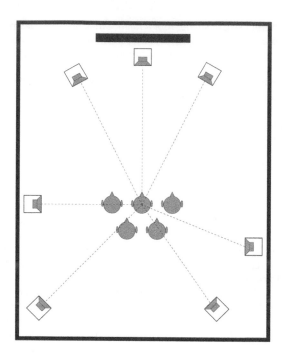

not ideal, it is still possible to achieve excellent sound quality with such an asymmetrical layout. In other cases, a room might require you to break up the symmetry of the loudspeaker arrangement itself. Beware though, as this can have a major impact on sound quality. Proper localization of sound images in the frontal soundstage is only possible with a symmetrical arrangement of the front loudspeakers. Keep them symmetrical with respect to the audience. If your placement options are rather limited, it is always better to compromise on the symmetry of the surround loudspeakers and keep the front symmetrical. An example of such a layout is shown in figure 4.4.

> **Recommendation 4.2** *Never compromise on the symmetry of the front loudspeaker arrangement. The prime listening position should form a triangle with the front left and right loudspeakers. Place the center loudspeaker and the video display centered between the left and right loudspeakers.*

The symmetry of the loudspeaker layout requires you to sit in the middle between the left and right loudspeakers. Audience members seated towards the sides are not optimally placed. They will experience a slight degradation of sound quality. It is a well-known fact that not all seats in the audience are equally good. Achieving the best possible sound quality for the entire audience is best tackled in two steps. First, you optimize the loudspeaker layout such that it provides the best possible listening experience at the prime seat. Next, you make some small adjustments such that excessive degradation at the other seats is avoided.

Symmetry with respect to the left and right room boundaries, as in figure 4.1, ensures that the left and right loudspeakers interact with the room in the same way (ITU-R BS.1116, 2015; Jones, 2015b). This interaction has two different aspects. First, the distance between the loudspeaker and the nearest sidewalls influences the spectral balance below about 200 Hz, because of the adjacent-boundary effect (see section 2.3.4). In the asymmetrical example shown in figure 4.3 the right front loudspeaker is much farther away from the nearest sidewall than the left loudspeaker is. Hence, the adjacent-boundary effect will not be the same on

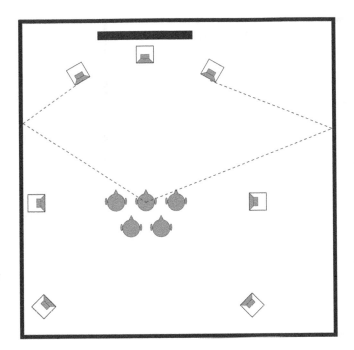

Figure 4.5
Symmetrical loudspeaker arrangement placed asymmetrically in the room. Due to this asymmetrical placement the reflected sound from the left and right sidewalls travels different distances before reaching the listener in the prime seat.

both sides and the loudspeakers will sound different for frequencies below 200 Hz. Typically, the loudspeaker that is the farthest away from the wall, in this case the right loudspeaker, will appear to be lacking in bass output. Second, the distance to the nearest sidewall influences the intensity and timing of the reflections of mid and high frequencies. In the asymmetrical example shown in figure 4.3 the reflected sound from the right loudspeaker has to travel a larger distance than the reflected sound from the left loudspeaker. Figure 4.5 clearly shows the difference in path length. As a result, the reflected sound from the right arrives later at the prime seat and is softer than the reflected sound from the left. Fortunately, our hearing system has quite some tolerance for such differences, but in extreme cases the sound quality may deteriorate and spectral balance, clarity, and localization may suffer.

Recommendation 4.3 *Ideally all your main loudspeakers should be placed symmetrical with respect to the left and right room boundaries. Beyond that, symmetry for the front loudspeakers is more important than symmetry for the surround loudspeakers.*

In some rooms perfect symmetry of the loudspeaker layout with respect to the left and right room boundaries is simply not possible. In such a case, always try to keep the front part of the room and the front loudspeaker arrangement symmetrical. If you have to choose between a symmetrical layout of the front loudspeakers or symmetrical surrounds, always choose a symmetrical layout at the front. Figure 4.6 shows an application of this rule of thumb for a loudspeaker layout in an L-shaped room.

The placement of furniture in the room, especially the larger pieces, can influence sound quality too. In a symmetrical layout (figure 4.1) careless placement of large pieces of furniture can distort the symmetry of the room. Again, try to keep the front part of the room symmetrical with respect to the loudspeakers and the sidewalls. On the other hand, in an asymmetrical loudspeaker layout, strategically placed furniture can help to recover some symmetry. Figure 4.7 revisits the asymmetrical example of figure 4.5 where the sound reflected off the sidewalls travels different distances. In figure 4.7 a large cupboard has been strategically placed

Figure 4.6
Symmetrical front
loudspeaker arrangement
placed in an L-shaped
room such that the front
loudspeaker arrangement is
symmetrical with respect to
the room boundaries.

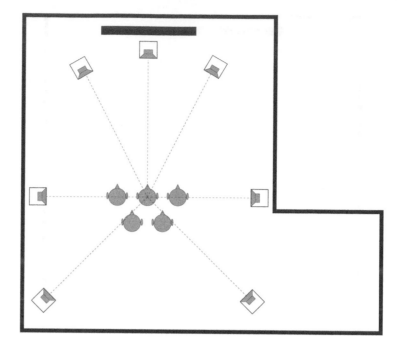

Figure 4.7
Symmetrical loudspeaker
arrangement placed
asymmetrically in the room.
A large cupboard has been
placed at the right sidewall
such that the reflected
sound from the left and
right sides of the room
travel the same distances.

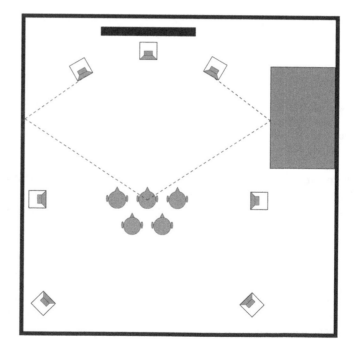

at the right sidewall such that the reflected sound from both the left and right sides of the room now travel approximately the same distance. This situation improves the symmetry with respect to the early reflections of the high and mid frequencies. However, due to the adjacent-boundary effect, the front left loudspeaker which is closer to the corner, still appears to reproduce more bass than the right loudspeaker. This difference in low-frequency output can be reduced by installing a bass absorber in the corner behind the left front loudspeaker.

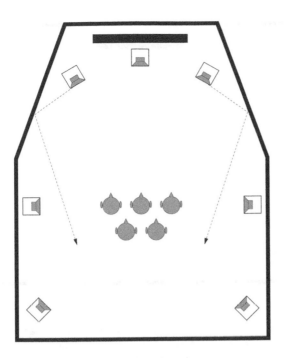

Figure 4.8
Loudspeaker arrangement
in a room with nonparallel
walls.

Bass absorbers and other acoustical devices are dealt with in Chapter 8. For now, remember that such devices can be used to improve the sound quality when you are forced to use an asymmetrical loudspeaker or room layout.

In rooms with nonparallel surfaces, I recommend that you place the front loudspeakers at the narrow end of the room (Harley, 2015; Winer, 2018). An example of such a setup is shown in figure 4.8. It shows that such a setup ensures that the sidewall reflections are directed away from the listener. Many studio control rooms are built in this fashion. The idea behind it is that strong sidewall reflections might deteriorate sound quality. In section 8.2.1 I explain in more detail the effect of reflections and the possibilities to control them.

Recommendation 4.4 *If your room has nonparallel surfaces (sidewalls or ceiling) position your front loudspeakers at the narrow end of the room.*

In all examples presented thus far, the front loudspeakers are positioned along the shortest wall of the room. Figure 4.1 clearly illustrates such a short-wall placement. The alternative would be to place your front loudspeakers along the longest wall of the room as in figure 4.9. Long-wall placement increases the distance between the front loudspeakers and the sidewalls and this weakens the sidewall reflections. Long-wall placement can be beneficial if you have a small room, because ideally your front loudspeakers should be at least 1 meter away from the sidewalls. If the loudspeakers are positioned closer to the sidewalls, the reflections of these walls may deteriorate soundstaging and localization. Another benefit of long-wall placement in a small room is that it allows you to increase the distance to the surround loudspeakers (compare figure 4.1 with figure 4.9). When you end up sitting too close to the surround loudspeakers, the illusion of envelopment (see section 3.4.5) might break down, especially when you are not exactly centered between them; the sound might appear to originate only from the loudspeaker closest to you.

Whether you choose short-wall or long-wall placement, you should never sit too close to the back wall (Briere and Hurley, 2009; Harley, 2015; Rushing, 2004; Winer, 2018). You need enough space behind you to optimally place the surround loudspeakers such that their sound

Figure 4.9
Front loudspeakers placed
along the longest wall of
the room.

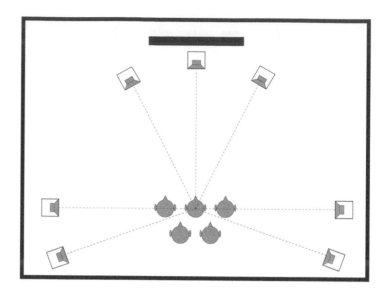

can truly envelop you. Another reason why you shouldn't sit too close to the back wall is that excessive bass buildup occurs near this wall. Recall from section 2.3.4 that certain low frequencies can resonate throughout the room. At the room boundaries, including the back wall, the energy of these resonances is at a maximum. As a result, the bass response is highly irregular close to a room boundary: the room resonance frequencies sound much louder than the other frequencies. The bass may sound like a low rumble, lacking clarity and having no transient impact. A third and final reason why you shouldn't sit too close to the back wall is that the reflections of this wall deteriorate soundstaging and localization (see section 3.4.4). While these early reflections can be reduced by covering the back wall in thick sound-absorbing acoustical panels, this is still far from ideal. You are better off moving the seating area away from the back wall. Ideally, keep a distance of at least 1.5 m (5 ft) from this wall (EBU-Tech 3276, 1998).

Recommendation 4.5 *Never put your prime listening position against the back wall. Sit at least 1.5 m (5 ft) from the back wall.*

Where you sit in the room very much determines the audibility of detrimental room resonances. The position of the loudspeakers, the acoustical properties of the room boundaries, and the furnishing of the room also influence the extent of the resonances. It is a complicated mix. The best way to optimize low-frequency sound quality is by measuring and experimenting with different layouts (more details follow in sections 4.2.6 and 8.1.3). Nevertheless, it helps if you sit in a location that is a good starting point for optimizing bass response. An often used rule of thumb states that the listening position should be at one-third (33%) or two-thirds (67%) of the length of the room between the front and back wall (Harley, 2015). Another rule of thumb states that the listening position should be 38% of the length (Winer, 2018). The exact choice is not critical as long as you avoid sitting in the exact middle between the front and back wall and stay away at least 1.5 m (5 ft) from the back wall.

The prime seat is the listening position that gives you the best possible sound quality. It is the seat that is positioned straight ahead of the center loudspeaker such that the loudspeaker arrangement on the left of this seat is the mirror image of the arrangement on the right. The other seats should be organized around the prime seat such that the sound quality at these locations does not deteriorate too much. In most rooms it is indeed possible to achieve good-quality sound for multiple seats. Obviously, the bigger your audience, the bigger your listening room needs to be. Try to keep all the seats at least 1.5 m (5 ft) away from back and sidewalls (EBU-Tech 3276, 1998).

> **Recommendation 4.6** *Make sure you arrange your seating such that it always includes the prime listening location that has the same distance to the left and right front loudspeakers.*
> **Recommendation 4.7** *Position all seats at least 1.5 m (5 ft) away from the front, side, and back walls.*

The seats next to the prime seat will be less than ideal (Rushing, 2004; Toole, 2018). Next to the prime seat, you are sitting closer to either the left or right front loudspeaker. This means that the frontal soundstage that you experience will be shifted a bit towards the side where you are sitting. The soundstage will only be centered when you sit exactly between the left and right front loudspeakers. The presence of a center loudspeaker will help to keep the sound more or less centered at the screen, but the front loudspeaker that is the farthest away from you will sound softer. A similar thing happens with the surround loudspeakers. The sensation of envelopment deteriorates as you move away from the prime seat towards the sidewalls. If you are too close to one of the surround loudspeakers the sound might appear to emerge only from that side of the room. Therefore, do not put too many seats in a row. In most rooms, three seats in a row shouldn't cause any trouble. A larger room allows you to sit farther away from the loudspeakers. This means that the relative level difference between the loudspeakers decreases. For example, if one loudspeaker is at 2 m (6.6 ft) from the listener and the other at 3 m (9.8 ft) their level difference will be 3.5 dB (see box 4.1). If you move the loudspeakers another 2 m (6.6 ft) away from the listener, the distances become $2 + 2 = 4$ m (13.1 ft) and $3 + 2 = 5$ m (16.4 ft) and the level difference will be reduced to 1.9 dB. This is the inverse-square law at work (see section 2.3.3).

The seats in front or behind the prime seat will also be less than ideal (Toole, 2018). The sound level balance between front loudspeakers and surround loudspeakers changes when you move towards the front or the back of the room. If you sit more towards the front, then the surround loudspeakers sound softer; if you sit more towards the back they sound louder. In both cases the front-back balance will be distorted. Again, the farther away you sit from all the loudspeakers (fronts and surrounds), the smaller the level difference becomes.

If you sit too close to a loudspeaker the sound quality suffers. Close to the loudspeaker you hear a different timbral quality than farther away. The sound of the individual loudspeaker drivers needs to travel a certain distance to be able to integrate into a pleasing whole. In the so-called *far field* the timbral quality has become independent of the listening distance. Typically, the far field starts at three to ten times the size of the sound source (Toole, 2018). Make sure your listening position is situated in the far field. In practice, this means that all your seats should be at least 2 m (6.6 ft) away from any loudspeaker in your room. If the tonal quality changes a lot when you move a bit farther away, then you are still sitting too close (Harley, 2015).

BOX 4.1 RELATIVE LEVEL DIFFERENCE

The level difference L in dB between two conventional loudspeakers placed at two different distances from the listener is given by:

$$L = 20 \log_{10} \left(\frac{d_2}{d_1} \right)$$

where d_1 and d_2 are the distances from the listener to the loudspeakers.

Recommendation 4.8 *Make sure the distance between you and your main loudspeakers is large enough for the sound of the different drivers to integrate. About three to ten times the height of the front of the loudspeaker cabinet should be sufficient.*

Ideally, you should be seated the same distance from each of the loudspeakers when you are at the prime listening location. The loudspeakers should be positioned around you on the circumference of a circle, as depicted in figure 4.10, but in practice you often end up sitting closer to the surround loudspeakers. Your AV-controller is equipped with delay settings to compensate for such a situation. In fact, most controllers allow you to input the distances between your prime listening position and each of the loudspeakers. The signals that the controller sends to the loudspeaker are then delayed appropriately such that the sound from all the loudspeakers arrive at the prime seat at exactly the same time instant (see also section 7.5.3).

4.2.2 Position of the Video Display

Where you sit in the room not only influences how you experience the sound from your loudspeakers, it also very much influences how you experience your video screen. The closer you sit to the screen, the larger it appears. Size does matter, because a large screen gives you a heightened sense of reality (Hatada et al., 1980; ITU-R BT.1769, 2008; Sugawara et al., 2007). When you look at a large screen, you are less aware of the physical boundaries of it and you feel more immersed in the movie. A large screen allows you to use head and eye movements to select parts of the image to focus on and to track moving objects across the screen. This increases the feeling of naturalness but may sometimes result in motion sickness if the screen is too large. By contrast, a small screen is tiring to look at, because watching it does not involve any head or eye movements.

Figure 4.10
Loudspeakers on the circumference of a circle in a 7.1-channel system.

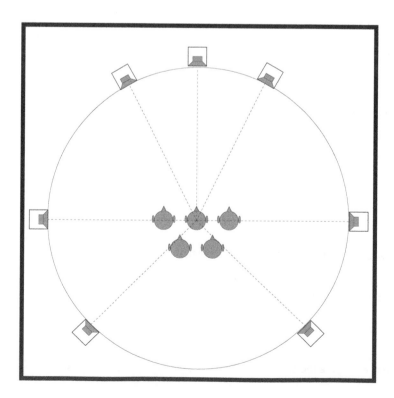

How big should your screen be? Well, that obviously depends on how far away you sit from it. Figure 4.11 shows that the horizontal viewing angle α decreases as you sit farther away from the screen (see box 4.2 for the mathematical relation). To give you that heightened sense of reality, the horizontal viewing angle should span a significant part of your entire visual field. Research has shown that horizontal viewing angles ranging from 30° to 80° produce the required psychological effects. The larger the angle the greater the effect, but when the viewing angle exceeds 80° the sensation of reality doesn't increase any further (Hatada et al., 1980).

The closer you sit to the screen, the higher the sense of reality. But there is a limit on how close you can sit. When you are too close to the screen you will start to see the individual pixels; they no longer blend together and the image becomes pixelated, which of course destroys the feeling of reality. Therefore, the optimal distance to your screen not only depends on its size, but also on its resolution. The higher its resolution, the more pixels it has, and the closer you can sit. Most people with normal vision can discern a pixel if it spans an angle of about 1/60 degrees or 1 arcmin (Bodrogi and Khanh, 2012; Drewery and Salmon, 2004). Based on this limit on the acuity of the human visual system, we can determine the optimal viewing distance. Table 4.1 shows the optimal viewing distances (v) for several

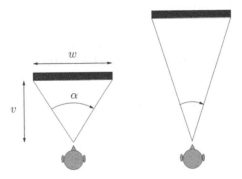

Figure 4.11
When you sit farther away from your video screen as shown on the *right*, the viewing distance v increases and the horizontal viewing angle α decreases. As a result, the video screen looks smaller, because it occupies a smaller area of your entire visual field.

BOX 4.2 VIEWING ANGLE AND DISTANCE

The relation between the viewing distance v and the viewing angle α in figure 4.11 is given by:

$$\alpha = 2\tan^{-1}\left(\frac{w}{2v}\right) \tag{4.1}$$

where w is the width of the screen.

Table 4.1 Optimal viewing distances and horizontal viewing angles based on visual acuity. The viewing distance is expressed as a multiple of the screen diagonal d. The viewing angles are based on a 16:9 screen.

Video Format	Number of Horizontal Lines n	Viewing Distance v	Horizontal Viewing Angle α
DVD NTSC	480	3.5 d	14°
DVD PAL	576	2.9 d	17°
HD	720	2.3 d	21°
HD/Blu-ray disc	1080	1.6 d	31°
UHD / 4K	2160	0.8 d	58°
8K	4320	0.4 d	96°

popular video resolutions. The optimal distances are expressed in terms of the diagonal (*d*) of the video screen. They have been calculated using the formula presented in box 4.3. For example, the optimal viewing distance for an HD video screen with 1080 horizontal lines equals 1.6 times the diagonal of the screen. If you sit farther away from the screen, you will not be able to discern all of the 1080 lines and you will miss a lot of detail. If you sit closer than 1.6 times the diagonal, you will start to see the individual pixels, and the image starts to look like a grid of colored squares. These squares become bigger the closer you sit resulting in an artificial-looking image.

Choose your viewing distance based on the resolution of your video display, not on the resolution of the video material that you are playing, the reason being that your display will convert the video material to its native resolution. It will either downscale or upscale the video material to make sure that the entire video image is visible and that it fills the entire display area. For example, if you are viewing a DVD on an HD video display, the display has more lines (1080) than the video material on the DVD (480). Your display will use upscaling to convert the 480 lines from the DVD to the 1080 lines needed to fill the entire display area. The additional lines are computed from the existing ones using a mathematical technique called *interpolation* (see section 6.3.1). The resulting upscaled picture will not be as sharp as an original image with 1080 lines; it looks a bit soft in comparison, but at least it will not look awfully pixelated as a display with only 480 lines would look.

BOX 4.3 OPTIMAL VIEWING DISTANCE BASED ON VISUAL ACUITY

The optimal viewing distance is the distance to the video screen at which you can just discern the different pixels. One row of pixels is called a *scan line*. The height *s* of one scan line equals the height of the screen *h* divided by the number of scan lines *n*, that is, $s = h/n$. At a viewing distance *v* one scan line spans an angle θ which equals

$$\theta = 2 \tan^{-1}\left(\frac{s}{2v}\right)$$

Note the similar to equation (4.1) in box 4.2. A person with normal vision can discern a θ as small as 1/60 degrees (Bodrogi and Khanh, 2012; Drewery and Salmon, 2004). Taking this value for θ, the optimal viewing distance *v* can be expressed as a multiple of the screen height *h*:

$$v = \frac{s}{2 \tan(\theta/2)} = \frac{h}{2n \tan(\theta/2)} \approx 3438 \frac{h}{n}$$

It can also be expressed as a multiple of the screen diagonal *d* of a 16:9 screen, because

$$d^2 = h^2 + w^2 = \left(1 + \left(\frac{16}{9}\right)^2\right) \cdot h^2$$

which yields

$$v \approx \frac{3438}{n} \frac{d}{\sqrt{(1 + (16/9)^2)}} \approx 1685 \frac{d}{n}$$

Table 4.1 shows this optimal viewing distance for different choices of the number of scan lines *n*. The corresponding horizontal viewing angles are obtained from equation (4.1).

Recommendation 4.9 *Determine the distance to your screen based on its resolution. Sit as close as possible to your screen to maximize the sense of reality, but don't sit any closer than the optimal viewing distances in table 4.1.*

The resolution of your display limits how close you can sit without seeing the pixels, and it also limits the maximum horizontal viewing angle that you can achieve. Table 4.1 shows that an HD video screen (1080 lines) spans a horizontal viewing angle of 31° at its optimal viewing distance. This viewing angle is just barely enough to create a weak sense of reality and immersion. The immersion effect is much stronger for larger angles between 30° and 80°. A UHD screen (2160 lines) gives you the possibility to sit much closer to the screen without seeing its pixels and thus allows you to achieve a viewing angle of 58°. To put these numbers into perspective, let's have a look at the typical horizontal viewing angles in a commercial movie theater. The prime seat in such a theater (about 2/3 into the room) is situated such that the horizontal viewing angle is between 45° and 50° degrees (Allen, 2000). Therefore, to ensure a minimal sensation of immersion in a home theater, it is recommended that you aim for a horizontal viewing angle of at least 40° (ITU, 2009a).

An immersive experience requires a really big video screen. To understand this conclusion, take a look at figure 4.12, which shows how the optimal viewing distance changes with screen size. This figure contains three lines that correspond to the optimal viewing angles of 21°, 31°, and 58°, which are the optimal viewing angles for 720, 1080 and 2160 lines of screen resolution, respectively. The figure shows that at a typical viewing distance of about 3 m (10 ft), you need a screen with a diagonal of more than 2 m (80 inch) to achieve a viewing angle larger than 31°. The most practical way to achieve such a large image is to use a front-projection system. Flat-panel displays of this size do exist, but they are quite expensive compared to a front-projection system.

Instead of buying a larger screen, you could of course also sit closer to your existing screen to increase the immersive experience. As long as your display has plenty of horizontal lines, you will be able to get closer without seeing any pixel structure. However, there is another limit on how close you can sit. This limit is determined by your loudspeakers. Most loudspeakers

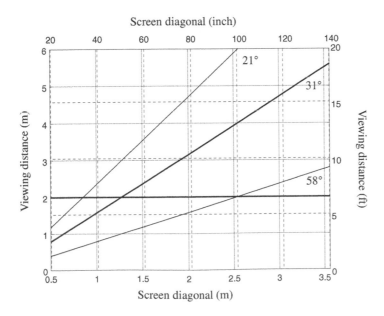

Figure 4.12
Viewing distance (in meters or feet) for the optimal viewing angles of 21°, 31°, and 58° as a function of the screen diagonal (in meters or inches). The *horizontal line* indicates that a minimum viewing distance of 2 m (6.6 ft) is recommended for all screen sizes.

sound their best if you do not sit any closer than 2 m (6.6 ft) to them. Since your video display will either be aligned with your center loudspeaker, or be slightly behind it, this lower limit of 2 m (6.6 ft) also applies to your viewing distance. In figure 4.12, the lower limit of 2 m (6.6 ft) is indicated by the horizontal line. This figure clearly shows that to achieve a minimum viewing angle of 31° at the closest distance to the screen (2 m or 6.6 ft), your screen needs to have a diagonal of at least 130 cm (50 inch).

In practice, most people are not really keen on installing large screens in their living room. As a result, most people have a flat-panel display that is too small for the distance from which they view it. Or, in other words, they sit too far away from the screen. Not only does this reduce the immersive experience, it also limits the amount of detail that can be discerned in the image. You should at least aim for a viewing angle of 21° for an HD video display with 1080 lines. At this viewing angle your eyes can discern about 720 different lines on the screen, despite the fact that the display has many more lines. Similarly, for a UHD video display with 2160 lines you should aim for a viewing angle of at least 31°, which allows you to discern about 1080 different lines.

> **Recommendation 4.10** *Choose your screen size and viewing distance using figure 4.12 such that you achieve a viewing angle of at least 21° with an HD flat-panel display and an angle of at least 31° with a UHD flat-panel display. For a more immersive viewing experience use a large projection system and aim for a viewing angle of at least 31° with an HD projector, and at least 40° with a UHD projector.*

Position your prime seat centered in front of the screen at the optimal viewing distance and arrange the other seats around it. Add multiple seats in a row and if necessary add multiple rows of seats. To provide everyone with an optimal experience you shouldn't put too many seats in a row. The seats to the left and right of the prime seat are not centered with respect to the display, and as a result the people sitting in those seats view the screen at an oblique angle. If this angle becomes too big, they need to turn their heads sideways too much and viewing becomes uncomfortable. Therefore, it is recommended keeping all the seats within 10° from the sides of the screen as illustrated in figure 4.13 (ITU, 2009a). Another factor to take into account, especially with LCD flat-panel displays, is that the contrast and color of the display may shift if the seats are positioned too far off to the side. This factor also limits the acceptable width of your seating area, as I explain in more detail in section 6.1.1. In most cases, keeping the seats within 10° from the sides of the LCD display will keep you out of trouble.

Figure 4.13
All the audience members should be seated within 10° from the sides of the screen.

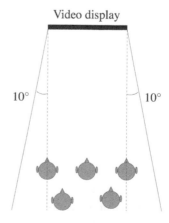

Recommendation 4.11 *Keep all seats within 10° from the sides of the screen as illustrated in figure 4.13.*

Each seat should have a clear line of sight to the screen. When you have multiple rows you need to take special care to ensure that everyone has an unobstructed view of the screen. With two rows you have the option to shift the seats in the second row by half a seat, such that the viewers can see the screen between the heads of the people in front of them, as illustrated in figure 4.14. To achieve this alternate viewing, you need to carefully adjust the spacing between the rows and the spacing between the seats in each row. Typically, the spacing between the rows is about 90 cm (3 ft), measured from the front of one seat to the front of the next row's seat (ITU, 2009a; Rushing, 2004).

In a typical living room having more than two rows of seating is not very practical. Even positioning two rows may be a challenge if the room is used for other activities. One solution would be to make the second row of seats removable. For example, using some lightweight chairs that can be easily moved and arranged differently for other uses of the room. In a dedicated home theater room, you could use massive reclining theater-style chairs, if you have the space for them. Avoid chairs that have a high back, because they will obstruct the sound coming from your surround loudspeakers (Rushing, 2004). Dedicated home theater seating often has a back that reclines about 15° from the vertical to compensate for the fact that the normal line of sight for a sitting person is about 15° below the horizontal line. These reclined seats make viewing more comfortable (CinemaSource, 2001a).

If you need more than two rows of seating in a dedicated home theater room, you can put the seating on risers such that the people in each row look over the heads of the people in the previous rows. Such a vertical arrangement of seats is shown in figure 4.15. It is called *every-row vision*. Since the average distance between the eyes of a viewer and the top of the viewer's head is about 12 cm (5 inch), you should aim for a head clearance of at least 15 cm (6 inch) between rows (ITU, 2009a). Use 30 cm (12 inch) seat risers if you want to be on the safe side. If you have a relatively low ceiling, you probably want to keep the height of the seat risers as low as possible. In this case you could instead use *every-other-row vision* where the

Video display

Figure 4.14
Horizontally staggered seats allow the people in the second row an unobstructed view of the video display.

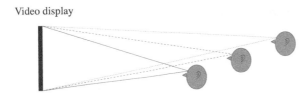

Video display

Figure 4.15
Vertically staggered seats allow the people in every row an unobstructed view of the video display.

rows are alternately shifted horizontally by half a seat such that you look between the heads of the people right in front of you and over the heads of the people two rows ahead of you, as in figure 4.14 (CinemaSource, 2001a).

Recommendation 4.12 *Make sure every seat has a clear line of sight to the screen.*

Viewers in all rows should feel comfortable watching the screen without having to look too far up. Therefore, as a general guideline, place the screen such that your audience does not have to look up more than 30° to see the top of the screen (ITU, 2009a).

Recommendation 4.13 *Place the video screen such that viewers don't have to look up more than 30° to see the top of the screen.*

4.2.3 Positions of the Front Loudspeakers

The front loudspeakers are arranged around your video screen to ensure that their sound lines up with the action on the screen. As its name implies, the front center loudspeaker is centered with the screen, and the left and right front loudspeakers are on the left and right sides of the screen, respectively. No surprises here. However, besides lining up the sound with the picture, the front loudspeakers also have another job to accomplish: they have to create a believable frontal soundstage that has a certain width and depth. The amount in which they succeed very much depends on their exact positions with respect to the video screen, the listener, and the nearest walls of the room.

The front loudspeakers should either be free standing or flush mounted in the front wall (Toole, 2018). In movie theaters the loudspeakers are almost always built into the front wall and placed behind a perforated theater screen. The small holes in this screen allow the sound to pass while being small enough to stay invisible for the audience. In your living room, free-standing loudspeaker placement is probably the most convenient option. It not only saves you from the hassles of remodeling your room, but it also allows you to freely experiment with loudspeaker position to optimize your sound. However, positioning your loudspeakers with respect to the nearest walls becomes a critical factor in achieving the best possible sound. Remember from section 2.3.4 that the reflections off the nearby walls alter the frequency response of the loudspeaker in the region below 200 Hz. By contrast, if you build the loudspeakers into the wall, there will obviously be no reflection from that wall. Still, you are better off using free-standing loudspeakers; you just need to experiment to find their best positions. I will give you detailed advice on fine-tuning the distances to the nearest walls in section 4.2.5, but right now let's start with the basics.

The recommended front loudspeaker arrangement is shown in figure 4.16a: the listener and the left and right front loudspeakers form a triangle with equal sides, such that the distances between the listener and the loudspeakers and the distance between the left and right loudspeakers are all equal. The center loudspeaker is placed at the same distance from the listener, such that it is situated slightly behind the left and right loudspeakers and forms a gentle arc with them (EBU-Tech 3276-s1, 2004; ITU-R BS.775, 2012). The video screen is either flush with the front of the center loudspeaker or placed behind the center loudspeaker, but never in front of it.

Recommendation 4.14 *Put the video screen flush with the front of the center loudspeaker, or put it behind the center loudspeaker, but never in front of it.*

In a small room where space is limited, the recommended placement for the center loudspeaker might not be the most convenient. The center loudspeaker might end up being too close to the wall, or the left and right loudspeakers might intrude too far into the room. In

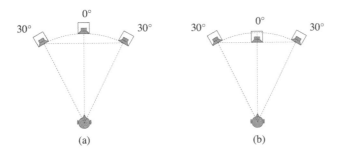

Figure 4.16
Front loudspeaker
arrangements:
(a) recommended
arrangement with all
loudspeakers at the
same distance from the
listener and *(b)* alternative
arrangement with all
loudspeakers lined up.

such a case you could bring the center loudspeaker a bit more towards the listener, up to the point where the three front loudspeakers are exactly lined up as shown in figure 4.16b. In fact, you can place the center loudspeaker anywhere on the line from the center point on the arc to the center point on the line connecting the left and right loudspeakers. However, when the center loudspeaker isn't positioned on the arc, it is closer to the listener than the other two loudspeakers are and therefore its sound will arrive earlier in time and will be a bit louder. Hence, the center loudspeaker will stand out as sounds will bunch together in the center of the soundstage, and the overall soundstage width will be reduced. To avoid this undesirable side effect, you need to slightly reduce the level of the center loudspeaker and introduce an electronic delay (Holman, 2008; ITU-R BS.775, 2012). Every A/V controller and receiver has a built-in facility to dial in such settings in the configuration menu. I will show you how to do this, along with the delay and level settings for your other loudspeakers, in sections 7.5.3 and 7.5.4.

> **Recommendation 4.15** *Put the center loudspeaker at the same distance from the prime listening position as the left and right front loudspeakers. The three front loudspeakers should form a gentle arc with the center loudspeaker exactly centered on the arc between the left and right loudspeakers. Alternatively, place the center loudspeaker between this position on the arc and the line connecting the left and right loudspeakers and use an appropriate time delay and level setting in your A/V controller or receiver.*

Ideally, all front loudspeakers are positioned such that their high-frequency drivers are approximately at the same height as the ears of the seated listener at the prime seat. Typically, this height is about 1.2 m (4 ft) from the floor (EBU-Tech 3276, 1998; ITU-R BS.775, 2012; ITU-R BS.1116, 2015). The height at which you place your loudspeakers will affect the spectral balance in the mid and especially the high frequencies. The main reason is that at high frequencies a loudspeaker radiates sound in a tightly focused beam (see section 2.3.3) and if your ears are outside of this beam the loudspeaker will sound less bright.

The optimal height of the center loudspeaker, positioned at ear height, conflicts with the optimal position of the video screen. A compromise has to be made. If you put your center loudspeaker at ear height and position your screen above it, the screen will be uncomfortable to watch, because it is way too far up. A better compromise is to place the video screen at a comfortable viewing height (see recommendation 4.13) and place the center loudspeaker either below or above the screen and at the same time try to keep the height of the center loudspeaker as close as possible to the heights of the other two front loudspeakers. Especially with a larger screen the position below seems to be better than the one above (Steinke, 2004). Try to keep the deviation in height within 6° (Toole, 2018), that is, not more than 10 cm for each meter of distance between the loudspeaker and the listener (or equivalently 1.26 inch for each foot). Figure 4.17 illustrates this recommendation. Never put the center loudspeaker high up, above the screen and tilt it down towards the listeners. The tilt will affect the spectral

Figure 4.17
Front loudspeaker
arrangement with the
center loudspeaker
positioned beneath the
video display. The height
difference between the
high-frequency drivers
should be less than 10 cm
for each meter of distance
between the center
loudspeaker and the listener
(1.26 inch for each foot).

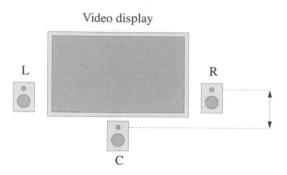

balance of the loudspeaker and this balance will audibly shift if you move just slightly forward or backward (Newell and Holland, 2007; Winer, 2018).

Recommendation 4.16 *Place the left and right front loudspeakers at the same height, such that the high-frequency driver is at ear height when you sit in the prime listening location (about 1.2 m or 4 ft from the floor).*
Recommendation 4.17 *Place the center loudspeaker beneath the video screen such that its height is as close as possible to the heights of the left and right front loudspeakers. The height difference between the high-frequency drivers should be less than 10 cm for each meter of distance between the center loudspeaker and the listener (1.26 inch for each foot).*

Large floor-standing loudspeakers usually do not need to be raised in height, because the designer already took the height of the listener's ears into account. Smaller loudspeakers, including dedicated center loudspeakers, need to be raised such that their high-frequency driver is positioned at the appropriate height. It is best to put these smaller loudspeakers on a dedicated loudspeaker stand. A good stand is sturdy and does not vibrate in response to the sound produced by the loudspeaker. What makes a good stand is explained in more depth in section 7.1.7. Try to avoid putting loudspeakers on top of shelves and cabinets, or even worse, inside a cabinet (Rushing, 2004; Toole, 2018). Shelves and cabinet walls will vibrate in response to the sound produced by the loudspeaker, and they will audibly resonate at certain frequencies. These audible resonances will negatively impact the overall sound quality by changing the overall spectral balance and reducing clarity. Nevertheless, a lot of people put their center loudspeaker in front of their video display on top of a low cabinet that houses some other audio and video equipment. This appears to be a convenient place to put such a speaker. Keep in mind that the sturdier and heavier the cabinet, the less prone it is to resonate. Also keep in mind that it is better to put the loudspeaker on top of the cabinet instead of inside an otherwise empty resonating cavity. Finally, align the front of the loudspeaker absolutely flush with the edge of the cabinet. If the cabinet protrudes in front of the loudspeaker, it reflects and diffracts the sound and in the process deteriorates the clarity of the sound (Harley, 2015).

The distance between the left and right loudspeakers has a substantial effect on the spatial quality of the frontal soundstage. Two factors are at play. First, to create a believable experience, the frontal soundstage should have a considerable width that extends beyond the boundaries of the video screen and sometimes even beyond the boundaries of the room. This requires a wide spacing of the left and right front loudspeakers. In a typical living room a distance of 2 to 3 meters (6.6 to 10 ft) is recommended; in larger rooms a distance of up to 5 m (16 ft) might be acceptable (EBU-Tech 3276, 1998; ITU-R BS.1116, 2015).

Recommendation 4.18 *Place the left and right front loudspeakers 2 to 3 meters apart (6.6 to 10 ft) or up to 5 m apart (16 ft) in larger rooms.*

Second, the sounds that accompany the action on the video screen should seem to emerge from a believable location in the frontal soundstage. When the left and right loudspeakers are placed too far from the edges of the video screen, the sounds may not seem to align and track the on-screen action. Such a discrepancy can be confusing and distracting (Harley, 2015; Holman, 2008). As long as the location of the sound and the action on the screen do not differ by more than 4°, the displacement is not noticeable. If the difference becomes more than 15°, it starts to annoy the majority of people (Komiyama, 1989; Theile, 1991). Therefore, it is recommended placing the left and right loudspeakers within 15° from the sides of the screen, as indicated in figure 4.18. Since the recommended front loudspeaker arrangement is based on a triangle that has equal sides, the left loudspeaker will be positioned 30° to the left of the center loudspeaker and the right loudspeaker 30° to the right. A simple calculation reveals that the horizontal viewing angle of the screen (α) needs to be at least 30° to keep the left and right loudspeakers within the recommended 15° from the sides of the screen.

Recommendation 4.19 *Place the left and right front loudspeakers within 15° from the sides of the screen.*

Only the larger video screens will allow you to place the left and right loudspeakers at 30° to the left and right, while at the same time keep them within 15° from the sides of the screen. For smaller screens, compromises have to be made. Try to keep the loudspeakers within the 15° limits, but never place them closer than 22° to the center loudspeaker. In other words, make sure that the angle β in figure 4.18 between the center and the other front loudspeakers stays between 22° and 30°. Keep in mind that the larger the angle β, the larger the soundstage and the more engaging the experience (Bech, 1998; Toole, 2018). With the loudspeakers at $\beta = 30°$, the triangle formed by the listener and the left and right loudspeakers has equal sides. Thus, at $\beta = 30°$ the distance between the loudspeakers b equals the listening distance ℓ. The loudspeakers can be placed at a smaller angle β by slightly increasing the listening

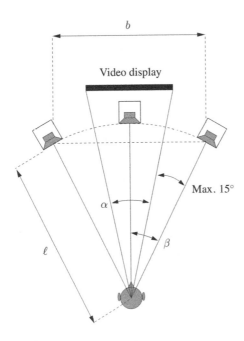

Figure 4.18
The left and right front loudspeakers should be placed within 15° from the sides of the video display. This means that $\beta - \alpha / 2$ should be less than 15° where β is the angle at which the left and right loudspeakers are placed and α is the viewing angle of the display.

Figure 4.19
Relation between the
listening distance and the
distance between the left
and right loudspeakers
(in meters and feet). Each
diagonal line corresponds
to a different angular
placement; their spacing
is 1°. Always keep the
listening distance between
the *thick lines* that
correspond to the extremes
of 22° and 30°.

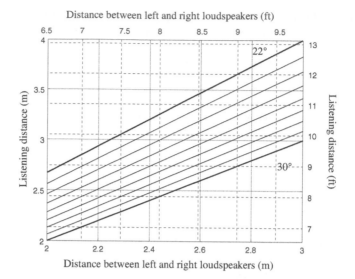

BOX 4.4 LISTENING DISTANCE AND ANGULAR POSITION OF THE FRONT LOUDSPEAKERS

The relation between the listening distance ℓ and the angle β at which the left and right loudspeakers are placed is given by:

$$\beta = \sin^{-1}\left(\frac{b}{2\ell}\right)$$

where b is the distance between the left and right loudspeaker.

Table 4.2 Recommendations for positioning the left and right front loudspeakers (EBU-Tech 3276, 1998; EBU-Tech 3276-s1, 2004; ITU-R BS.775, 2012; Toole, 2018).

	Target	Tolerance
Angular position L and R (β)	30°	22°–30°
Distance between L and R (b)	2–3 m	2–5 m
Listening distance (ℓ)	b	between b and $1.33\,b$

distance ℓ. You can use figure 4.19 to determine the listening distance that corresponds to a certain angle β between 22° and 30° (see also box 4.4). For easy reference, I have summarized the above recommendations for placing the left and right front loudspeakers in table 4.2.

Recommendation 4.20 *Place the left and right front loudspeakers at 30° from the center loudspeaker. If they end up too far away from the sides of the video screen, you can slightly reduce this angle up to a minimum of 22°.*

It is important to take the viewing angle of your display into account when you determine the angular position of the left and right loudspeakers. For example, if you have a relatively small video screen and aim to achieve the minimum recommended viewing angle of 21° (recommendation 4.10), you should place the left and right loudspeakers about 25° from the center loudspeaker in order to keep them within 15° of the sides of the screen.

The three front loudspeakers should be aimed at the audience and everyone in the audience should have a clear line of sight of the high- and mid-frequency drivers. The center loudspeaker is always aimed straight ahead, but the left and right loudspeakers should be angled in to a certain extent. If not, the reproduction of the high and mid frequencies will suffer. Recall that loudspeakers radiate high- and mid-frequency sounds in tightly focused beams (see section 2.3.3). Listeners outside of these beams will receive less high- and mid-frequency sound. They will hear a less bright sound with a distorted tonal balance. High-quality loudspeakers maintain the correct tonal balance within a beam that extends about 15° to either side of the loudspeaker (Toole, 2018). Therefore, make sure that your front loudspeakers are aimed such that all your audience members are seated within these 30° beams, as in figure 4.20. Also make sure that there are no objects between the loudspeakers and the audience that obstruct these beams. At high and mid frequencies the wavelengths of the sound range from a few millimeters to a few centimeters, so even objects that are quite small can hinder the propagation of the sound (see section 2.3.3).

Recommendation 4.21 *Aim the front left and right loudspeakers at the audience and position the audience such that everyone is covered by the 30° beams from each of the three front loudspeakers.*

Audience coverage is not the only reason for aiming the left and right front loudspeakers. Equally important, aiming strongly influences the spatial quality of the frontal soundstage, especially presence, soundstage width, and image focus. When the loudspeakers are properly aimed, you get a wide and continuous soundstage that allows you to exactly pinpoint the location of the sound sources. Therefore, I recommend that you not only aim the loudspeakers to completely cover your audience, but also try to achieve the best possible soundstaging at your prime listening seat. The amount at which the loudspeakers are angled in is commonly called *toe-in*. Too little toe-in results in a wide and spacious soundstage where it

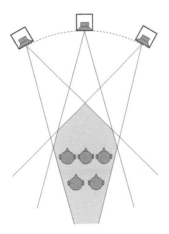

Figure 4.20
Front loudspeaker arrangement where the front left and right loudspeakers are aimed at the audience such that everyone is within the 30° beams from each of the three front loudspeakers.

is difficult to precisely pinpoint the location of the individual images. Or, it may result in a discontinuous soundstage where sounds that move from one side to the other might suddenly jump as if there are gaps in the soundstage. On the contrary, too much toe-in creates a tightly focused soundstage with greatly reduced width that sounds closed, narrow, and less spacious. In addition, image separation may suffer as images start to overlap. Besides these spatial quality aspects, toe-in also affects the tonal balance, in particular the amount of high-frequency sounds. Furthermore, toe-in alters the ratio of direct to reflected sounds, because as the loudspeakers are angled in they will direct more sound at the listener and less towards the sidewalls. The optimal amount of toe-in strongly depends on the type of loudspeaker that you have, its position in the room, and the acoustics of the room. The only way to determine the right amount is to experiment with different angles and listen to the results (Harley, 2015).

The left and right loudspeakers must always have the same amount of toe-in. The easiest way to check this is to measure the distance from the rear wall to the two rear corners of the loudspeaker as indicated in figure 4.21 (Harley, 2015). Alternatively, you could hold a laser pointer against the side of the loudspeaker to see where it points (McLaughlin, 2005). When the loudspeakers have the same amount of toe-in you will see the same amount of their sidewalls when you are seated in the prime seat.

Recommendation 4.22 *Ensure that the front left and right loudspeakers have the same amount of toe-in.*

The optimum amount of toe-in varies with the distance b between the left and right loudspeakers. In general, the smaller this distance, the smaller the toe-in. To achieve the best possible frontal soundstage, you need to experiment with changing both the toe-in and the distance between the loudspeakers at the same time (Rives, 2005). The larger the distance between the loudspeakers, the larger the soundstage, but if the loudspeakers are too far apart, they will start to sound as two separate sound sources and no longer create a continuous soundstage across the front of the room. Use the following procedure to find the optimal setup (Harley, 2015; Rives, 2005):

1. Put your A/V controller/receiver into stereo (two-channel) mode to disable the center and surround loudspeakers. Play a high-quality recording of a female singer where the voice is well-defined and positioned in the center of the frontal soundstage.
2. Start with the loudspeakers close together with a minimum separation of $b = 2$ m (6.6 ft). Aim the loudspeakers at a point that is $\ell/2$ meter behind the listener, where ℓ is the distance between the listener and the loudspeakers (see figure 4.18).
3. Gradually move the loudspeakers further apart and adjust their angle such that they remain aimed at the point $\ell/2$ meter behind the listener. Increase the distance approximately 10 cm (4 inch) at a time and listen to how the soundstage changes. The voice should stay well-defined between the two loudspeakers. As you move the loudspeakers

Figure 4.21
The left and right loudspeakers should have the same amount of toe-in. An easy way to check this is to measure the distance from the wall behind the loudspeakers to the two rear corners of each loudspeaker. The loudspeakers have the same amount of toe-in when $x_L = x_R$ and $y_L = y_R$.

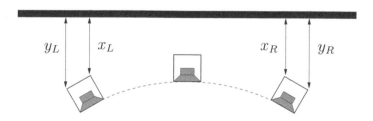

farther apart, you will eventually reach a point where the vocal will sound more diffuse, sound less focused, and appear to be coming from two separate loudspeakers. At this point, the loudspeakers are too far apart and you should slightly reduce their separation. Remember not to position the loudspeakers too far out; keep them within 15° from the sides of your video screen.

4. When you have reached the optimal distance, it is time to adjust the toe-in. Point the loudspeakers right at the listening position, and gradually reduce their toe-in by pointing them slightly outward each time. Again listen to the female voice and compare the results. You will reach a point where the vocal ceases to come from a well-defined position centered between the loudspeakers. At this point, the angle is too far out, and you should turn the loudspeakers slightly inward.

5. When the angle found in step 4 differs a lot from the angle that you started with in step 2, it is a good idea to revisit step 3 and check whether you need to adjust the distance between the loudspeakers. Usually, two iterations is all that is needed.

The toe-in of the left and right loudspeakers also influences the spatial quality of the frontal soundstage when the center loudspeaker is active. All three loudspeakers should play nicely together, such that when sounds are panned across the front they move seamlessly from one side to the other. Sometimes you need to slightly adjust the toe-in to reach the optimal balance. Again a listening test is the best way to check this out. Before attempting such a test it is important to set the correct delay and level for the center loudspeaker. If the center loudspeaker is slightly ahead in time or slightly louder than the other two loudspeakers, it will always stand out regardless of the toe-in of the other loudspeakers. In that case, the sound will have a tendency to bunch up in the middle around the center loudspeaker. So make sure you have the correct settings (see sections 7.5.3 and 7.5.4). I recommend listening to artificially generated noise signals that travel between the three front loudspeakers. You can find such noisy test signals for example on the DVE HD Basics Blu-ray disc (see appendix A). One such test signal moves in discrete steps from left to right and back again. It stops at the left, center, and right loudspeaker positions and at the positions halfway between these loudspeakers. These in-between positions should sound as if they come from the exact midpoints between the loudspeakers (see figure 4.22). If they sound too close to the center loudspeaker you should try to move them further out by reducing the toe-in of the left and right front loudspeakers. Another useful signal to listen to is a noise source that moves smoothly from one side to the other. With the correct loudspeaker setup, it should move smoothly at a regular speed from one side to the other without being pulled to any particular loudspeaker and without leaving any gaps in the soundstage. Another useful noisy test signal for evaluating toe-in is the so-called LEDR (Listening Environment Diagnostic Recording) test on the Audio Check website www.audiocheck.net. This test consists of pinna-filtered noise signals that appear to float around your loudspeakers and have been specially crafted to evaluate the accuracy of stereo image reproduction (Jones et al., 1986).

Recommendation 4.23 *Use a listening test to fine-tune the amount of toe-in of the left and right front loudspeakers. Listen with a disabled center loudspeaker to check for two-channel imaging and listen with an active center loudspeaker to check for three-channel imaging.*

Reflective objects near your front loudspeakers can significantly degrade imaging in the frontal soundstage. It is best to avoid any reflective objects, such as equipment racks and

Figure 4.22
A useful signal to test imaging in the frontal soundstage consists of noise that moves from left to right and back again in discrete steps. It stops at the left, center, and right loudspeaker positions and at the positions halfway between these loudspeakers indicated by the two *large crosses*.

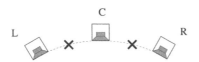

furniture in the direct vicinity of the loudspeakers, especially in the area between the left and right front loudspeakers. Moving reflective objects from between the loudspeakers to the sides of the room often results in a huge improvement in image focus and soundstage depth. If this is not possible, it is best to move these objects as far behind the loudspeakers as possible. Thus, if you have a flat-panel video display, it is better to hang it on the wall some distance behind the center loudspeaker than to align it flush with this loudspeaker. If the wall behind the loudspeakers contains any windows, it is best to hang drapes in front of them to absorb some of the reflected sound (see section 8.2.2). The LEDR test mentioned above is very sensitive to reflections that degrade imaging. Use this test to check whether imaging is degraded by reflective objects near your loudspeakers (Harley, 2015).

> **Recommendation 4.24** *Avoid putting reflective objects in the direct vicinity of your front loudspeakers, especially in the area between the left and right loudspeakers.*

4.2.4 Positions of the Surround Loudspeakers

The surround loudspeakers are placed at the sides and at the back of the room. Their main purpose is to immerse you in sound. They create the illusion that you are present in a real acoustical space: you are there at the concert, and you are in the middle of the action taking place in the movie. In addition, the surround loudspeakers also occasionally deliver directional sounds that originate off-screen from the sides and the back of the room. However, the direction of these sounds is much less well-defined than the direction of sounds in the frontal soundstage. Despite the fact that there are only three loudspeakers at the front, the directions from which sounds appear to come are not limited to these three positions. Varying the intensity by which the sound is reproduced by each of the front loudspeakers allows the sound to appear anywhere between the three front loudspeakers. This mechanism is called *phantom imaging.* It appears that phantom imaging only works reliably in the front and back, not at the sides of the listeners. At the sides it results in a large uncertainty with either very diffuse or jumping images (Holman, 2008; Ratliff, 1974; Theile and Plenge, 1977). Because of this uncertainty, filmmakers usually limit directional sounds from the surround loudspeakers to momentary sound effects for which no real precision is demanded. A typical example is the flyover of an airplane.

Since directional sounds from the surround loudspeakers are rare and also ambiguously located, the optimal placement of these loudspeakers is governed by their ability to immerse the listeners in sound. It may come as a surprise, but there is no need for the sound to come from all directions to feel enveloped by it. In fact, just two surround loudspeakers are sufficient to give you the illusion of being completely enveloped in surround sound. All that matters is that the sound from these two loudspeakers arrives from appropriate angles. Maximum envelopment occurs when the direct sound from the surround loudspeakers arrives from angles between 60° and 135° in the horizontal plane (Hiyama et al., 2002). In a multi-channel recording the illusion of envelopment can be created by using multiple surround loudspeakers at these angles to imitate reflections that arrive later than 80 ms and have a significant frequency content between 100 Hz and 1000 Hz.

In a 5.1-channel sound system the two surround loudspeakers (LS and RS) are typically placed at angles in the range of 110°–120° (EBU-Tech 3276-s1, 2004; ITU-R BS.775, 2012). Figure 4.23 shows such a typical arrangement. At these angles the surround loudspeakers are optimally placed to create enveloping sound and they also contribute slightly to the sense of image broadening in the frontal soundstage. Furthermore, they are placed far enough behind the listener to provide the occasional off-screen sounds that need to emerge from the back of the room. If your room and its furnishings do not allow you to place the surround loudspeakers at these angles, do not despair because there is quite some flexibility in their placement. If you place the surrounds somewhere between 90° and 135° they are still very capable of creating

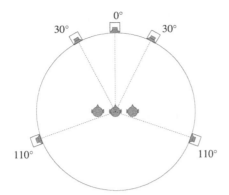

Figure 4.23
Loudspeaker positions in a
5.1-channel system.

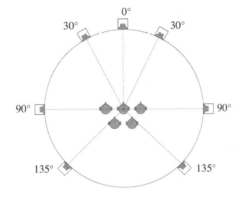

Figure 4.24
Loudspeaker positions in a
7.1-channel system.

a sense of envelopment. A front-back symmetrical arrangement with the surrounds placed at 150° opposite to the front loudspeakers is not a good idea; it strongly reduces the sense of envelopment (Hiyama et al., 2002; Toole, 2018).

In a 7.1-channel sound system the four surround loudspeakers are typically placed at both ends of the 90°–135° range: LS and RS at 90° and LB and RB at 135°, as depicted in figure 4.24. Again there is some flexibility in placement. The addition of two extra surrounds strengthens the sense of envelopment in larger rooms. Having four surround loudspeakers is also a big advantage when you have multiple rows of listeners, because it is easier to cover a larger listening area such that the sense of envelopment becomes less dependent on the listener's position within the listening area (AES, 2001; Theile, 1991). Of course, it doesn't harm to have four surround loudspeakers in a small room, but it can be more of a challenge to integrate them harmoniously in the room. Furthermore, it increases the costs of your system. You should bear in mind that in a small room it is always better to have two high-quality surround loudspeakers than four mediocre ones (Harley, 2015).

In the largest of rooms, you could consider setting up a 9.1-channel system that has three pairs of surround loudspeakers. In such a system the two additional surrounds are placed in front of the listener. They are called the *front wide loudspeakers* (LW and RW), because not only do they contribute to the sense of envelopment, they also provide a strong support for the image broadening effect in the frontal soundstage. In general, if more than two surround loudspeakers are used, they should be distributed at equal intervals along an arc between 60° and 150°. Table 4.3 summarizes all the placement options for the 5.1-, 7.1-, and 9.1-channel systems.

Table 4.3 Recommendations for the angular position of the surround loudspeakers (EBU-Tech 3276-s1, 2004; ITU-R BS.775, 2012; Toole, 2018).

Configuration		Target	Tolerance
5.1 channels	LS and RS	110°	90°–135°
7.1 channels	LS and RS	90°	70°–110°
	LB and RB	135°	120°–150°
9.1 channels	LW and RW	60°	60°–75°
	LS and RS	90°	90°–120°
	LB and RB	135°	135°–150°

Recommendation 4.25 *Put the surround loudspeakers at the angular positions recommended in table 4.3, such that they are placed symmetrically on the left and right sides of the audience.*

Ideally, all front and surround loudspeakers are at the same distance from the prime listening seat. This means that if you draw a circle around this seat, all loudspeakers are placed on the circumference of this circle as in figures 4.23 and 4.24. The idea is that the sound emerging from each loudspeaker arrives at exactly the same time instant at the prime seat. However, you can place the surround loudspeakers at different distances as long as you compensate for the path difference by using appropriate delays in your A/V controller or receiver. A left-right symmetrical placement of the surround loudspeakers is preferred, but not strictly necessary (see for example figure 4.4).

As with the front loudspeakers, everyone in the audience should have a clear line of sight to all surround loudspeakers. The surround loudspeakers should also be aimed towards the prime seat, such that all audience members are within the 30° beams of high-frequency sound emerging from these loudspeakers (Toole, 2018). Figure 4.25 shows such a setup. The amount of toe-in of the surround loudspeakers is less critical than it is with the front loudspeakers, but it doesn't hurt to fine-tune it using a listening test. Such a listening test should focus on the enveloping qualities of the sound. Useful test material consists of movie scenes with heavy rain all around you and scenes that take place in a hollow-sounding acoustical environment, such as inside a freeway tunnel or a large cathedral. Fine-tune the toe-in of the surround loudspeakers such that all audience members experience an enveloping surround sound. Often the seats closest to the surround loudspeakers are the most difficult to get right. Since they are closer to one of the surround loudspeakers, the sound is being pulled towards this nearest loudspeaker and the illusion of being enveloped by sound is destroyed. It helps to place the surround loudspeakers as far away from the audience as possible, because increasing the absolute distance minimizes the relative level difference between the nearest and farthest surround loudspeakers.

Recommendation 4.26 *Aim the surround loudspeakers at the listening positions such that all audience members are within the 30° beams emerging from the surround loudspeakers. Use a listening test to fine-tune the toe-in of the surround loudspeakers to ensure that all audience members experience an enveloping surround sound.*

The height of the surround loudspeakers should be chosen such that their high-frequency drivers are approximately level with the ears of a seated listener. This is about 1.2 m (4 ft) from the floor (EBU-Tech 3276-s1, 2004). The surrounds may also be installed a bit higher to facilitate covering the entire audience, but never higher than about 1.5–1.8 m (5–6 ft) from the floor (Toole, 2018). In a lot of rooms, the most convenient place to put the surround loudspeakers is

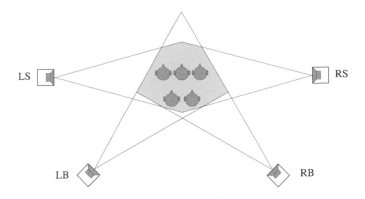

Figure 4.25
Surround loudspeaker
arrangement where the
loudspeakers are aimed
at the audience such that
everyone is within the 30°
beams from each of the
loudspeakers.

on the walls. Wall-mounted loudspeakers are often placed a bit higher than seated ear height, while free-standing surround loudspeakers are mostly positioned at ear height. Both options are perfectly fine as long as the angular positions in table 4.3 are being observed.

Recommendation 4.27 *Place all the surround loudspeakers at the same height, such that the high-frequency driver is about 1.2 m (4 ft) from the floor. Alternatively place them a bit higher to facilitate covering the entire audience, but never more than 1.5–1.8 m (5–6 ft) from the floor.*

In a Dolby Atmos or DTS:X surround system you have additional overhead surround loudspeakers installed in the ceiling. Although there can be up to ten overhead speakers in such a system, in practical home theater systems there are either two or four overhead loudspeakers. These loudspeakers are always combined with at least two ordinary surround loudspeakers at ear height. The most commonly used configurations are 5.1.2, 5.1.4, 7.1.2, and 7.1.4. It is important that the overhead loudspeakers have a wide dispersion pattern, preferably 45° on either side from 100 Hz to 10 kHz, such that all audience members are within the area that they can cover. Localization of sound in the vertical plane only works if the frequencies above 6 kHz are able to reach the ears of the listeners (Toole, 2018). Wide dispersion overhead loudspeakers can be conveniently installed facing directly downward. Loudspeakers with narrow dispersion patterns can still be used, but they have to be angled towards the audience to achieve appropriate high-frequency coverage. Install your overhead loudspeakers on the left and right sides of the audience at the same distance from the sidewalls as the front loudspeakers, as illustrated in figures 4.26 and 4.27. If you use only one pair of overhead loudspeakers (configurations 5.1.2 and 7.1.2), install them such that the line from the loudspeaker to the prime listener makes an 80° angle with the horizontal line as shown in figure 4.28. Again there is some flexibility in placement: as long as you position the overhead loudspeakers within the 65°–100° range, you should be fine. If you use two pairs of overhead loudspeakers (configurations 5.1.4 and 7.1.4), install one pair slightly in front of the prime listener and the other pair slightly behind this listener. As shown in figure 4.29, their preferred position is at 45°, but installing them within the range of 30°–55° is also possible (www.dolby.com).

Recommendation 4.28 *Place the overhead surround loudspeakers on the left and right sides of the audience at the same distance from the sidewalls as the front loudspeakers.*

Recommendation 4.29 *Mount the overhead loudspeakers in a 5.1.2 and 7.1.2 system at an angle of 80° as shown in figure 4.28, or alternatively within the 65°–100° range. Install the overhead loudspeakers in a 5.1.4 and 7.1.4 system towards the front and the back of the listener at an angle of 45° as shown in figure 4.29, or alternatively within the 30°–55° range.*

Figure 4.26
Top-down view of a room showing the overhead loudspeaker positions in a 5.1.2 Dolby Atmos system. The overhead loudspeakers are installed on the left and right sides of the audience at the same distance from the sidewalls as the front loudspeakers.

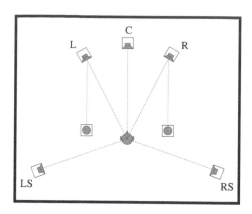

Figure 4.27
Top-down view of a room showing the overhead loudspeaker positions in a 5.1.4 Dolby Atmos system. The overhead loudspeakers are installed on the left and right sides of the audience at the same distance from the sidewalls as the front loudspeakers.

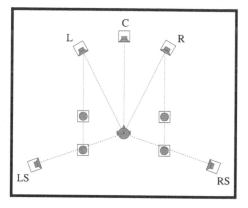

Figure 4.28
Side view showing the overhead loudspeaker positions in a 5.1.2 or 7.1.2 Dolby Atmos system. The overhead loudspeakers are preferably installed at an angle of 80° with the *horizontal line* passing through the ears of the listener, or alternatively within the 65°–100° range indicated by the *broken lines*.

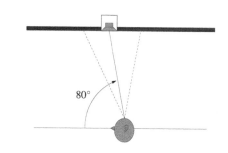

Figure 4.29
Side view showing the overhead loudspeaker positions in a 5.1.4 or 7.1.4 Dolby Atmos system. The overhead loudspeakers are preferably installed at an angle of 45° with the *horizontal line* passing through the ears of the listener, or alternatively within the 30°–55° range indicated by the *broken lines*.

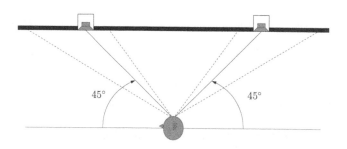

For the best experience the overhead surround loudspeakers should not be too close to the listener. They should be mounted at a minimum height of 2.4 m (8 ft) from the floor, and in case of vertically staggered seating (as in figure 4.15) at least 1.2 m (4 ft) above the ears of the highest seated listener. Mounting them closer to the listeners will destroy the sense of being enveloped in overhead sound, because you start to hear exactly which loudspeakers produce the sounds (www.dolby.com).

Recommendation 4.30 *Mount the overhead surround loudspeakers at least 1.2 m (4 ft) above the ears of the highest seated listener.*

If you are unable to install overhead loudspeakers in your ceiling, you could use so-called *Dolby Atmos enabled speakers* as an alternative. Dolby Atmos enabled speakers are mounted in the top of a loudspeaker cabinet and are tilted towards the ceiling such that their sound is reflected off the ceiling towards the listener, as illustrated in figure 4.30. They can be integrated in the cabinet of a standard front or surround loudspeaker or come as a separate enclosure that is placed on top of the front and surround loudspeakers. In a 5.1.2 or 7.1.2 system the Dolby Atmos enabled loudspeakers are the left (L) and right (R) front loudspeakers. In a 5.1.4 system also the two surround loudspeakers (LS and RS) are Dolby Atmos enabled, while in a 7.1.4 system the two back surround loudspeakers (LB and RB) are Dolby Atmos enabled, but the other two surrounds (LS and RS) are not. Table 4.4 summarizes the recommended configurations.

The use of Dolby Atmos enabled loudspeakers requires that the part of the ceiling between the listener and these loudspeakers is highly reflective. Ceilings made of drywall, plaster, concrete, hardwood, or some other rigid material work well, while ceilings made of softer sound-absorbing materials don't. The ceiling should also be completely flat, neither angled nor vaulted. The Dolby Atmos enabled loudspeakers should be aimed towards the prime seat, in exactly the same way as the front and surround loudspeakers on which they are placed. They should be placed slightly above ear level to minimize any direct sound reaching the listeners. Finally, the ceiling height should be between 2.4 and 4 m (between 8 and 13 ft) such that the reflections are able to reach the listeners at an appropriate angle (www.dolby.com).

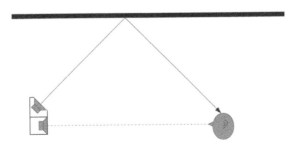

Figure 4.30
Side view showing how a Dolby Atmos enabled loudspeaker bounces the sound off the ceiling towards the listener. The angled loudspeaker is either a separate unit that is placed on top of an ordinary front or surround loudspeaker, or it is built into the top of the cabinet of a dedicated Dolby Atmos enabled loudspeaker that provides both overhead reflected sound and direct sound.

Table 4.4 Recommended configuration for Dolby Atmos enabled loudspeakers. Ordinary loudspeakers are indicated by the letter *O* and Dolby Atmos enabled ones by the letter *A* (www.dolby.com).

Configuration	L	C	R	LS	RS	LB	RB
5.1.2	A	O	A	O	O		
5.1.4	A	O	A	A	A		
7.1.2	A	O	A	O	O	O	O
7.1.4	A	O	A	O	O	A	A

Recommendation 4.31 *Dolby Atmos enabled loudspeakers can be used as an alternative to overhead loudspeakers in the configurations listed in table 4.4. Ensure that the area on the ceiling that lies between the listener and these loudspeakers is reflective and that the height of the ceiling is between 2.4 and 4 m (between 8 and 13 ft).*

4.2.5 Distance Between Loudspeakers and Nearby Walls

One more important factor to consider when placing the front and surround loudspeakers is their distances to the three nearest walls. These distances have a profound effect on the reproduction of low and mid frequencies below about 500 Hz as they strongly influence the presence or absence of unwanted coloration at these frequencies. Therefore, the particular choice of the distances to the nearby walls is another important factor to consider in addition to all the placement recommendations given thus far. Things are getting a little complex, because in setting the distances to the nearest walls, you also have to take into account all the other recommendations. For example, the distance to the floor is mainly governed by the fact that the high-frequency drivers of the front and surround loudspeakers should be positioned at ear height. You can only deviate slightly from this requirement: as we saw in section 4.2.3 the center loudspeaker is often placed a bit lower beneath the video display, and as we saw in section 4.2.4 the surround loudspeakers can be placed a bit higher than ear height. But, that is it. Another example is choosing the distance between the front left and right loudspeakers (*b* in figure 4.18). Setting the distance between these two loudspeakers also determines the distances of these loudspeakers to the nearest sidewalls. Furthermore, once the left and right fronts are brought into position, the distances between the sidewalls and the center loudspeaker are also fixed, because the center loudspeaker needs to be centered between the other front loudspeakers. The remaining factor to play with is the distance between the front loudspeakers and the wall behind them. Despite all the dependencies between the different recommendations, it is worthwhile to fine-tune the distances to the nearest room boundaries in order to get the best possible reproduction of low- and mid-frequency sound.

The quality of low-frequency sound reproduction mainly depends on the position of the loudspeakers in the room, in particular on their distances to the walls. Two factors come into play: 1) the adjacent-boundary effect that results from reflections from the nearby walls and 2) the room resonances that develop between opposing walls (see section 2.3.4). As I discussed in section 4.1 on the need for subwoofers, loudspeaker placement becomes a bit easier when the lowest frequencies, below about 80 Hz, are reproduced by subwoofers instead of the front and surround loudspeakers. When you use subwoofers, you can focus on the frequency range above 80 Hz when placing the front and surround loudspeakers.

Room resonances above 80 Hz are not much to worry about. As the frequency rises the number of resonances increases and they become more closely packed, making the frequency response of the loudspeakers much smoother (Everest and Pohlmann, 2014). Figure 4.31 illustrates this effect for a typical room of 6 × 5 × 2.5 m (20 × 16 × 8 ft). In fact, in most rooms the resonances above 150 Hz are so numerous and closely packed that they have a negligible influence on the frequency response. Hence, the resonances in the range of 80–150 Hz are the ones that influence the low-frequency response of the front and surround loudspeakers the most. Recall from section 2.3.4 that every resonance generates a particular pattern of pressure maximums and nulls between opposing walls. In theory, you should avoid placing the loudspeakers at either a maximum or a null. For a rectangular room you can easily calculate the positions of these maximums and nulls. For an example refer to figure 2.35. However, in practice it turns out that such simple calculations do not always work. The walls of the room may absorb some energy at certain frequencies as may some of the furniture, or the shape of the room diverts from a simple rectangle. Fortunately, the negative influence of resonance frequencies between 80 and 150 Hz can easily be reduced by absorbing these resonances using simple acoustical devices to be discussed in section 8.1.2. Therefore, there is no need

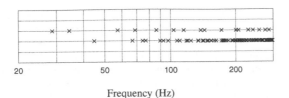

Frequency (Hz)

Figure 4.31
The room resonance frequencies in a rectangular room of 6 × 5 × 2.5 m (20 × 16 × 8 ft). Every *cross*
corresponds to a resonance frequency. The *top row* **shows the resonances from reflections between**
two opposing walls. The *bottom row* **shows the resonances from reflections involving four walls. As the**
frequency increases the number of resonances increases and they get closer together. I will explain in
section 8.1.1 how to calculate these room resonance frequencies.

taking these resonances into account when determining the optimal distances of the front
and surround loudspeakers to the nearby walls. If you use subwoofers there is also no need
to take into account the room resonances below 80 Hz when placing the front and surround
loudspeakers. This is most fortunate, because the resonances below 80 Hz are much harder to
deal with. Below this frequency the response of the loudspeaker is dominated by a few strong
resonances that clearly stand out, because of their relatively large separation in frequency
(see figure 4.31). As a result, bass notes that coincide with a resonance will either sound too
loud or will sound way too soft, depending on where you sit.

The adjacent-boundary effect is what matters most when you are fine-tuning the distances
between each loudspeaker and the three nearest walls. The goal is to minimize the detrimental effect of reflections from these walls. These reflections cause unwanted variations in the
low and mid frequencies, and the resulting peaks and dips can cause ugly coloration of the
sound, especially in the mid frequencies where the ear is most sensitive. The peaks and dips in
the frequency response can be predicted using the equation in box 4.5. This equation, which
describes the adjacent-boundary effect, is a bit difficult to analyze, therefore we will look at
some examples to get a better feel for what is happening.

BOX 4.5 ADJACENT-BOUNDARY EFFECT

The sound power output of a loudspeaker changes due to the reflection from nearby
walls. The amount depends on the distance to these boundaries and changes with frequency according to the following equation. The relative power output is given by
(Allison, 1974; Waterhouse, 1958):

$$W(f) = 1 + \frac{\sin(kfx)}{kfx} + \frac{\sin(kfy)}{kfy} + \frac{\sin(kfz)}{kfz}$$

$$+ \frac{\sin(kf\sqrt{x^2+y^2})}{kf\sqrt{x^2+y^2}} + \frac{\sin(kf\sqrt{x^2+z^2})}{kf\sqrt{x^2+z^2}}$$

$$+ \frac{\sin(kf\sqrt{y^2+z^2})}{kf\sqrt{y^2+z^2}} + \frac{\sin(kf\sqrt{x^2+y^2+z^2})}{kf\sqrt{x^2+y^2+z^2}}$$

where $k = 4\pi/c$ with $c = 344$ m/s the speed of sound, and x, y, and z are the distances
to the nearby walls as measured in meters from the center of the low-/mid-frequency
driver.

Figure 4.32 shows an example of the variation in the frequency response of a loudspeaker when the low-/mid-frequency driver is positioned at a distance of 1 m (3.3 ft) from each of the three nearest walls. The thin line shows the ideal frequency response for a front or surround loudspeaker. This response is flat at the high frequencies and rolls slightly off towards the lower frequencies, because the A/V controller is redirecting the lowest frequencies to the sub-woofers. The subwoofer *crossover frequency* in this example is 80 Hz. The crossover frequency is defined as the frequency at which the response has been reduced to –3 dB. The broken line in figure 4.32 shows the adjacent-boundary effect, that is, the change in the frequency response due to the reflections from the three nearest walls. The sound you hear is a combination of the adjacent-boundary effect (broken line) and the response of the loudspeaker (thin line); it is indicated in the figure by the thick line. Two things are noteworthy. First, around 95 Hz the response has a large and wide dip of more than 10 dB. The low frequencies around this region are strongly suppressed. Second, above 200 Hz several smaller peaks and dips occur that result in coloration of the mid-frequency sound. The big suppression around 95 Hz is the result of the reflections from all three walls arriving out of phase with the direct sound produced by the loudspeaker. The frequency at which the maximum suppression occurs depends on the distance to the walls. Figure 4.33 shows that if we move the loudspeaker closer to the walls, the dip moves upwards in frequency. Placing the loudspeaker further away from the walls will move the low-frequency dip to lower frequencies and with a large enough distance it will eventually cease to affect the frequencies above 80 Hz. However, this means that the loud-speaker needs to be placed at a distance of more than 1 m (3.3 ft) from the floor and the nearest walls. In most rooms this is not a viable option.

Fortunately, there is another way to alter the adjacent-boundary effect: when the distances to the nearest room boundaries are different the reflections no longer reinforce each other and the response becomes smoother. Figure 4.34 shows an example where the loudspeaker is placed 0.75 m (2.5 ft) from the floor and 0.5 m (1.6 ft) from both walls. The dip is much

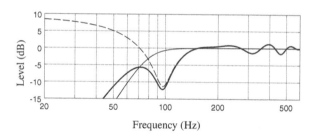

Figure 4.32
The change in the low-frequency response of a loudspeaker when the center of its low-frequency driver is placed at a distance of 1 m (3.3 ft) from all three room boundaries. The *thick line* is a combination of the adjacent-boundary effect *(broken line)* and the response of the loudspeaker *(thin line)*.

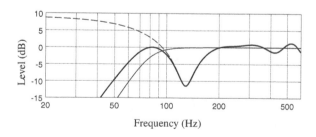

Figure 4.33
The change in the low-frequency response of a loudspeaker when the center of its low-frequency driver is placed at a distance of 0.75 m (2.5 ft) from all three room boundaries. The *thick line* is a combination of the adjacent-boundary effect *(broken line)* and the response of the loudspeaker *(thin line)*.

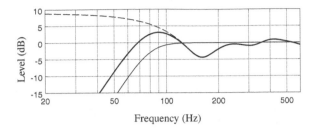

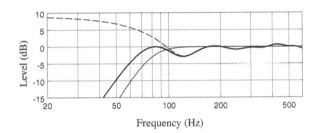

Figure 4.34
The change in the low-frequency response of a loudspeaker when the center of its low-frequency driver is placed at a distance of 0.75 m (2.5 ft) from the floor and 0.5 m (1.6 ft) from both walls. The *thick line* is a combination of the adjacent-boundary effect *(broken line)* and the response of the loudspeaker *(thin line)*.

Figure 4.35
The change in the low-frequency response of a loudspeaker when the center of its low-frequency driver is placed at a distance of 0.75 m (2.5 ft) from the floor, 0.5 m (1.6 ft) from one wall, and 1 m (3.3 ft) from the other wall. The *thick line* is a combination of the adjacent-boundary effect *(broken line)* and the response of the loudspeaker *(thin line)*.

smaller now, only 5 dB, and the overall response becomes much smoother. Figure 4.35 shows the response when the distances to the wall also differ: 0.5 m (1.6 ft) from one wall, and 1 m (3.3 ft) from the other wall. The response is even smoother in this case. In conclusion, the smoothest low-frequency response is obtained when the distances to the walls are as different as possible, while the worst response is obtained when the loudspeaker is placed at exactly the same distance from each of the walls (Allison, 1974; Everest and Pohlmann, 2014).

It should come as no surprise that the goal is to tune the distances to the nearby walls such that the front and surround loudspeakers have the smoothest possible frequency response without any large peaks and dips above 80 Hz. (Allison, 1974; Ballagh, 1983). Since the height of the front and surround loudspeakers is already determined (see recommendations 4.16, 4.17, and 4.27), all that is left is to fine-tune the distances to the nearest sidewall and the wall behind the loudspeakers. You can use figure 4.36 or 4.37 to make an informed choice. The gray areas in this figure show the maximum difference between the ideal loudspeaker response (the thin lines in figures 4.32–4.35) and the altered response due to the adjacent-boundary effects (the thick lines in figures 4.32–4.35) for frequencies above 80 Hz. The lighter the area the smaller the difference, and the smoother the response. There are four different figures corresponding to different heights of the low-/mid-frequency driver. Pick the one that is closest to your situation and use it to fine-tune the distances to the sidewall and the wall behind the loudspeakers.

Recommendation 4.32 *Fine-tune the distances to the walls for the front and surround loudspeakers to obtain the smoothest possible frequency response. Use figure 4.36 or 4.37 as a guide and make the distances to the nearby walls as different as possible.*

Figure 4.36
Loudspeaker boundary effects for different distances to the nearest room boundaries in meters. The height (m) of the low-frequency driver is indicated below the graphs. The *gray areas* in the graphs show the maximum difference between the ideal loudspeaker response and the altered response due the adjacent-boundary effect for frequencies above 80 Hz. The lighter the area, the smaller this difference, and hence the most favorable loudspeaker positions are indicated by the light areas. The darkest area in each graph corresponds to the worst case response that occurs when all three distances are equal.

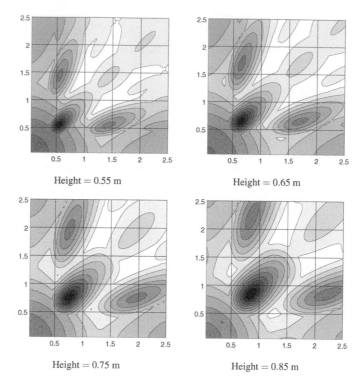

Height = 0.55 m   Height = 0.65 m

Height = 0.75 m   Height = 0.85 m

Figure 4.37
Loudspeaker boundary effects for different distances to the nearest room boundaries in feet. The height (ft) of the low-frequency driver is indicated below the graphs. The *gray areas* in the graphs show the maximum difference between the ideal loudspeaker response and the altered response due the adjacent-boundary effect for frequencies above 80 Hz. The lighter the area, the smaller this difference, and hence the most favorable loudspeaker positions are indicated by the light areas. The darkest area in each graph corresponds to the worst case response that occurs when all three distances are equal.

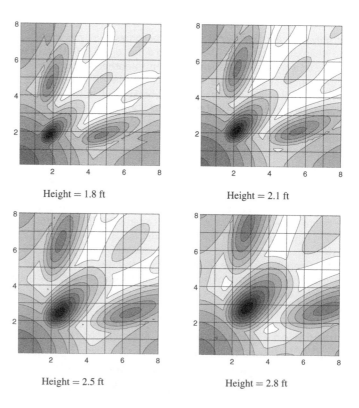

Height = 1.8 ft   Height = 2.1 ft

Height = 2.5 ft   Height = 2.8 ft

Figures 4.36 and 4.37 show that it can be beneficial to make the distance to the sidewall and the wall behind the loudspeaker larger than 1 m (3.3 ft): the lightest areas, which represent the smoothest responses, all lie beyond the 1 m (3.3 ft) mark. In fact, it is often recommended placing the loudspeakers at least 1 m (3.3 ft) from the walls (AES, 2008; EBU-Tech 3276, 1998; ITU-R BS.1116, 2015). However, as figures 4.36 and 4.37 show there are still some darker areas beyond the 1 m (3.3 ft) mark. Therefore, it is not so much the minimum distance to the walls, but the mutual relation between these two distances that determines the overall smoothness of the response.

Fine-tuning the distances to the walls for the three front loudspeakers can really make a significant difference in the quality of their low- and mid-frequency sound. So, by all means, go ahead and experiment to find the best possible placement with respect to the nearby room boundaries. However, keep in mind that you should never violate the other placement constraints for the front loudspeakers imposed by the recommendations in section 4.2.3. For the surround loudspeakers I recommend a different approach. While it is true that their frequency response can also be improved by fine-tuning the distances to the walls, it is more important to place them such that they provide a realistic sense of immersion for all audience members. This means that they should be placed as far as possible from the audience, and as a consequence in most small rooms they will end up close to the walls or even in the corners of the room. If you are able to achieve both a good sense of immersion and a smooth frequency response, go ahead and do it, but if you have to make a compromise always give priority to the immersive qualities of the surround loudspeakers (Rives, 2005).

Some loudspeaker manufacturers provide placement guidelines in their user manuals. They often recommend placing the loudspeakers at a certain distance or range from the rear and sidewalls. Such recommendations are given to provide the best possible low-frequency response when the loudspeakers are used as full-range loudspeakers that reproduce the whole frequency range, including the lowest frequencies below 80 Hz. Thus, these recommendations are only useful in case you don't use subwoofers to reproduce those lowest frequencies. When you don't use subwoofers, the position of the front and surround loudspeakers with respect to the walls will determine the lowest frequency that you can produce with your audio system. Some loudspeakers sound thin and lean when placed too far away from the walls, while others may start to sound heavy and weighty when placed too close to the walls (Harley, 2015). All loudspeakers are limited in the amount and extent of low frequencies they can produce. The loudspeaker's frequency response will gradually decrease when the frequency gets lower. Figure 4.38 shows a typical example of a loudspeaker that gradually rolls off below 40 Hz. As illustrated in this figure, the adjacent-boundary effect can partly compensate for this roll-off: it boosts the lowest frequencies and thus extends the low-frequency response towards lower frequencies (approximately 30 Hz in this case). When the boost doesn't exactly match the inverse of the low-frequency roll-off of the loudspeaker, the

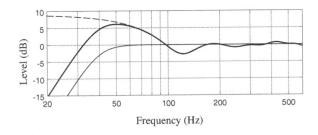

Figure 4.38
An example of the adjacent-boundary effect boosting the lowest frequencies of this full-range loudspeaker when the center of its low-frequency driver is placed at a distance of 0.75 m (2.5 ft) from the floor, 0.5 m (1.6 ft) from one wall, and 1 m (3.3 ft) from the other wall. The *thick line* is a combination of the adjacent-boundary effect *(broken line)* and the response of the loudspeaker *(thin line)*.

bass may sound too heavy (Newell and Holland, 2007). To facilitate tuning the low-frequency response, some loudspeaker manufacturers provide foam plugs or foam linings that can be inserted into the bass-reflex ports of their loudspeakers. Once inserted, the foam reduces the low-frequency output of the loudspeaker and thus allows you to adapt the response of a full-range loudspeaker to its local environment (Colloms, 2018).

An additional complicating factor with full-range front and surround loudspeakers is that you also have to take into account all the room resonances, not only those above 80 Hz. These lowest resonances often introduce quite heavy colorations in the bass. As I explained in section 4.1, if you remove the lowest frequencies from the front and surround loudspeakers and redirect them to the subwoofers, you do not need to worry about the bass extension and the lowest room resonances. This use of subwoofers greatly simplifies the search for the optimal positions of your front and surround loudspeakers. As stated before, I strongly believe that the best low-frequency reproduction requires the use of subwoofers.

Finding the optimal distances to the walls is relatively easy when the lowest frequencies are directed to the subwoofer, but it can become quite complicated with full-range loudspeakers. There are a lot of different factors to consider like room resonances, bass extension, and a smooth response throughout the entire low- and mid-frequency range. Computer software exists that can be used to make a prediction of the frequency response of a loudspeaker when placed at a certain position in the room (Chéenne, 2015; Everest and Pohlmann, 2014). CARA (Computer Aided Room Acoustics) is an example of such a program (www.cara.de). It uses a simple acoustical model of the room to calculate a quite accurate prediction of the sound field in the room. All you have to do is specify the shape, dimensions, and acoustical properties of the surfaces in the room. You can even specify the positions of doors and windows, and create models of the major pieces of furniture. Obviously, the program is limited in presenting the real world and can only approximate what is really happening in the room. Nevertheless, the results turn out to be quite useful in practice, so you might want to give it a try.

Another interesting method to determine the optimal distances from the walls was developed by Wilson Audio (Harley, 2015; Wilson Audio, 2012). It is based on a simple listening test in which you determine a so-called *zone of neutrality* where your loudspeakers interact the least with the adjacent boundaries. To determine this zone you position yourself at the prime listening seat and ask an assistant to walk around in the room while speaking. Your assistant should speak in a moderately loud voice and at a constant volume. Ask your assistant to stand against the front wall (behind the loudspeakers) and walk in a straight line parallel to the sidewall. Listen to how the assistant's voice changes in timbre from overly heavy and chesty towards a more neutral sound. At the point where your assistant's voice starts to sound more clear and uncolored, you have reached the outer edge of the zone of neutrality. Use some tape to mark this distance from the front wall on the floor. In most rooms this point is slightly less than 1 m (3.3 ft) from the wall. Next, position your assistant against the sidewall at the distance from the front wall that you have just determined. Ask your assistant to walk along the line that runs parallel to the front wall. The best position for your loudspeaker is at the point on this line where your assistant's voice sounds the most neutral.

4.2.6 Positions of the Subwoofers

The position of your subwoofers in the room can make or break the sound quality at low frequencies. Contrary to popular belief, the principal determinant of low-frequency sound quality is not the subwoofer itself, but the room and the positions of the subwoofer and listeners in this room. Careful positioning is crucial if you want to achieve the best possible low-frequency sound. Nevertheless, it is not easy. Many factors influence the end result: in addition to the positions of the subwoofers and listeners in the room, the result also depends on the shape and dimension of the room; the type of wall construction; and the positions of doors, windows, and the larger pieces of furniture. Every room is different and imposes its own sonic signature on the low-frequency sound (Toole, 2018).

Properly set up, the subwoofers only reproduce the lowest frequencies up to the subwoofer crossover frequency, which typically lies somewhere between 80 Hz and 120 Hz. At these low frequencies the human hearing system cannot determine the location of the sound source. In other words, these low frequencies don't contain any spatial information. This means that your subwoofers could be placed anywhere in the room and that they are not limited, as the front and surround loudspeakers are, to specific front-back, left-right, or angular positions with respect to the prime seat.

Subwoofers radiate their sound in all directions, and therefore the driver doesn't need to be aimed at the listeners; it can be turned in any direction. Some subwoofers even have their driver installed in the bottom of the cabinet firing down. One thing that matters though is that the driver must have enough clearance to the nearest surface to prevent compression of the air in front of the driver. The same holds true for the bass-reflex port (the hollow tube) if your subwoofer has one. Keep the nearest surface about 10 cm (4 inch) away from the port and the driver and everything should be fine (McLaughlin, 2005). Also make sure that the back panel of the subwoofer is at least 15 cm (6 inch) away from the nearest surface. Most subwoofers are equipped with heat radiating fins at the back to cool the integrated amplifier. You want to provide ample ventilation around these fins to avoid overheating. It is also a good idea to orient the bass-reflex port away from the listener (Holman, 2008). The air that flows through this port can sometimes create a low volume noisy sound, especially with large excursion of the loudspeaker's diaphragm. This unwanted sound has frequency components above 120 Hz that might attract your attention to the location of the subwoofer (AES, 2001).

It is important that the sound produced by the subwoofers integrates well with the sound produced by the other main loudspeakers, especially the front loudspeakers. To achieve a well-integrated sound you need to make some adjustments to the bass management settings of your A/V controller. Use the bass management feature to direct all low frequencies to the subwoofers. In most controllers this can be achieved by setting the front and surround loudspeakers to 'small', and setting the crossover frequency to a value between 80 and 120 Hz (see section 7.5.2). Around the crossover frequency the sound from the main loudspeakers gradually fades out and is taken over by the subwoofers, as shown in figure 4.39. Thus, around the crossover frequency both the main loudspeakers and the subwoofers produce the same low-frequency sound. It is important to ensure that the sound from all these loudspeakers arrives in phase at the prime listener seat, because a phase difference between the sound from the main loudspeakers and the sound from the subwoofers can significantly reduce the amount of bass around the crossover frequency. In the extreme case, where the main loudspeakers are 180° out of phase with the subwoofers, the mains and subs work against each

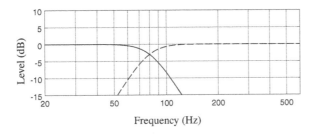

Figure 4.39
Example of how the frequency range is divided between the front/surround loudspeakers and the subwoofer when the crossover frequency is set to 80 Hz. At the crossover frequency, the response of the subwoofer *(solid line)* is 3 dB down and gradually diminishes for higher frequencies. The response of the front/surround loudspeakers *(broken line)* complements the subwoofer and starts to dominate above the crossover frequency.

other, canceling each other's sound (as in the example in figure 2.20). To control the phase, most subwoofers have either a switch that inverts the phase or a knob that controls the delay of the sound, and hence its phase. Some A/V controllers also offer subwoofer delay options. The delay controls should be set such that the sound from the mains and the sound from the subs arrive in phase at the prime listening seat. I explain how to configure the loudspeaker delay settings in more detail in section 7.5.3.

Typically, subwoofers are placed on the floor and they often end up along one of the walls. Positioning them close to a wall is convenient, because most subwoofers need to be plugged into an electrical power outlet. However, don't let the convenience of a nearby power outlet be the decisive factor on where you place your subwoofers. Instead, choose the position that yields the best low-frequency sound quality. Finding that position requires some experimentation. Make sure that both the power cord and the cable from the subwoofer to the A/V controller are long enough to experiment with different subwoofer positions.

The best positions for your subwoofers are those that provide a smooth and even reproduction of low frequencies, where all bass notes are clearly articulated with well-defined attacks and decays. Two factors conspire to prevent you from achieving such a frequency response: 1) the adjacent-boundary effect and 2) the room resonances. The resonances are the more troublesome. At the low-frequency range in which the subwoofer operates, the relative sparsity of the room resonances (see the example in figure 4.31) will result in certain frequencies being emphasized or deemphasized. Resonances that are spaced more than 20 Hz apart or resonances that coincide are frequent sources of coloration (Everest and Pohlmann, 2014; Fielder and Benjamin, 1988). The amount to which these resonances are perceived depends in part on the positions of the subwoofers and the listeners. That is why it is important to carefully position your subwoofers.

If you use one single subwoofer the best position is often on the floor between the left and right loudspeakers close to the front wall and away from the sidewalls (AES, 2001; Genelec, 2009). Keeping the subwoofer close to the floor and the front wall increases its output, because the adjacent-boundary effect boosts the lowest frequencies. A position close to the wall also avoids the adjacent-boundary effect causing severe irregularities in the subwoofer's response. It is recommended that you place the subwoofer not more than 60 cm (2 ft) from the front wall (Genelec, 2009). Figure 4.40 shows an example of the adjacent-boundary effect on the response of a subwoofer when it is placed at a distance of 0.2 m (0.7 ft) from the floor, 0.6 m (2 ft) from the front wall, and 2 m (6.6 ft) from the sidewall. The thin line in this figure shows the subwoofer's frequency response without any influence of the room. At 30 Hz the response of the subwoofer has been reduced from 0 dB to −3 dB; this is the subwoofer's *roll-off frequency*, commonly thought of as the lowest frequency it can produce. The subwoofer produces low-frequency sound up to the subwoofer crossover frequency, which is 120 Hz in this

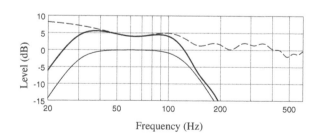

Figure 4.40
The change in the low-frequency response of a subwoofer when the center of its low-frequency driver is placed at a distance of 0.2 m (0.7 ft) from the floor, 0.6 m (2 ft) from the front wall, and 2 m (6.6 ft) from the sidewall. The *thick line* is a combination of the adjacent-boundary effect *(broken line)* and the response of the subwoofer *(thin line)*.

example. The thick line shows how the response of the subwoofer is modified when placed in the room. Note that there is only a slight ripple in the response, and more important, the overall response is lifted by about 5 dB. This means that the room is reinforcing the sound of the subwoofer and therefore the subwoofer will be able to play louder. Another interesting effect is that the room is lowering the subwoofer's roll-off frequency from 30 Hz to about 25 Hz. This lowering can be seen in figure 4.40, but you need to take into account that the room has boosted the nominal response of the subwoofer by 5 dB. As a result, the roll-off frequency where the nominal response is 3 dB down now occurs at the frequency where the response of the subwoofer has been reduced from 5 dB to 2 dB. As you can see in figure 4.40, the response drops to about 2 dB at a frequency of approximately 25 Hz. Thus, the room not only reinforces the sound of the subwoofer, it also enables the subwoofer to go deeper and lower. In practice, the adjacent-boundary effect that boosts the low frequencies strongly depends on the rigidity of the room's walls. The largest bass boost and extension is obtained with walls made of brick or concrete. With less rigid structures like timber frame walls the boost and extension will be far less, because such walls will absorb bass energy (Allison, 1974; Colloms, 2018).

With the subwoofer positioned on the floor along the front wall, what remains is choosing an appropriate distance to the sidewalls such that the detrimental effects of the room resonances are minimized. It is not advisable to position the subwoofer in a corner. In that position it will excite all room resonances and the resulting bass response will be far from smooth (Fazenda et al., 2012). The corner corresponds to a pressure maximum for all resonances of the room. To smooth out the bass response you should avoid placing the subwoofer at the pressure maximums and pressure minimums of the room resonances that occur between the two sidewalls. You only have to take into account those resonances that have a resonance frequency that lies below the subwoofer crossover frequency. Placing the subwoofer at a pressure maximum will strongly stimulate the associated resonance frequency such that this frequency starts to sound annoyingly louder than all the other bass frequencies. Placing the subwoofer at a pressure minimum will also result in stimulating the associated resonance frequency, but now the opposite happens, the corresponding frequency almost disappears from the reproduced sound. Figure 4.41 revisits the example of room resonances presented in figure 2.35. It shows the pressure distribution between two sidewalls that are 4 m (13 ft) apart for the two lowest resonance frequencies of 43 Hz and 86 Hz. The positions to avoid are the pressure maximums at 0, 2, and 4 m; as well as the pressure minimums at 1, 2, and 3 m. You should also avoid placing the subwoofer too close to these positions, because in real-life situations their effect often extends a bit beyond the theoretically calculated position. In the example shown in 4.41, a good position for your subwoofer would be at 1.5 m.

Recommendation 4.33 *If you use a single subwoofer place it on the floor between the left and right loudspeakers close to the front wall and away from the sidewalls. Do not place it further than 60 cm (2 ft) away from the front wall, and do not place it at the pressure maximums and minimums of the room resonances that occur between the two sidewalls.*

Figure 4.41
Pressure distribution for standing waves between two walls that are 4 m (13 ft) apart. *At the top,* the distribution of the first room resonance at 43 Hz. *At the bottom,* the distribution of the second resonance at 86 Hz. The + and – indicate that at each pressure minimum (null) the pressure variation changes polarity.

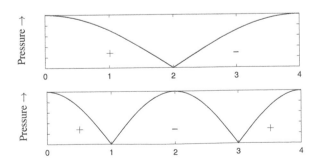

The subwoofer positioning advice given thus far takes into account the adjacent-boundary effect and the room resonances between the two sidewalls. Of course, the resonances between the front and the back wall, and the resonances between the floor and the ceiling also influence the low-frequency reproduction. However, it is difficult to do something about them. At the recommended position, on the floor and not further than 60 cm (2 ft) from the front wall, the subwoofer will be positioned near the pressure maximums of all the room resonances between the front and back wall and all the resonances between the floor and the ceiling. These resonances can alter the low-frequency sound considerably, and there is not much that can be done about it in terms of subwoofer positioning. For example, raising the subwoofer off the floor is not very practical and will alter the adjacent-boundary effect. Similarly, moving the subwoofer further away from the front wall will also alter the adjacent-boundary effect. The only thing that might help to tame some of the resonances is to carefully position the prime listening seat. Avoid placing it in a pressure maximum or minimum of one of the room resonances between the front and the back wall.

In practice, the theory that predicts the pressure distribution of the room resonances often breaks down. Rooms are not always rectangular, nor symmetrical, and the walls might be less rigid. Therefore, the amount to which resonances influence the sound is far less predictable from theory and has to be experienced in real life. The only way to know for sure what is happening at low frequencies is to perform acoustical measurements (Toole, 2018; Welti and Devantier, 2006; Winer, 2018). Figure 4.42 shows such a measurement for a subwoofer positioned on the floor near the front wall at a distance of 110 cm (3.6 ft) from the left sidewall. As you can see there are some large peaks and dips in the response. This is quite typical. Almost every room has an uneven response in the low-frequency region. Figure 4.43 shows what happens if the subwoofer is moved to the other side of the room at a distance of 110 cm (3.6 ft) from the right sidewall. The response changes a lot and actually gets even less smooth. Clearly, at this alternative position the room resonances are not excited in the same way. Thus, the pressure distribution of the room resonances is not symmetrical in this room.

Performing acoustical measurements is essential if you want to find the best positions for your subwoofers. In section 8.1.3 I will explain it in more detail. If you are not able to or if you don't want to perform such measurements, you could alternatively use your ears to

Figure 4.42
Response at the prime listening position with one subwoofer located at the front wall, 110 cm (3.6 ft) from the left sidewall.

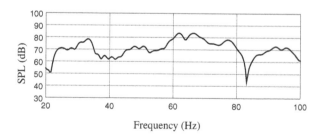

Figure 4.43
Response at the prime listening position with one subwoofer located at the front wall, 110 cm (3.6 ft) from the right sidewall.

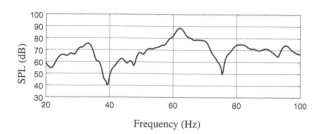

listen to the response of the subwoofer at different positions in the room (Harley, 2015; McLaughlin, 2005). Be aware though that this is much less accurate and can be time consuming. Before you start listening, disconnect all the other loudspeakers. Next, play a bass-heavy piece of music and sit in the prime listening seat to listen to the sound of the subwoofer. Try out different positions for the subwoofer to find the position at which it sounds the smoothest.

Using multiple subwoofers instead of just one will greatly improve the sound quality in the low-frequency region. Multiple subwoofers that are positioned at different locations in the room will excite the room resonances differently, and as a result the low-frequency response gets smoothed out and often becomes substantially more even (Harley, 2015; Holman, 2008; Toole, 2018; Welti and Devantier, 2006). For example, when two subwoofers are placed to the left and the right of a pressure minimum, they will drive the corresponding room resonance at opposite phases and thus effectively cancel it out (Welti, 2002). Have a look at the top of figure 4.41 to see that placing subwoofers at 1 m and 3 m from the left wall will indeed drive the first resonance of 43 Hz at opposite phases. Now let's look at a real-life example. Recall that figure 4.42 shows the measured response of only one subwoofer placed 110 cm (3.6 ft) from the left wall, and that figure 4.43 shows the response when that single subwoofer is placed 110 cm (3.6 ft) from the right wall. Now, if you place two sub-woofers in this room at these same positions (one 110 cm from the left wall and the other 110 cm from the right wall), you get the smoothed-out response shown in figure 4.44. Note that the large dips around 39 Hz, 75 Hz, and 82 Hz are all gone, making the overall response much more even.

Using multiple subwoofers will not only smooth out the bass at the prime listening seat, but throughout the entire listening area. This is a major advantage. Multiple subwoofers at different locations are likely to excite the room resonances in a more balanced matter, reducing the seat-to-seat variations in the low-frequency response. Intuition would suggest that the more subwoofers you use the better, but this is not the case. Extensive experiments with different subwoofer configurations in a rectangular room have shown that two or four subwoofers are sufficient (Welti, 2002; Welti and Devantier, 2006). The optimal configurations are shown in figures 4.45 and 4.46. There happens to be no advantage of having three subwoofers; the optimal subwoofer configurations all use an even number of subwoofers. However, bear in mind that these configurations are only optimal in a symmetrical rectangular room. Again, in real life, rooms often differ, because there are windows, doors, alcoves, openings to other rooms, and large pieces of furniture (Toole, 2018). Nevertheless, it remains advantageous to use multiple subwoofers in a room to smooth out the bass response. It is just that the configurations in figures 4.45 and 4.46 might not be the most optimal ones. Finding the optimal subwoofer positions requires some experimentation, preferably using your own acoustical measurements.

Regardless of the type of room you have, I recommend using at least two subwoofers. When you have more than one row of seating, you could consider using four subwoofers. With four subwoofers you will be able to make the bass response more consistent among the different rows of seating, and this might justify the additional cost of two extra subwoofers.

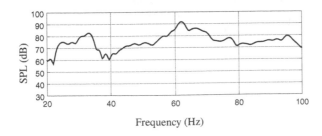

Figure 4.44
Response at the prime listening position with two subwoofers located at the front wall; one subwoofer 110 cm (3.6 ft) from the left sidewall and the other 110 cm (3.6 ft) from the right sidewall.

Figure 4.45
Best arrangements
for two subwoofers.
The subwoofers are
either placed at the wall
midpoints or at a quarter
distance from the sidewalls.

Based on data from Welti and
Devantier (2006).

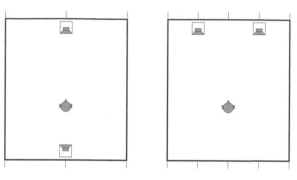

Figure 4.46
Best arrangements for
four subwoofers. The
subwoofers are placed at
the wall midpoints, at a
quarter distance from the
sidewalls, or in the corners
of the room.

Based on data from Welti and
Devantier (2006).

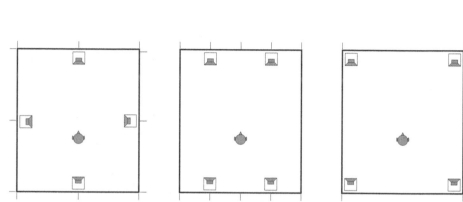

Recommendation 4.34 *Use at least two subwoofers placed at two different positions in the room to get the smoothest bass response. Use the configurations from figure 4.45 as a starting point. If you have multiple rows of seats, consider using four subwoofers distributed around the room to minimize the variations from row to row. Use the configurations from figure 4.46 as a starting point.*

It is absolutely essential that all subwoofers receive the same low-frequency signal from the A/V controller. They should all be playing the same monoaural bass that consists of the bass from all the channels (front, surround, and LFE) summed together. If not, the results become unpredictably bad, just like the case with full-range loudspeakers discussed in section 4.1. Ideally, all subwoofers should be identical models for the best overall performance. Mixing different models is possible but requires some care. If the subwoofers have different roll-off frequencies, they will be in phase for almost the entire frequency band, but at or below the roll-off frequency they may be out of phase. If that happens, the subwoofer with the highest roll-off frequency reduces the low-frequency output of the other sub below this roll-off frequency. Hence, the highest roll-off frequency of the two subwoofers becomes the roll-off frequency of the entire system (DellaSala, 2010).

4.3 Room Requirements

Now that we know how to position the loudspeakers and the video display in the room, let's have a look at the room itself. The room in which you watch and listen should be free from anything that distracts you from having the optimal audio/video experience. The room should provide a comfortable environment that is neither too hot nor too cold, lacks any drafts, and has comfortable chairs. To have the best possible experience your room should be devoid of extraneous light and sound. Strong ambient light needs to be avoided, because it washes out the contrast and colors of your video screen. Loud background noise also needs to be avoided as it can easily distract you and obscure the subtle nuances in the reproduced sound (Noxon, 1994).

4.3.1 Ambient Light

Too much ambient light is a bad thing. It reflects off the video screen, making the blacks less black and the colors less saturated. A window or a room light in the wrong place can create an annoying glare on the screen. Therefore, close the window shades and turn off the lights. Your display looks best in a dark environment. Furthermore, in this darkness, the room itself disappears from your view, enabling you to get more involved with the movie and eliminating unwanted visual distractions from your environment. Being able to control the amount of light in your room makes a huge difference on the perceived video quality (Briere and Hurley, 2009; Rushing, 2004).

Install window treatments that can block out the incoming sunlight. Obviously, the fewer windows a room has and the smaller these windows are the better. Use window treatments that are opaque and have a blackout feature that prevents light from entering. Use dark and medium hues and stay away from white, bright, and light colors, because these will reflect the light from the video screen. For the same reason, choose matte finishes instead of shiny fabrics (Rushing, 2004). Thick heavy curtains not only block out the light, they also reduce sounds coming in from outside and can be used as sound absorbers in taming room acoustics as will be explained in section 8.2.2.

> **Recommendation 4.35** *Install window treatments such that you can darken the room.*

Controlling the light in your home theater room is important. You do not want to stumble around in the dark looking for your remote control. Besides being of practical use, lighting helps to create the right mood and atmosphere in your room. Three types of lighting can be distinguished: 1) the room's main light source that provides the necessary ambient light before and after the movie; 2) focused accent lighting that can be used to enhance certain features of the room, such as artwork, sculptures, or even your loudspeakers; and 3) task lighting that is focused on certain areas of the room to facilitate activities such as fetching a drink or loading a Blu-ray disc. Consider investing in a lighting control system that allows you to switch the lights on and off without getting up from your seat. Even a simple control system has its uses, because it is not very convenient to get up from your seat, walk to the light switch, and then find your way back in the dark (Briere and Hurley, 2009; Rushing, 2004).

Should your room be completely dark while you are watching a movie? Well, it depends. Large projection displays can be viewed in a completely dark room, but small and very bright displays may cause eye strain when being watched in a completely dark environment. Flat-panel displays are typically much brighter than projection displays (see table 3.2) and can easily cause eye strain when viewed in a dark room (Bodrogi and Khanh, 2012). It is therefore recommended that you view them with a bit of low-level ambient light. By contrast, projection displays are best viewed in a completely dark room, because they are not as bright and their contrast is reduced when ambient light hits the projection screen. Let's look into these two different situations in more detail.

Bright flat-panel displays are best viewed in a dimly lit environment. It is best to illuminate the wall behind the display such that no direct light can reflect off the video screen and such that you cannot see any direct light from your viewing position. Often the best place to put your ambient light source is behind the video screen. The light source should be a neutral white, ideally with a color temperature of 6500 K to match the color of white on your video display. The wall behind the video display should also be neutrally colored. The smaller the display the more ambient light you need. Typically, you need an amount that equals 10% to 15% of the maximum light output of the display, as specified in table 3.2 (ITU-R BT.710, 1998). If your flat-panel display is rather large you might prefer less than 10% ambient light, because the display itself will easily illuminate its surroundings. The exact amount of ambient light can be chosen based on personal preference but should never exceed the recommended maximum of 15%. For watching HDR material it is recommended that the ambient light level lies below 5 cd/m² (ITU-R BT.2100, 2017).

There are two ways that you can determine the right amount of ambient light. The first method only requires the use of a video test pattern. The AVS HD 709, DVE HD Basics, and HD Benchmark video test patterns (see appendix A) contain test patterns to visually estimate the optimal amount of ambient light (see table 4.5). These patterns consist of a gray rectangle that represents the desired amount of ambient light. You should fine-tune the amount of ambient light such that the light that falls on the wall behind the display matches the intensity of the gray rectangle on the video display. You could either try out light bulbs of different intensity, use a dimmer, or partially mask the light source. The second way to determine the right amount of ambient light requires the use of a light level meter. You use this meter to measure the amount of light that is emitted by the video display when it displays a 100% white field from a video test disc. All the discs mentioned above contain such a pattern. Point the light level meter towards the video screen, just a few centimeters from the surface, and measure the amount of light in lux (see box 4.6). Next, in a similar fashion measure the amount of light (again in lux) that reflects off the wall. This reflected light should be 10% to 15% of the light that you measured off the screen.

Recommendation 4.36 *If you use a flat-panel display, light the wall behind it such that the intensity of the light on this wall is 10% to 15% of the maximum light output of the display. For watching HDR content a maximum of only 5 cd/m² is recommended.*

Projection screens are best viewed in a completely dark environment. Eye strain is not a problem, because the display is large and not as bright as a flat-panel display. The problem with projection screens is that ambient light degrades the achievable contrast, the reason being that the blackest black that can be displayed depends on the ambient light. The screen can

Table 4.5 Test patterns for determining the optimal amount of ambient light.

AVS HD 709	Misc. Patterns—Various	Backlight Comparison
DVE HD Basics	Basic Patterns	Maximum Ambient Light
HD Benchmark	Advanced Video—Setup	Bias Light 10%
		Bias Light 15%

BOX 4.6 ILLUMINANCE AND SCREEN LUMINANCE

Illuminance in lux (lx) is the quantity measured by a light meter. The formal definition of illuminance (E_v) is the luminous flux per unit area arriving at a surface throughout all directions in a hemisphere (Poynton, 2012). The luminance (L_v) of the screen in cd/m² is obtained from an illuminance measurement (E_v) as follows:

$$L_v = \frac{E_v R}{\pi} \approx 0.3183\, E_v R$$

where R is the reflectivity of the area.

Table 4.6 Test patterns for measuring sequential and simultaneous contrast.

AVS HD 709	ChromaPure/HCFR	Contrast 0% Black Contrast 100% White ANSI Contrast
DVE HD Basics	Advanced Patterns	Window 0% with PLUGE Window 100% with PLUGE
DVS UHD	Miscellaneous Setup Patterns	Contrast Ratio ANSI Contrast
HD Benchmark	Advanced Video—Contrast Ratio	Black 0% White 100% 4×4 Checkerboard

never be blacker than the reflection of the ambient light. Consequently, the less ambient light there is, the blacker the screen can become and the higher the achievable contrast. In a well-lit room the sequential contrast can be as low as 2:1 (Brennesholtz and Stupp, 2008). Recall that sequential contrast is the ratio of the luminance of a completely white field (100% white) divided by the luminance of a completely black field (see section 3.2.1). If you have a projection system, I highly recommend that you check whether your room is dark enough to achieve a sequential contrast of at least 500:1 (see table 3.2). You can do this using a light level meter, or alternatively using a colorimeter as I explain in section 6.2.3. Hold the light meter close to the projection screen and aim it at the projector (not at the screen). Measure the light level in lux for a completely white field and a completely black field. You obtain the sequential contrast by dividing the measured lux value for white by the value for black. Test patterns for a completely white and black field are available as a video test disc or as downloadable files; see table 4.6 and appendix A. In the unfortunate case that your sequential contrast is on the low side, you could either make the room darker to lessen the light emitted from a black field, or alternatively increase the light emitted from a completely white field by using a smaller projection screen or a more powerful projector. Given this choice, most people will understandably go for the darker room.

Achieving sufficient simultaneous contrast on a projection screen is often much harder than achieving sufficient sequential contrast. Recall from section 3.2.1 that simultaneous contrast is measured using a checkerboard pattern of white and black squares. The reason that it is difficult to achieve good simultaneous contrast is that the light from the white squares will spill over into the black squares. The light from the white squares will reflect off the walls, floor, and ceiling back to the screen. When this reflected light hits the black squares they will become less black and will look more like dark gray squares. Ideally, the squares remain dark enough such that simultaneous contrast is at least 100:1 (see table 3.2). Maximum simultaneous contrast can only be achieved when all the walls, the floor, and the ceiling have a black matte finish. This is why such a finish is used in most dedicated home theater rooms. If you are using a multipurpose room, painting everything black will probably not appeal to you. Still, it is a good idea to choose dark subdued colors for the room's surfaces and decorations. At the bare minimum, you should avoid glossy light colored surfaces, because they reflect a lot of light and will show up as distracting blotches of light in the room. Be especially careful with mirrors and picture frames with a glass surface. Also avoid brightly colored walls as they create an unwanted color cast on your projection screen. Pick a neutral subdued color scheme for your room and stick with matte and textured finishes (Briere and Hurley, 2009; Harley, 2002; McLaughlin, 2005; Rushing, 2004).

Recommendation 4.37 *If you use a projection screen make the room as dark as possible to achieve maximum contrast. Also pick neutral, matte, and subdued colors for your room's surfaces, furnishing, and decorations. Consider painting the walls and ceiling a dark color to minimize the amount of light that is reflected back to the screen.*

The appearance of brightness and color on the video display is influenced by the brightness and color of the wall surrounding your display. This holds true for both flat-panel displays as well as projection screens. Because of simultaneous brightness contrast (see figure 2.11) the brightness of the wall influences the apparent brightness of the display. With a dark surrounding wall the video display looks bright, while with a bright wall it looks dim. Similarly, the color of the surrounding wall influences the perception of the colors on the display. A correct color appearance requires a neutral or nearly neutral colored wall (Bodrogi and Khanh, 2012; McLaughlin, 2005; Poynton, 2012). For an optimal viewing experience it is recommended that the background luminance and color be controlled in an area around the video display that is 53° high by 83° wide. If that is not possible then aim to control luminance and color in an area that is at least 28° high by 48° wide (ITU-R BT.710, 1998). Since this surrounding area is specified as a viewing angle, the actual size in meters or feet depends on the viewing distance. You can look up the height and width in meters or feet in figure 4.47. I explain how the values in these figures were calculated in box 4.7 and figure 4.48. As a final note, you should also avoid distracting lights in this area surrounding the display. Indicator lights and displays on your audio and video equipment easily distract your eyes. Therefore, either move your equipment off to the side or place it behind a closed cabinet door.

Figure 4.47
Size of the area surrounding the display that should have medium brightness and a neutral color. The size is measured in meters or feet from the center of the display. The width of the area (*x* in figure 4.48) is indicated by the two solid *diagonal lines*. The *upper solid line* corresponds to the recommended width (83°), the *lower solid line* to the minimum width (48°). The height of the area (*y* in figure 4.48) is indicated by the two *broken diagonal lines*. The *upper broken line* corresponds to the recommended height (53°), the *lower broken line* to the minimum height (28°).

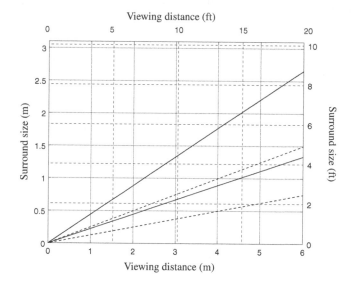

BOX 4.7 AREA SURROUNDING THE DISPLAY

Luminance and color should be controlled in an area around the display that is 53° high by 83° wide. Since this area is specified as a viewing angle, the actual size in meters or feet depends on the viewing distance (compare the computations of viewing distance in box 4.2). The width *x* and height *y* of the area surrounding the display can be computed as

$$x = v \tan\left(\frac{83°}{2}\right), \quad y = v \tan\left(\frac{53°}{2}\right)$$

where *v* is the viewing distance.

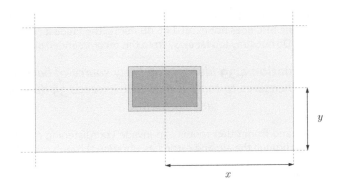

Figure 4.48
Width *x* and height *y* of the
area surrounding the video
display as measured from
the center of the display.

Recommendation 4.38 *Control the background luminance and choose a neutral color in an area around the video display that is 53° high by 83° wide. Also avoid having any bright lights or reflecting objects in this area.*

4.3.2 Background Noise

Hearing sound that is not part of the movie or recording that you are playing is distracting and annoying. Therefore, the aim is to avoid such distractions as much as possible. Extraneous sounds can originate from within the room itself or penetrate the room from outside. Examples are numerous: the fridge softly humming away in the corner, your projector's fan spinning, the rain hitting your roof hard, cars whooshing by your house, or people shouting in the street. Not only does such background noise distract you from the movie or recording, it also interferes with your ability to hear details. The background noise may mask certain subtle nuances in the reproduced sound, reducing the overall clarity and obscuring low-level spatial cues (Harley, 2015). Furthermore, any sound in the source material that is softer than the background noise will get completely lost in it.

It is recommended that the background noise in your listening room not exceed an SPL of 35 dBA or 50 dBC (AES, 2008). The 'A' and 'C' in this SPL specification indicate the use of different frequency weighting curves (IEC 61672, 2013). The A-weighting mimics the sensitivity of the ear at moderate sound levels and emphasizes the contribution of the mid and high frequencies, while the C-weighting takes into account the entire frequency spectrum including the low frequencies. You need an SPL meter to check the average sound level of the background noise in your own room. Such a meter can be a good investment, because you can also use it to accurately calibrate the levels of the different loudspeakers in a surround sound system (as explained in section 7.5.4) and you can use it to calibrate and equalize your subwoofers (as explained in section 8.1.3). To measure the amount of background noise with your SPL meter, position the meter at ear level at the prime listening position, set the meter to a slow response, and take a measurement using A-weighting. The lower the background noise the better. A rating below 40 dBA is considered to be quiet and a rating below 35 dBA is excellent. Ratings below 30 dBA are difficult to achieve in an urban environment and often require extensive and expensive remodeling to reduce the penetration of outdoor sounds (Jones, 2015a).

It is not a good idea to increase the playback volume to counter too much background noise. You might be tempted to increase the overall volume to make the softest sounds in the source material louder than the background noise level. However, this typically results in the loudest sounds in the source material becoming too loud. So loud that you might no longer enjoy it. And so loud that your neighbors start to complain, because you have become the source of their background noise. Furthermore, remember from section 2.3.2 that if you play the source material at too loud (or too low) a volume the frequency balance will change due to the frequency selective sensitivity of the human hearing system. The recommended listening level at home is such that the SPL does not exceed 96 dB during the loudest peaks in the source material (see box 3.2). Do not stray too far away from this recommended playback level.

Recommendation 4.39 *Identify the likely source of background noise and take measures to tame it if its SPL exceeds 35 dBA or 50 dBC in your room.*

Sounds from outside and from other rooms can invade your listening room in two ways: transmitted as sound through the air; and carried as vibration through solid structures, such as the floor, walls, and doors. Isolating a room from outside noise requires thick walls and the elimination of all air leaks. The amount of noise that can enter a room through small cracks is astonishing. Close doors and windows, and seal their perimeters with weather stripping. Even a small crack under a door can radiate lots of outside noise into your room. Solid structures like walls, doors, and window panes vibrate in response to the sound hitting them. The heavier they are the less sound they will transmit. Solid core doors, double glass window panes, and rigid walls help to isolate the room from unwanted sounds. The heavier the wall the better the isolation. Double walls with an unbridged air cavity between them work even better. But the ultimate sound isolation requires a room within a room, such that all the inner walls, including the floor and the ceiling, are mechanically isolated from the outside structure of the room with air cavities in between (Everest and Pohlmann, 2014; Jones, 2015a). Such a room within a room requires serious effort and commitment. It is a highly specialized undertaking that is best entrusted to an acoustical consultant who is an expert in such matters. Fortunately, in most situations such extreme measures are not necessary.

Appliances in your room can also generate distracting background noises. Sit down in the prime listening position and listen for distracting sounds, such as a clock ticking or the humming of a refrigerator. Remove noise-generating devices from the room, create a sound-insulating barrier between your prime seat and the device, or move the device farther away from your seat (Briere and Hurley, 2009; Rushing, 2004). Projectors, amplifiers, and home theater PCs have built-in fans that also generate some background noise. Do not position such equipment too close to your prime listening seat.

The low-frequency sound from your subwoofers causes vibrations that propagate through your room. These vibrations can cause certain objects and surfaces in the room to produce a rattling or buzzing sound. These extraneous sounds are annoying and distracting and should therefore be eliminated (EBU-Tech 3276, 1998). Make sure that all resonating structures and objects in the room are sufficiently damped. Rattling windows, picture frames, and lamp shades can often be silenced by using a small piece of caulk or double sided-tape. Sticking a small block of felt to them might also help. The easiest way to check for rattling and buzzing objects in your room is to play a low-frequency sweep and listen for any problems. The sweep should cover at least the frequencies between 15 Hz and 200 Hz. A suitable low-frequency sweep is included on several audio test discs (see appendix A). You can also download a low-frequency sweep from the Audio Check website www.audiocheck.net.

Recommendation 4.40 *Listen to a low-frequency sweep that ranges from 15 Hz to 200 Hz to identify and locate any objects in the room that rattle or buzz. Take measures to silence these objects: remove, move, dampen, or fixate them.*

4.4 Installing a Front-Projection System

Installing a flat-panel display in your room is rather straightforward. You either place it on its own stand or hang it on the front wall. Installing a front-projection system is a bit more involved, because you need to carefully align the projector and the screen. The projector can be mounted on the ceiling, the floor, or the back wall. In most situations, the ceiling is your best option. If you place a projector on the ceiling, the light reflecting off the screen bounces towards the viewers. This makes the projected image appear brighter. In contrast, if you place a projector on the floor, the light reflecting off the screen will bounce towards the ceiling, resulting in a dimmer image (Rushing, 2004). Mounting the projector high on the back wall is another option, but in most cases the projector might be too far from the screen. Make sure you have sufficient clearance under the projection beam, such that your audience is not shadowing the projection beam with their heads.

It is important to mount the projector at the appropriate distance from the projection screen. The distance between the projector and the screen is called the *throw distance*. It determines the size of the image on the screen. To make the size of the projected image fit the size of the screen you need to carefully determine the throw distance. To make things easier, most projectors have a variable zoom lens that allows you to vary the image size without changing the throw distance. This gives you a little more leeway as you can position the projector at a certain throw distance and use the zoom lens to make the image fit the screen (see figure 4.49). Most projector manuals contain a table listing the screen size at a number of throw distances for the maximum and minimum zoom setting of the lens. From such a table you can determine the *throw ratio* of your zoom lens at its minimum and maximum setting. The larger the throw ratio, the more the lens has been zoomed in. The throw ratio allows you to compute the screen size at any throw distance, as explained in box 4.8. Alternatively, you can use the online calculator from Projector Central (www.projectorcentral.com) to determine the appropriate throw distance for your screen size. Sometimes the manufacturer of the projector has a calculator available online (Brennesholtz and Stupp, 2008).

The closer the projector is to the screen, the brighter the projected image will be. Although a zoom lens gives you some flexibility in projector placement, it is not always a good idea to

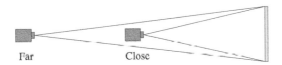

Far   Close

Figure 4.49
When the projector is farther away from the projection screen, you need to zoom in with the projector's lens to make the projected image fit the screen. If you do not zoom in at such a far distance, the projected image will be much larger than the projection screen.

BOX 4.8 PROJECTOR THROW DISTANCE

The size of a projected image depends on the throw distance (the distance between the projector and the screen) and on the zoom setting of the projector's lens. The width (W) of a projected image can be calculated as follows:

$$W = \frac{D}{R}$$

where D is the throw distance and R the throw ratio of the lens. The throw ratio R varies with the zoom setting of the lens. The throw ratios for the minimum and maximum zoom settings of the lens can be easily calculated from the mounting examples given in the projector's user manual. Once you have determined these two values, you can calculate the variation in image size that the zoom lens allows at any given throw distance.

use it. The more you zoom in to achieve the same image size, the farther away you need to place the projector and the dimmer the projected image becomes. For example, with a zoom lens that has a zoom ratio of 2:1, the maximum luminance of the screen might be halved when the lens is fully zoomed in. By zooming in all the way you lose quite some luminance output from the projector. You can use the online calculator from Projector Central (www.projector-central.com) to get an estimate of the image brightness at different zoom settings and throw distances for many different projector models.

Recommendation 4.41 *For maximum image brightness, do not zoom the projector's lens too far in, but instead place the projector closer to the screen. Also mount the projector on the ceiling and not on the floor.*

The screen and projector need to be aligned to each other properly; otherwise, the image becomes geometrically distorted. The front of the projector should be parallel to the projection screen in both the vertical and the horizontal direction. Never tilt the projector as shown in figure 4.50, because it results in trapezoidal distortion of the image where the bottom will be wider than the top. Most projectors have a so-called *keystone correction* in their setup menu to counteract trapezoidal distortion. When you activate it, an inverted trapezoidal distortion will be digitally applied to your image to counteract the optical distortion. Never use such a digital 'keystone correction', because it reduces the resolution in your image, destroying fine details and making the image softer on one side. Tilting the projector vertically as shown in figure 4.50 will also cause focus problems, because the top part of the screen will be closer to the projector than the bottom. As a result, you cannot get both the top and bottom part of the screen in focus at the same time. Trapezoidal distortion and focus problems can occur in both the horizontal and vertical direction. Therefore, always keep your projector's front parallel to the projection screen in both directions, as shown in figure 4.51.

Figure 4.50
Ceiling-mounted projector tilted down to project an image on the screen. The front of the projector is not parallel to the surface of the screen and as a result the projected image is no longer a rectangle but a trapezoid that is wider at the bottom.

Figure 4.51
Ceiling-mounted projector with its front parallel to the surface of the screen. Vertical lens shift is used to shift the projected image down towards the screen. There is no trapezoidal distortion; the projected image is perfectly rectangular.

The projector is equipped with an adjustable vertical lens shift, so that you can move the image in the vertical direction up or down to align it to the screen. Without this lens shift, you would be limited to mounting the projector such that its lens lines up to the center of the projection screen—not very practical in most rooms, as the projector would be hovering in midair somewhere between the floor and the ceiling. The vertical lens shift allows you to mount the projector on the ceiling and shift its projected image downwards to where the screen is located. Most projectors are also equipped with a horizontal lens shift, which obviously allows you to move the image in the horizontal direction. You only need it when you are not able to line up the projector's lens to the horizontal center of the screen. Typically, the range of horizontal shift is less than the range of vertical shift. Figure 4.52 shows a typical example. For the best image quality, avoid using too large a lens shift. Without any lens shift, the image is projected through the center part of the lens. Vertical and horizontal shifts will move the projected image towards the edges of the lens, as illustrated in figure 4.53. Most lenses are less sharp at their edges and may introduce some optical aberrations when used close to their edges.

Recommendation 4.42 *Keep the front of the projector parallel to the projection screen. Use the lens shift controls to move the image up and down. Do not tilt the projector and do not use the digital 'keystone correction' feature from the projector's setup menu.*

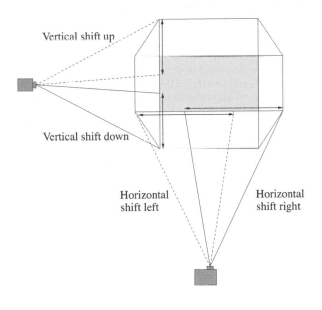

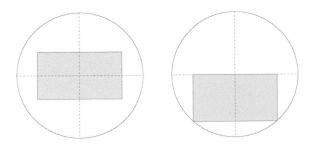

Figure 4.52
Vertical and horizontal lens shift. The *gray area* indicates the image from the projector without any vertical or horizontal lens shift applied. Maximum lens shift up, down, left, and right are indicated by the *arrows*. Typical maximum vertical lens shift equals 85% of the image height. Typical maximum horizontal lens shift equals 25% of the image width. Vertical and horizontal lens shift settings influence each other's range. When horizontal and vertical lens shifts are applied together, the image must be within the *octagonal shape.*

Figure 4.53
A lens shift will move the projected image towards the edges of the lens. *On the left,* the projected image is perfectly centered and uses the center part of the lens. *On the right,* the projected image is vertically shifted all the way up to the edge of the lens.

Table 4.7 Test patterns for focusing the lens of a projector.

AVS HD 709	Basic Settings	Sharpness and Overscan
DVE HD Basics	Basic Patterns	Overscan
DVS UHD	Basic Setups	Sharpness and Overscan
HD Benchmark	Video Calibration	Sharpness

After aligning the screen and the projector, you need to focus the lens of the projector. Without the proper focus the image won't be sharp. Use a test pattern from AVS HD 709, DVE HD Basics, DVS UHD, or HD Benchmark (see appendix A) that has a lot of small details (see table 4.7). Stand close to the screen such that you can see the individual pixels. If your projector has a motorized focus, you can use the remote control to focus the lens. Otherwise, ask somebody to help you out. Or, alternatively, stand next to the projector and use binoculars to look at the screen while you turn the focus dial. Carefully check the sharpness in all corners of the screen. When you are not able to get the entire screen in sharp focus, find a spot on the focus dial where all corners are approximately equally sharp. In such a case, you may want to check the alignment between your projector and the screen, and you may also want to check the flatness of the screen. When only some random parts of the screen are not as sharp as you would like, the reason might be a dirty lens.

Recommendation 4.43 *Focus the lens of the projector to achieve maximum sharpness. Use a test pattern with a lot of small details to check the result.*

When installing the projector, you also need to consider the air flow around it. Projectors run very hot, and unless they are properly ventilated they will overheat and malfunction. Do not cover the air inlet or air outlet ports, and keep at least 10–30 cm between the projector and the walls. Do not place the projector in a hot and humid environment, and keep it away from other heat radiating sources. If you want to conceal the projector when it is not in use, you could consider putting it in a box that matches the ceiling, but make sure that the projector is still properly ventilated. Alternatively, you could put it on a motorized lift that descends from the ceiling when you want to watch a movie, and that retracts into the ceiling to store the projector out of sight when you are not using it (Rushing, 2004).

5
DIGITAL SOURCES

Playing a mediocre recording on an excellent sound system won't make it sound better. In fact, it will sound worse. The high-quality sound system will ruthlessly reveal every flaw in the recording. The same holds true for mediocre video material. To enjoy high-quality audio and video reproduction you need to start with high-quality source material.

The quality of audio source material depends on two different factors. The first factor is the quality of the recording itself. The quality varies, because different sound engineers have different styles and quality standards. The recording quality also depends on its age. Recordings made in the past with equipment that is now considered obsolete often sound much worse than recordings made with the newest technology. For example, recordings made before the 1960s have a limited dynamic range and are lacking the highest and lowest parts of the audio frequency range. Back then, almost every recording was made in mono, so these recordings are also severely limited in the spatial cues that they provide. As technology for audio recording and delivery has improved over the years, the technical quality of the recordings has also improved. In recent years, a number of legacy recordings have been digitally remastered. It is worthwhile to check them out, because it often means that the sound quality has been improved.

Despite the technology advantages not all modern recordings and remastered tracks have improved in sound quality. A major spoiler has been the so-called *loudness war* (Deruty and Tardieu, 2014; Milner, 2009; Vickers, 2010). In this loudness war artists and producers use

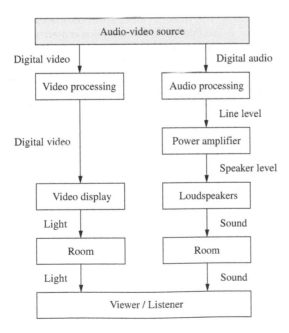

dynamic compression to make their recordings sound louder than those of the competitors (see figure 3.17 for an example of compression). The origin of this loudness envy lies in the fact that when two identical programs are presented at different loudness levels, the louder of the two appears to sound 'better' and attracts your attention. Applying too much compression will actually hurt the sound quality of the program. As the dynamic range is decreased the sound becomes uninvolving and flat. The quest for more loudness has even resulted in some recordings that were so compressed and raised in level that they became heavily distorted. The height of the loudness war took place between 1989 and 2011 (Deruty and Tardieu, 2014). Loudness normalization in music streaming services and media players has helped to put an end to the loudness war. This normalization reduces the loudness of every recording to a certain standardized level (more on that in section 5.3.4). With such normalization in place it doesn't make much sense anymore to aim for the loudest recording. The media player will reduce the level upon playback and all that remains is a heavily compressed and distorted sound. A widespread use of loudness normalization in the future might provide the right incentive for artists and producers to focus again on making their recordings sound more dynamic instead of competitively loud (Katz, 2015).

The second factor that determines the quality of audio source material is the delivery format. Today, you can access audio sources in many different ways, from streaming services on the Internet to the good old compact discs. A recording delivered to you as a compressed MP3 file sounds worse than an uncompressed high-resolution audio recording. In the MP3 encoded file, the sound quality has been traded for a small file size. Highly compressed audio files save disk space and Internet bandwidth and can have their use, but avoid them if you want the best possible sound quality. The audio quality of each particular delivery format depends for a large part on the digital coding technique that it employs. Different codecs give different results. And beyond that, there is the issue of how many bits and which sampling rate to use. All these topics will be addressed in this chapter, so that you can make an informed choice when looking for high-quality source material.

Video source material typically provides both sound and imagery. Regarding the sound quality, similar aspects play a role as with audio-only sources. The sound quality depends both on the recording itself and on the delivery format. The same holds true for the video part. The quality of the moving images strongly depends on the age of the equipment used to record them. During the 1980s lots of material intended for television broadcasts was shot with electronic video cameras. This material looks awful compared to today's high-quality digital HD and UHD video. Restoration of mediocre video material is possible, but in general it will not improve much. However, imagery that was originally shot using photographic film can be rescued and converted into high-quality digitized video. The original film can be scanned with a high-resolution scanner that retains all the fine details present in the original. Advanced digital restoration techniques can be used to clean up and improve each individual frame. These techniques can automatically remove scratches and dirt, get rid of color casts, restore faded blacks, and much more. It is an expensive process, but it often creates a stunning result. When you want to watch an older movie, I highly recommend that you go for the digitally restored version. You will certainly be amazed by the quality that can be extracted from the original film.

For modern high-quality digital video material, the video coding method used is the major factor that determines image quality. While at the time of its introduction the image quality of the DVD amazed everyone, it has now lost its former glory if you compare it to HD and UHD video. Different delivery formats use different video codecs. In this chapter I will show you their quality differences.

5.1 Video Coding

Digital video is just a sequence of still images where each image is made up of a large number of individual pixels. Each individual pixel is described by a collection of ones and zeros

Table 5.1 Common video delivery formats (BDA, 2011, 2015b; DCI, 2008; ITU-R BT.709, 2015; ITU-R BT.1543, 2015; ITU-R BT.2020, 2015; Pohlmann, 2011; SMPTE RP 431–2, 2011; Taylor et al., 2006). Anamorphic wide-screen is indicated as 16:9a. DVD NTSC and DVD PAL formats can also be used on Blu-ray disc (BD).

Media	Aspect Ratio	Resolution (pixels)	Frame Rate (Hz)	Bit Depth
VCD NTSC	4:3	352 × 240	30i	8
VCD PAL	4:3	352 × 288	25i	8
SVCD NTSC	4:3	480 × 480	30i	8
SVCD PAL	4:3	480 × 576	25i	8
DVD NTSC	4:3, 16:9a	720 × 480	30i	8
DVD PAL	4:3, 16:9a	720 × 576	25i	8
BD	16:9	1280 × 720	24p, 50p, 60p	8
	16:9a	1440 × 1080	24p, 25i, 30i	8
	16:9	1920 × 1080	24p, 25i, 30i	8
UHD BD	16:9	1920 × 1080	24p	10
	16:9	3840 × 2160	24p, 25p, 50p, 60p	10
SDTV	4:3	640 × 480	24p, 30i, 60p	8
HDTV	16:9	1280 × 720	24p, 50p, 60p	8
	16:9	1920 × 1080	24p, 25i, 30i, 50p, 60p	8
UHDTV	16:9	3840 × 2160	24p, 25p, 30p, 50p, 60p, 120p	10, 12
Digital Cinema	19:10	2048 × 1080	24p, 48p	12
	19:10	4096 × 2160	24p	12
LSDI	16:9	7680 × 4320	24p, 25p, 30p, 50p, 60p, 120p	10, 12

called *bits*. The larger the number of bits per pixel the greater the tonal definition as I already explained in section 3.2.1. The number of horizontal and vertical pixels is called the *resolution*. Different delivery formats have different resolutions. The most common formats are listed in table 5.1. The ratio of the width of the video image to its height is called the *aspect ratio*. Almost all modern displays have a ratio of 16:9, and therefore a lot of common delivery formats use this ratio. Before the introduction of wide-screen television, the most commonly used aspect ratio was 4:3. Table 5.1 also shows the frame rate of the different delivery formats. The frame rate is simply the frequency at which the different still images are displayed. Each still image is called a *frame*. The letter *i* indicates *interlaced scanning* and the letter *p* *progressive scanning*. These are two different ways in which the horizontal lines of pixels are being displayed. I explain these two scanning techniques in more detail in section 5.1.3.

5.1.1 Aspect Ratio

The aspect ratio of 16:9 is the standard for wide-screen displays in the home. It has been chosen as a reasonable compromise between the older television aspect ratio of 4:3 and the commonly used aspect ratios in cinema (Taylor et al., 2009). The *Academy Standard* or *Flat* aspect ratio of the cinema is almost equal to the 16:9 ratio. The widely used *CinemaScope* or *Panavision* aspect ratio is much wider. Table 5.2 compares the different ratios and shows the percentage of a 16:9 screen that is used when either the height or width is fitted. Black bars appear on the top and bottom of the picture when one of the wider aspect ratios is fitted within the 16:9 screen. The corresponding display mode is called *letterbox*. Black bars appear on the sides of the picture when a 4:3 aspect ratio is fitted within the 16:9 screen. This display mode is called *pillarbox*. Consult figure 5.1 to see how it looks.

Some people find the black bars annoying. For this reason video displays and source components such as Blu-ray disc players offer different display modes. One way to get rid of the

Table 5.2 Common aspect ratios (Briere and Hurley, 2009; CinemaSource, 2001b; Taylor et al., 2009).

Name	Aspect Ratio	Display Mode on 16:9 Display	Percentage Used on 16:9 Display
Standard (4:3)	1.33:1	Pillarbox	75%
Wide-screen (16:9)	1.78:1	Full-screen	100%
Academy, Flat	1.85:1	Letterbox	96%
CinemaScope, Panavision	2.35:1	Letterbox	76%
CinemaScope, Panavision	2.40:1	Letterbox	74%

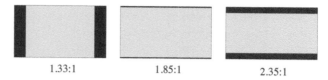

1.33:1   1.85:1   2.35:1

Figure 5.1
Common aspect ratios displayed on a 16:9 screen with black bars. The *gray area* is used to display the picture.

black bars is to zoom in. Such a display mode is often appropriately called 'zoom' or 'expand'. With 4:3 video material you will lose the top and bottom part of the picture, and with the CinemaScope aspect ratio you will lose parts on the left and right of the picture. Another option to get rid of the black bars is to stretch the picture in one dimension. This introduces all kinds of geometric distortions: circles become ellipses and people look either squeezed or too fat. The display mode that stretches the picture is often called 'wide'. Some manufacturers use a 'smart' stretch mode in which the center of the picture is less stretched than the sides. This tends to make the geometric distortion less noticeable. However, objects that move through the picture can look weird, because they change form when they move from the side to the center of the screen (Taylor et al., 2009).

It is best to leave all the fancy display modes alone and put your display and source components in full-screen mode. Yes, this will give you the ugly black bars, but that is way better than throwing away parts of the picture or changing its geometry. It is the only way to view the video material as it was originally intended without any unnatural distortions.

Recommendation 5.1 *Only use a full-screen mode to display letterbox material. Do not attempt to get rid of the black bars using another display mode, because it will result in a cropped or geometrically distorted picture.*

When you play some older material having the 4:3 aspect ratio, it is best to use the pillarbox format. Do not use a display mode that stretches the picture to fill the entire width, because it will not only distort the picture but also make it look fuzzy and less well-defined. The reason is that the original material does not have enough pixels to fill the entire width. So your display or source component is inventing additional pixels to fill in the gaps. This artificial increase of the number of pixels is called *interpolation* and will be covered in more detail in section 6.3.1. Although, in general, interpolation can work quite well, it often doesn't with older 4:3 material. Such material often has only 480 pixels on one horizontal line, which is not much if you consider that you need 1920 pixels to fill the entire width of an HD display.

Recommendation 5.2 *To avoid unnatural distortions of the picture, always use the pillarbox format to display material with a 4:3 aspect ratio on a 16:9 screen.*

5.1.2 Resolution

The most common resolution used today is 1920 × 1080, which is referred to as *high-definition* (HD) video. It is widely used on Blu-ray discs, the Internet, video streaming services, and digital television broadcasts. If you divide 1920 by 1080, you will get 1.78, so HD video has an aspect ratio that is exactly equal to 16:9. Video material that has a slightly lower resolution of 1280 × 720 is also called HD video. It is quite common on the Internet, because it saves some bandwidth. Again, this resolution has an aspect ratio of 16:9.

All the resolutions below 1280 × 720 are known as *standard-definition* (SD) video. Table 5.1 shows some typical examples, the most important ones being the DVD formats. There are two slightly different resolutions for DVDs. The resolution of 720 × 480 is used in North America and all the other countries that used the old NTSC (National Television Systems Committee) standard for analog television. The resolution of 720 × 576 is used in Europe and all the other countries that used the old PAL (Phase Alternating Line) standard for analog television. It is interesting to note that these resolutions do not have an aspect ratio of 16:9. DVDs use a so-called *anamorphic wide-screen* format, in which a wide-screen 16:9 image is squeezed into a smaller aspect ratio (approximately equal to 4:3). On playback, your video equipment restores the 16:9 aspect ratio by stretching the picture such that it fits the entire horizontal width of a 16:9 wide-screen display. As a result the pixels are no longer square; they are wider than they are tall (Taylor et al., 2006). Hence, a video image from a DVD has less detail in the horizontal dimension than it has in the vertical dimension. These nonsquare pixels are shown in figure 5.2 where they are compared to the pixels for the HD and UHD resolutions. Obviously, the smaller the pixels in the video format, the more detailed the video image will be.

UHD video has a resolution of 3840 × 2160, which is twice the resolution of HD video in both the horizontal and the vertical dimension. Or to put it differently, a UHD video image has exactly four times more pixels than an HD video image. As you can see in figure 5.2, you can fit exactly four UHD pixels into the area of an HD sized pixel. So a UHD video image has even more detail than an HD video image. As I already mentioned in section 4.2.2 in order to see all these details you need to sit quite close to a large video screen. UHD first appeared in digital cinema and other large screen installation and is now common in displays for domestic use.

The resolution used for digital cinema slightly differs from the HD and UHD ones. Digital cinema uses 2048 × 1080 and 4096 × 2160. This is related to the two most commonly used aspect ratios in digital cinema, which are 1.85:1 and 2.4:1. The 1.85:1 aspect ratio is fitted within the 2048 × 1080 and 4096 × 2160 resolutions by taking the maximum number of horizontal pixels (2048 × 858 and 4096 × 1716). While the 2.4:1 aspect ratio is fitted by taking the maximum number of vertical pixels (1998 × 1080 and 3996 × 2160) (DCI, 2008).

A resolution of 7680 × 4320 exists for super large screens such as those found at theme parks, sport venues, and concerts. This resolution is referred to as LSDI, which stands for *large screen digital imagery* (ITU-R BT.1769, 2008; ITU, 2009a). It has four times more pixels than UHD video. It is intended for large audiences at large venues and not for a typical home environment.

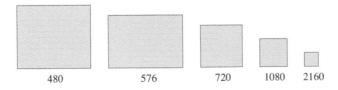

480	576	720	1080	2160

Figure 5.2
Comparison of the relative size of pixels for different resolutions displayed on a 16:9 screen of the same size. *From left to right:* 720 × 480 (SD), 720 × 576 (SD), 1280 × 720 (HD), 1920 × 1080 (HD), and 3840 × 2160 (UHD).

Recommendation 5.3 *Prefer high-resolution video sources over low-resolution ones. That is, if possible use HD or UHD video instead of SD video, choose HDTV broadcasts over standard television broadcasts, and Blu-ray discs over DVDs.*

On packaged media such as DVDs and Blu-ray discs the video resolution used is almost always listed on the backside. For streaming and other online video resources you often have the option to choose between different resolutions. It is quite common to use a shorthand notation to refer to the different video resolutions. This shorthand notation is based on the number of vertical pixels. Table 5.3 shows the most common ones. The letter *i* indicates interlaced scanning and the letter *p* progressive scanning. I will explain these different scanning methods in a moment. For the higher resolutions another shorthand notation has become popular. In this alternative notation the horizontal resolution is used and rounded to the nearest multiple of 1000: for example UHD video with a resolution of 3840 × 2160 is indicated as 4K, where the letter *K* stands for 'kilo' (one unit of 1000), similar to the letter *k* in km for kilometer.

5.1.3 Progressive and Interlaced Scanning

Modern digital video uses progressive scanning, which means that the video image is built from top to bottom by displaying each line of pixels subsequently. Interlaced scanning is an old technique stemming from the days of analog television. In interlaced scanning first all the odd lines are displayed from top to bottom and after that the even lines are shown. The video frame is split into two interweaved fields: one containing only the odd lines and another containing only the even lines. These two fields are displayed one after another. Figure 5.3 illustrates how this works.

Interlaced scanning was used in the days of analog television to cut the frame rate in half by alternately transmitting the odd and the even lines of the image. This reduced the required

Table 5.3 Commonly used shorthand notation for video resolutions.

Shorthand	Resolution	Scanning
480i	720 × 480	Interlaced
576i	720 × 576	Interlaced
720p	1280 × 720	Progressive
1080i	1920 × 1080	Interlaced
1080p	1920 × 1080	Progressive
2K	1920 × 1080	Progressive
4K	3840 × 2160	Progressive
8K	7680 × 4320	Progressive

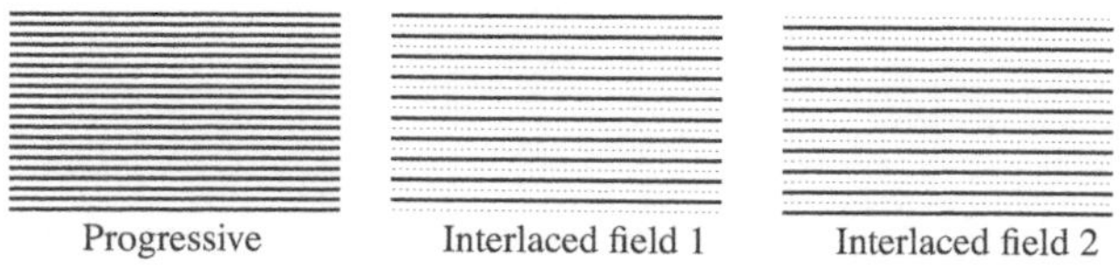

Progressive   Interlaced field 1   Interlaced field 2

Figure 5.3
Progressive and interlaced scanning. In progressive scanning the video image is built from top to bottom by displaying each line of pixels subsequently. In interlaced scanning first all the odd lines are displayed subsequently from top to bottom (field 1) and after that the even lines are shown (field 2).

bandwidth for transmitting the image, because the amount of information that had to be sent at the same time was cut in half (Jack, 2007; Taylor et al., 2009).

Progressive scanning provides superior picture quality compared to interlaced scanning. In interlaced scanning the resolution is effectively cut in half, reducing the amount of detail in the picture. Interlaced scanning can also cause unwanted flicker effects, because small details only appear in one of the two fields. For example a person wearing a striped shirt may start to flicker when these stripes are not present in both fields of the video frame (Taylor et al., 2009).

The type of scanning used, progressive or interlaced, is often denoted with a small letter *p* or *i* after the frame rate. The frame rate is given in frames per second (fps). The most common frame rate for movies is 24p. In interlaced scanning, each frame is made up of two fields. Therefore, you can distinguish between the frame rate and the field rate. To achieve a frame rate of 25 Hz you obviously need a field rate of 50 Hz. The frame rate and field rate are often carelessly interchanged. For example, it is customary to extend the shorthand notation for resolution shown in table 5.3 with the frame rate, such that you obtain expressions like 1080p24 and 1080i30. But the latter one is sometimes also written as 1080i60.

It is important to realize that a DVD always contains digital video with interlaced scanning, either 30i or 25i, the reason being that DVD video was designed during a time when analog television was still the norm and a DVD player would be hooked up to an NTSC or PAL television using analog video connections. The frame rate of 30i was derived from the 60 Hz frequency of the electrical power supply used in North America. Similarly, the frame rate of 25i stems from the 50 Hz frequency of the power supply used in Europe. Although not entirely correct, the 30i rate is associated with the NTSC standard and the 25i rate with the PAL standard. Strictly speaking, NTCS and PAL are color coding standards, not scanning standards (Poynton, 2012). To complicate matters further the NTSC system actually uses a frame rate that slightly differs from 30 fps. For some obscure technical reason during the conversion of the NTSC system from monochrome to color television the frame rate was changed to $30 \times (1000/10001) \approx 29.97$ fps. This corresponds to displaying 59.94 fields per second. To ease conversion between NTSC and the filmlike format of 24p, also the frame rate of $24 \times (1000/1001) \approx 23.976$ fps has been introduced (Poynton, 2012; Taylor et al., 2009).

Nowadays high-quality systems exist that use progressive scanning at twice the above frame rates: 48p, 50p, 60p, and even 120p. These increased display frame rates can significantly improve the perceived detail in a video sequence especially for moving objects (Taylor et al., 2009).

5.1.4 *Perceptual Coding*

The delivery format used for digital video can be characterized by its aspect ratio, resolution, frame rate, and bit depth. Despite these different features, in the end, digital video just consists of an extremely large collection of bits. The way in which these bits are stored or transmitted is called *video coding*. Different coding methods have been developed to store and transmit digital video in an efficient way. These methods differ in their quality and efficiency. And therefore, the video coding used is an another important feature of a particular delivery format.

Perceptual coding is a data compression technique that reduces the amount of video data without significantly reducing the quality of the picture. Compression of digital video is needed because storage and transmission capabilities are limited. Although these capabilities are increasing almost every year, at the moment compression is still very much needed. Let's have a look at a convincing example. Typical HD video has a resolution of 1920×1080, a color depth of 8 bits per color channel, and a frame rate of 24 fps. Such an HD video stream has a data rate of 1200 Mbit/s (mega bits per second). At this video data rate a single-layer Blu-ray disc can hold less than 3 minutes of HD video (see box 5.1 for the calculation). Furthermore, the electronics in a Blu-ray player can only process digital video at a maximum bit rate of 40 Mbit/s, and this is without taking into account the audio tracks. Clearly, something has to be done about the video data rate.

BOX 5.1 VIDEO DATA RATE

HD video has a resolution of 1920×1080 and a bit depth of 8 bits per color channel. This means that one frame takes up $1920 \times 1080 \times 8 \times 3 \approx 50000000$ bits. If you multiply this number of bits with a typical frame rate of 24 fps the data rate of the video stream amounts to approximately 1200 Mbit/s (mega bits per second). Since 8 bits equal a byte, this is equivalent to 150 MB/s (mega bytes per second). A single-layer Blu-ray disc holds about 25 GB of data, which at a rate of 150 MB/s fills up in less than 3 minutes ($25000/150 \approx 167$ seconds).

BOX 5.2 LUMA AND CHROMA

Luma Y' is constructed from a weighted sum of the gamma corrected RGB values. For HD video this weighted sum is specified in recommendation ITU-R BT.709 (2015) and equals:

$$Y' = 0.2126\,R' + 0.7152\,G' + 0.0722\,B'$$

where R', G', and B' are gamma corrected with the opto-electronic transfer function from figure 3.7. This weighted combination for Y' is slightly different for SD video (ITU-R BT.601, 2011) and UHD video (ITU-R BT.2020, 2015), but always roughly equal to a mix of 20% red, 70% green, and 10% blue. This mix is representative of the color sensitivity of the human eye: the eye is most sensitive to green and least sensitive to blue. The two chroma components are formed as

$$C_B = B' - Y' \text{ and } C_R = R' - Y'.$$

Poynton (2012) has pointed out that the terms luma Y' and luminance Y are used carelessly by many engineers and specialists. The prime that distinguishes luma from luminance is often omitted. This can lead to great confusion. Luma Y' is not equal to luminance Y; it is an engineering approximation. Luma is formed from the gamma corrected $R'G'B'$ signals, while luminance is formed from the uncorrected RGB signals. Note that applying the gamma correction to Y does not result in Y', because gamma correction is a nonlinear transformation and the order in which nonlinear transformations are applied does matter.

Perceptual coding is a so-called *lossy compression* method that can greatly reduce the video data rate. Part of the video data is permanently removed in order to achieve high compression ratios. These parts of the original data are permanently lost and cannot be retrieved upon decompression. The main idea is to use properties of human perception to discard as much data as possible without perceptually degrading the video image.

Lossy video compression techniques take advantage of the fact that humans are better at distinguishing fine details by differences in brightness than by differences in color (Poynton, 2012; Taylor et al., 2009). To exploit this fact, digital video is converted into a *luma* component (Y') that represents luminance (see box 5.2) and two *chroma* components (C_B and C_R) that represent color. Luma and chroma provide an alternative way of representing the RGB components of digital video. These two representations are completely equivalent; no information is lost by converting between them. Lossy compression is obtained by reducing the resolution of the two chroma components. This is called *chroma subsampling*. Typically, for every block of 4-by-4 pixels 16 Y' values are stored, one for each pixel; while only a total of 4 chroma values (two for C_B and two for C_R) are stored for the entire block. This reduces the total number of values that need to be stored from $3 \times 16 = 48$ to $16 + 4 = 20$ and thus approximately halves the video data rate in our example from 1200 Mbit/s to 500 Mbit/s.

To reduce the video rate even further, JPEG image compression (ISO/IEC 10918, 1994) is used on each frame. JPEG compression was designed by and named after the Joint Photographic Experts Group. It is an effective technique to compress continuous-tone color or grayscale still images and is widely used to compress digital photographs. It exploits the fact that the human eye is less sensitive to complex detail. Or in other words, the human eye is less sensitive to high-frequency spatial detail than to low-frequency spatial detail. It works roughly as follows: the image is divided into small blocks of pixels that are transformed with a discrete cosine transform (DCT). This transform effectively separates the low-frequency from the high-frequency spatial details. The details with the highest frequencies can be discarded without having an objectionable effect on the image quality. Typical JPEG compression ratios from 5:1 to 10:1 don't result in significant perceptual image degradation while at the same time reducing the amount of image data quite a bit (Poynton, 2012; Taylor et al., 2009).

Even more compression can be achieved by not only compressing each individual frame, but also taking into account the redundancy between different frames. For example, in a movie sequence of a person walking in front of a house the background remains fairly static from frame to frame. Instead of coding this static background for each individual frame, it is much more efficient to store it once and then reuse its data on the subsequent frames. This technique is called *interframe compression*. It is the method of choice, and many video codecs are based on it. A large family of video codecs has been jointly developed by the Moving Picture Experts Group (MPEG) and the International Telecommunications Union (ITU). The most commonly used ones are listed in table 5.4 along with the VC-1 codec of the Society of Motion Picture and Television Engineers (SMPTE ST 421, 2013), which is based on Microsoft's Windows Media Video 9 technology.

All video codecs can be operated at different quality settings. The quality is determined by the bit rate. The lower the bit rate, the more bits are being discarded and hence the lower the image quality. Modern codecs support variable bit rate (VBR) encoding: the bit rate is varied depending on the complexity of the scene. VBR encoding results in better quality as the data rate adapts itself to what is needed in a particular scene. Constant bit rate (CBR) encoders, such as the older MPEG-1 video codec, require a higher overall bit rate to achieve the same image quality (Jack, 2007; Poynton, 2012; Taylor et al., 2009). The image quality does not gracefully decrease with the bit rate. It rapidly deteriorates once the bit rate falls below a certain threshold. To give you an idea of what a good bit rate is, I have listed some typical bit rates in table 5.5 for the codecs used for DVD and Blu-ray disc.

Since each codec works differently it makes no sense to compare the bit rates among codecs. For example, the MPEG-2 codec always requires a much higher bit rate than the MPEG-4 AVC

Table 5.4 Compression ratios for MPEG and SMPTE video codecs (ISO/IEC 13818–2, 2013; ISO/IEC 14496–2, 2004; ISO/IEC 14496–10, 2014; ISO/IEC 23008–2, 2017; SMPTE ST 421, 2013). The MPEG codecs have an alternative name after the corresponding ITU standard (ITU-T H.262, 2012; ITU-T H.263, 2005; ITU-T H.264, 2017; ITU-T H.265, 2018).

Name	Compression Ratio
MPEG-2 / H.262	20:1
MPEG-4 / H.263	30:1
MPEG-4 AVC (Advanced Video Coding) / H.264	50:1
MPEG-H HEVC (High Efficiency Video Coding) / H.265	100:1
SMPTE VC-1	50:1

Table 5.5 Typical and maximum video bit rates for DVD and Blu-ray disc video codecs (BDA, 2015b; Taylor et al., 2006, 2009).

Medium	Video Codec	Resolution	Typical Rate (Mbit/s)	Maximum Rate (Mbit/s)
DVD	MPEG-2	720 × 480	4.5	9.8
BD	MPEG-2	1920 × 1080	24.0	40.0
BD	MPEG-4 AVC	1920 × 1080	16.0	40.0
BD	SMPTE VC-1	1920 × 1080	18.0	40.0
UHD BD	MPEG-H HEVC	3840 × 2160	35.0	100.0

Figure 5.4
Subjective video quality on a five-point scale as a function of bit rate for 1920 × 1080 video coded with the MPEG-4 AVC codec.

Based on data from Cermak et al. (2011).

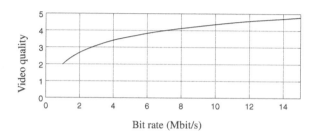

codec to achieve a comparable image quality. It only makes sense to compare different bit rates when the same codec is being used. Such a comparison is shown in figure 5.4 for the MPEG-4 AVC codec, which is widely used on the Internet with different bit rates. This comparison is based on extensive measurements done by the Video Quality Experts Group (VQEG) for HD video with a resolution of 1920 × 1080 (Cermak et al., 2011). The image quality is indicated using a five-point grading scale; the higher the score the better the image quality. You can see that the image quality degrades when the bit rate is lowered from the 16 Mbit/s used on Blu-ray discs. First, it degrades gradually and then below about 5 Mbit/s it deteriorates quite fast.

Recommendation 5.4 *When you are given a choice between high bit rate and low bit rate digital video streams or files (that are encoded with the same video codec), always choose the high bit rate ones.*

5.2 Audio Coding

Digital audio is simply a series of numbers that represent the magnitude of an audio signal at different moments in time. Digitizing an audio signal involves two fundamental processes: *sampling* and *quantization*. Sampling the audio signal is nothing more than selecting certain time instances at which the signal will be digitally represented. Only at the sampled time instances will the magnitude of the signal be digitized. An example of sampling a sine signal is shown in figure 5.5. The *sampling rate* or *sampling frequency* is the number of samples per second. Typical sampling rates used in digital audio are 44.1 kHz (CD) and 48 kHz (DVD and Blu-ray). Multiples of these rates are also used for high-resolution audio.

The next step is quantization. In this step the range of continuous values that the magnitude can take is converted into a finite number of fixed levels. The magnitudes at the sampled time instances are rounded to the nearest quantization level. This process is similar to the quantization used in digital video, as explained in figures 3.4 and 3.5. Recall that due to the rounding, quantization introduces a *quantization error.* The larger the number of quantization levels, the smaller the error and thus the more accurate the digital representation resembles the original analog audio signal. Similar to the coding used in digital video the different quantization levels are coded using bits. The number of bits used is called the *bit depth*. Thus, the larger the bit

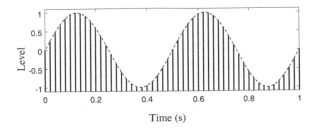

Figure 5.5
Sampling a sine signal.
The *thick vertical lines*
indicate the sampling
instances. Only at these
time instances will the
magnitude of the sine
signal be digitized.

Table 5.6 Common audio delivery formats used on optical discs (BDA, 2011, 2015b; Holman, 2008; Pohlmann, 2011; Taylor et al., 2006, 2009).

Media	Coding Method	Lossless	Max. Channels	Sampling Rate (kHz)	Max. Bit Depth
CD	LPCM	yes	2.0	44.1	16
VCD	MPEG-1 Layer II	no	2.0	44.1	16
SVCD	MPEG-2 Layer II	no	5.1	44.1	16
SACD	DSD	yes	5.1	2822.4	1
DVD-Audio	LPCM-MLP	yes	5.1	48, 96, 192[a]	24
DVD-Video	LPCM	yes	5.1	48, 96	24
	MPEG-2 Layer II	no	7.1	48	20
	Dolby Digital	no	5.1	48	24
	DTS Digital Surround	no	6.1	48, 96[b]	24
BD/UHD BD	LPCM	yes	7.1	48, 96, 192[c]	24
	Dolby Digital	no	5.1	48	24
	Dolby Digital Plus	no	7.1	48	24
	Dolby TrueHD	yes	7.1	48, 96, 192[c]	24
	Dolby Atmos	yes	7.1.4	48, 96	24
	DTS Digital Surround	no	6.1	48, 96[b]	24
	DTS-HD High Resolution	no	7.1	48, 96	24
	DTS-HD Master Audio	yes	7.1	48, 96, 192[c]	24
	DTS-X	yes	9.1.6	48, 96	24

[a] Sampling rates 44.1, 88.2, 176.4 kHz are also supported
[b] Sampling rate 96 kHz only with the X96 extension also known as DTS 96/24
[c] Sampling rate 192 kHz only with a maximum of 5.1 channels

depth, the smaller the quantization error. Typical bit depths used for digital audio are 16, 20, and 24 bits. The CD uses a bit depth of 16, while in high-resolution audio the bit depth is 24 bits.

In the end, digital audio is a sequence of bits. The way in which this sequence is stored or transmitted is called *audio coding*. The most common way to code digital audio is *linear pulse code modulation* (LPCM) (Pohlmann, 2011). In LPCM the signal is sampled at uniform time periods and also the quantization levels are uniformly spaced across the entire magnitude range. Over the years several other coding formats have been developed and used for different delivery formats. Different delivery formats use different audio codecs, support different number of audio channels, and utilize different sampling rates and bit depths. Some audio codecs are lossless, preserving the original digital LPCM signal, while others are lossy and based on perceptual coding techniques. In the early years of digital audio and video it was quite a challenge to fit the enormous amount of bits on a single DVD. This challenge resulted in several lossy compression codecs of which Dolby Digital and DTS Digital Surround are the most widely recognized ones. In table 5.6 you find some common delivery formats used on optical discs.

5.2.1 Sampling Rate

What sampling rate should you use for digital audio? The answer is deceptively simple: at least twice the highest frequency in the audio signal. When you adhere to this rule no information is lost due to sampling. At a glance, this might seem strange, since you only take into account the magnitude of the signal at the sampled time instances and discard all the information in between these samples. Even so, it appears that the information between the samples is not needed, as long as the sampling rate is at least twice the highest frequency in the signal. It can be proven mathematically that in that case the entire original signal can be reconstructed from just a few samples. This fact is known as the *sampling theorem*. This theorem is usually attributed to either Shannon or Nyquist (Pohlmann, 2011).

To understand the sampling theorem better, take a second look at figure 5.5. The theorem states that you can reconstruct this sine signal if you sample at least twice during the time it takes the sine to complete one period. Obviously, if you only take one sample during its period, you cannot reconstruct the sine. With only one sample, you have no idea how the sine progresses over time. You need at least two samples. Figure 5.6 shows that you can fit other sines through the two sample points, but these sines will all have a higher frequency than the original one. Therefore, these sines do not comply with the sampling theorem: they are not sampled at least twice during their period. Now recall from section 2.3.1 that based on Fourier analysis any audio signal can be expressed as a sum of sines of different magnitudes, phases, and frequencies. Therefore, any audio signal can be exactly reconstructed if you sample it at a sampling rate that is at least twice its highest frequency.

Before an analog audio signal is sampled it is low-pass filtered. The low-pass filter removes all the frequencies above the frequency that corresponds to half the sampling rate. It ensures that the sampling theorem is not violated. For example, if the sampling rate is 48 Hz the filter removes all the frequencies above 24 kHz from the audio signal. Without this low-pass filter, the sampling process would map all frequencies above 24 kHz to frequencies below 24 kHz. This is called *aliasing*. Aliasing must be avoided, because it distorts the digital signal such that it is no longer possible to reconstruct the original analog signal from it (Pohlmann, 2011).

The choice of the sampling rate is one of the most important design criteria of a delivery format, because it determines the highest frequency that can be reproduced. The CD uses a sampling rate of 44.1 kHz and is thus limited to reproducing frequencies up to 22.05 kHz. Audio codecs used for video delivery formats often use a sampling rate of 48 kHz, because that number can be evenly divided by the common frame rates of 24, 25, and 30 fps. This makes

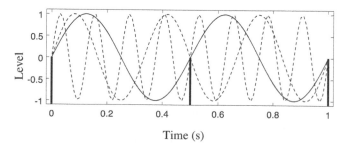

Figure 5.6
Sampling a sine signal. A sine wave with a frequency of 2 Hz *(solid line)* is sampled only once per period. The *thick vertical lines* indicate the sampling instances. It is possible to fit other sine waves through these sampling instances as long as they have a higher frequency. The *broken lines* show that a sine with a frequency of 3 Hz and a sine with a frequency of 4 Hz can be fitted through the same sampling points.

it easier to keep the audio tracks synchronized to the video images (Winer, 2018). Consequently, the audio tracks of these video delivery formats are limited to reproducing frequencies up to 24 kHz. Both 22.05 kHz and 24 kHz lie slightly above the upper limit of 20 kHz for
normal human hearing. Recall from section 2.3.1 that for most people the upper limit is lower
than 20 kHz and that under controlled conditions only a minority of people can hear a very
subtle difference between music with and without frequencies above 20 kHz. Based on these
facts the sampling rates of 44.1 kHz and 48 kHz should be more than sufficient. Nevertheless,
sampling rates of 88.2 kHz, 96 kHz, and sometimes even 176.4 kHz, 192 kHz, or 384 kHz are
being used in high-resolution audio. These higher sampling rates should presumably result in
superlative sound quality, but not everyone is convinced of the merits.

High-resolution audio (HRA) or high-definition (HD) audio is defined as lossless audio that
is capable of reproducing the full range of sound from recordings that have been mastered
from better than CD-quality music sources (DEG, 2014). Strictly speaking, this means that in
HRA the sampling rate is higher than 44.1 kHz and the bit depth larger than 16 bits. However,
audio with a sampling rate of 48 kHz is not generally considered to be high-resolution audio.
Typically, in HRA the sampling rate needs to be a multiple of 44.1 kHz or a multiple of 48 kHz:
the most common sampling rates for HRA are 88.2, 96, 176.4, and 192 kHz.

The question arises if HRA sounds distinguishably different from CD-quality audio. The
ultrahigh frequencies that are included in HRA lie far above the upper frequency limit
of human hearing, which is at most 20 kHz and much lower when you are getting older.
In addition to this audibility issue, another factor to consider is the ability of your loudspeakers and room to reproduce audio frequencies higher than 20 kHz (Colloms, 2006).
The higher the frequency, the more directional a loudspeaker becomes (recall figure 2.28).
At these ultrahigh frequencies, your loudspeakers will radiate the sound in an unusually
tiny beam. The chances of this beam reaching your ears directly are small. In addition, the
higher the frequency, the more sound is absorbed by the room. Therefore, the chances that
reflections of these ultrahigh frequencies reach your ears are also slim. Furthermore, many
loudspeakers cannot reproduce frequencies above 45 kHz. In fact, many audio devices
include a filter that removes all frequency content above 50 kHz as a safety measure to
avoid the high-power high-frequency signals that damage amplifiers and loudspeakers. For
this reason, it does not make sense to increase the sampling rates of 88.2 and 96 to 176.4
and 192 kHz. The increased high-frequency range above 45 kHz will not be reproduced by
the loudspeakers.

There exists some controversy about the usefulness of high sampling rates. Some listening
tests have shown that listeners fail to observe any differences between CD-quality recordings
with 44.1 kHz sampling rate and HRA with much higher sampling rates (Meyer and Moran,
2007) while other listening tests have shown that expert listeners can in fact detect subtle
differences between them (Pras and Guastavino, 2010). These subtle differences resulted in
improved spatial reproduction and increased clarity. However, the differences are difficult
to detect and strongly depend on the musical excerpt used for testing. A systematic meta-
analysis of 18 different controlled listening tests showed a small but statistically significant
ability of listeners to discriminate HRA (Reiss, 2016). Listeners were able to correctly distinguish between HRA and standard audio in about 55% of the cases. When the listeners were
trained this increased to about 60%. This is only slightly better than taking a random guess,
because that would result in 50% of the cases being correctly identified. Further research is
needed to determine the cause of the perceived effect of HRA.

Interesting enough, one factor that might play a role in distinguishing HRA from standard
audio has nothing to do with sampling rate and bit depth. It is the quality of the mastering of
the recording (Waldrep, 2017). In one listening test it was found that virtually all HRA material
(SACD and DVD-audio) sounded better compared to the same material released on CD, even
when the HRA material was downsampled to 44.1 kHz (Meyer and Moran, 2007). This difference is due to the production quality. CD-quality recordings are aimed at the mass consumer

market and are often heavily compressed in their dynamic range during the mastering process. Compressing the dynamic range ensures that these recordings sound reasonably well on lesser-quality audio systems and suit most casual listening conditions (see figure 3.17 for an example of a compressed audio signal). By contrast, in the production of HRA recordings the engineers are given the freedom to produce recordings as good as they can possibly make them. These recordings are made with great care and they sound like it.

Recommendation 5.5 *When you are given a choice between HRA with 88.2 kHz or 96 kHz sampling and standard audio with 44.1 kHz or 48 kHz sampling, consider choosing the HRA version. It might sound better, if it has been mastered with greater care and has a larger dynamic range.*

While HRA audio might provide you with better sound quality, increasing the sampling rate beyond 96 kHz is of no use (Waldrep, 2017). A sampling rate of 96 kHz ensures that all information up to 48 kHz is present in the recording. Musical instruments do not produce any frequencies higher than 48 kHz, and contemporary microphones are not able to record beyond this frequency limit. Furthermore, the loudspeakers and other audio equipment that you use to reproduce the sound are also limited in their frequency range: rarely do their capabilities extend beyond 50 kHz.

Recommendation 5.6 *There is absolutely no compelling reason to use a sampling frequency higher than 96 kHz for digital audio.*

Simply increasing the sampling rate and bit depth does not automatically result in improved audio quality. This is especially true for older material. In the 1980s and 1990s lots of digital recordings were made and mastered using 48 kHz 16-bit digital recorders that were prevalent during that era (Harley, 2015). Simply upsampling this material to a higher sampling rate and increasing the bit depth is useless. The frequency range and dynamic range were already limited during the original recording and mixing process and will not magically get larger. Similarly, lots of older analog recordings have been made with microphones and recording equipment that was limited in their frequency and dynamic range compared to today's HRA standards. However, this does not stop certain companies from trying to sell you older material that is simply upconverted material as genuine HRA (Waldrep, 2017). Beware of this marketing ploy! That being said, remastering older material as HRA, even when it has limited frequency and dynamic range, might still result in a quality improvement, especially when the sound engineers were able to go back to the original source material and use advanced digital signal processing to clean it up. Just remember that the difference in sound quality is not the result of the higher sampling rate or bit depth, but due to careful remastering.

5.2.2 Bit Depth

Quantization introduces errors, because the magnitude of the analog audio signal is rounded to the nearest quantization level. The rounding error is called the *quantization error*. A higher bit depth results in more quantization levels and a lower quantization error. This is illustrated in figure 5.7. In the example shown, the amplitude of the audio signal is normalized such that it varies between −1 and 1. As the number of quantization levels between −1 and 1 increases, the distance between these levels decreases. As a result, the digital audio signal can more accurately represent the analog audio signal. In other words, the quantization error, which is simply the difference between the analog and digital signal levels at the sample points, becomes smaller. Note that the maximum quantization error occurs when the analog signal level falls exactly between two quantization levels.

Digital audio comes in three different bit depths: 16, 20, and 24 bits. The corresponding number of quantization levels is shown in table 5.7 (a bit depth of n bits results in 2^n quantization

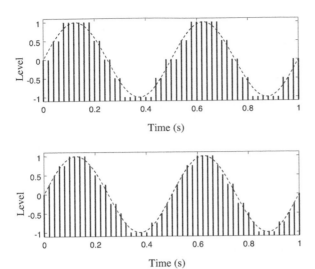

Figure 5.7
Sampling and quantization
of a sine signal. *At the top*,
a sampled sine wave that
is quantized with only five
levels. *At the bottom*, a
sampled sine wave that
is quantized with nine
levels. Compare these two
approximations to the
sampled sine wave shown
in figure 5.5.

Table 5.7 Number of bits, number of levels, and dynamic range in quantization of audio signals.

Number of Bits	Number of Levels	Dynamic Range
16	65 536	98 dB
20	1 048 576	122 dB
24	16 777 216	146 dB

BOX 5.3 SIGNAL-TO-ERROR RATIO

The signal-to-error ratio (SER) is similar, but not equal to the signal-to-noise ratio (SNR).
It is defined as the ratio of the root-mean-square (rms) values of the signal and the quantization error. For a quantization system with n bits the SER in dB equals (Pohlmann,
2011):

$$SER = 20 \log_{10} \left[\left(\frac{3}{2} \right)^{1/2} (2^n) \right] \approx 1.76 + 6.02n$$

where the value of 1.76 is based on the peak-to-rms ratio of a sine wave signal and it is
assumed that the quantization error is uniformly distributed and uncorrelated with the
signal that is quantized.

---

levels). The table also lists the corresponding dynamic range, that is the range in dB between
the highest and the lowest signal level that can be represented using a certain number of bits.
The higher the bit depth, the larger the dynamic range. The dynamic range can be calculated
from the signal-to-error ratio (see box 5.3). Figure 5.8 puts these dynamic ranges into perspective. The 0 dBFS in this figure indicates the maximum signal level, where FS stands for full
scale. It is customary in digital audio to relate the digital signal level to the maximum possible
level at 0 dBFS. At the maximum signal level all the bits have been used to code the signal.
All the other signal levels lie below 0 dBFS and use fewer bits. The lowest signal level that can
be coded uses only 1 bit and is slightly larger than the quantization error. All signal levels that
are smaller than the quantization error will be lost, because they are all rounded to the lowest

Figure 5.8
Dynamic range perspective.
The dynamic ranges
corresponding to 16, 20,
and 24 bits are compared
to the 135 dB dynamic
range of the human
hearing system and the
96 dB dynamic range
that corresponds to the
recommended peak
listening level.

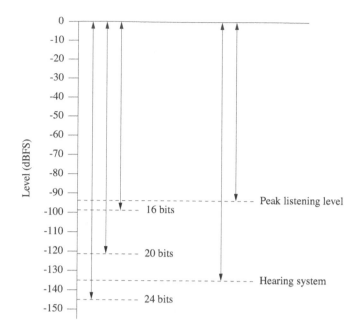

quantization level in the system. When a bit depth of 16 bits is used, the lowest possible signal level lies at –98 dBFS. For 20 and 24 bits the lowest possible levels are –122 dBFS and –146 dBFS, respectively. Usually 16 or 20 bits provide sufficient accuracy and dynamic range. Recall from section 3.5 that at the recommended listening level for home theater the loudest peaks of the program are reproduced at 96 dB SPL (EBU-Tech 3276, 1998; ITU-R BS.1116, 2015). Since these peaks correspond with the maximum digital level of 0 dBFS, the lowest signal levels in the program will end up at –96 dBFS. This illustrates that the dynamic range provided by a bit depth of only 16 bits should be sufficient, and this is without taking into account the background noise level in a typical home theater. Recall from section 4.3.2 that the recommended maximum background noise level for home theater equals 35 dBA (A-weighted) and that it is difficult to obtain background noise levels below 30 dBA. Most of the sound reproduced at a volume lower than 30 dB will be difficult to distinguish from the background noise. This would suggest that the useful dynamic range is reduced from 96 dB to about 66 dB. But, this is not the whole story. It turns out that the human ear is quite good at resolving sounds below the noise floor, if these sounds have a narrow frequency content (Pohlmann, 2011). For example, a pure sine wave can easily be detected even if its magnitude is much smaller than the background noise level. Therefore, it is safe to say that the required dynamic range is approximately 96 dB.

The dynamic range of 16 bits is already so large that you only hear the effects of quantization errors during the softest parts of the program when you play this program at a volume much louder than the recommended peak listening level of 96 dB. At normal listening levels, you are much more likely to hear either the background noise in your own listening room, or noise that is present in the recording. Older recordings made with analog master tapes often have some amount of analog noise in them. But, noise in the recording can also be the result of background noise in the recording environment that has been picked up by the microphones when the recording was being made (Winer, 2018). In the unlikely event that quantization error is audible, it will sound horrible. It creates an effect called *granulation noise* that sounds rough and gritty and is especially noticeable during the low-level parts of the program (Pohlmann, 2011).

The maximum bit depth used in digital audio is 24 bits. At this bit depth the quantization error is extremely small and lies at a signal level of –146 dBFS. It is difficult to take advantage of a bit depth of 24 bits, because the thermal noise in the electronic circuits of a digital-to-analog converter will be larger than –146 dBFS. There are currently no digital-to-analog converters available that can achieve the full dynamic range of 24 bits. Even the best 24-bit converters are limited by electronic noise in their circuits and can only achieve a dynamic range that is equivalent to about 21 bits (Pohlmann, 2011; Winer, 2018). Furthermore, the dynamic range of our hearing system equals 135 dB, from the threshold of audibility to the threshold of pain (see figure 2.23). So no matter how loud you play a 24-bit audio signal, you will never be able to hear all the details. Consequently, there is not much point in having 24 bits of dynamic range when listening to digital audio; a dynamic range of 20 bits is more than sufficient and often 16 bits will be just as fine. However, it still pays to have a 24 bits dynamic range during recording and mixing. The extra dynamic range that the 24 bits provide is called *headroom*. Headroom reduces the risk of introducing errors in the digital signals when they are manipulated during the mixing and mastering process. As a consequence, most mastered and finalized HRA recordings are delivered with 24 bits. While of course this doesn't hurt, it is not strictly necessary for high-quality playback.

To summarize, 16 or 20 bits will be more than sufficient for a high-quality reproduction of sound. Still, it is possible that a 24-bit HRA recording sounds better than the CD-quality version that has only 16 bits. Similar to the case with higher sampling rates, such a difference in sound quality is not due to the increased bit depth, but more likely the result of a different mastering of the recording.

> **Recommendation 5.7** *When you are given a choice between HRA with 24 bits and standard audio with 16 bits, consider choosing the HRA version. It might sound better, if it has been mastered with greater care and has a larger dynamic range.*

5.2.3 Direct Stream Digital (DSD)

Direct Stream Digital (DSD) is an alternative coding method for digital audio. DSD is completely different from LPCM: it uses only 1 bit and a very high sampling rate of 2.8224 MHz or higher. Instead of coding the magnitude of the audio signal using several bits, the change of this magnitude is coded using 1-bit pulses. This coding process is called *pulse-width modulation* (PWM). When the magnitude is increasing a binary one is recorded and when it is decreasing a binary zero. The width of the resulting pulses represents the magnitude of the audio signal.

DSD is the method used in Super Audio CD (SACD), a disc-based format introduced in 1999 that never really took off and is now considered a failure. Nowadays, DSD music releases are mostly distributed as downloadable media files.

Since DSD uses only 1-bit quantization, the quantization error is quite large, much larger than the error obtained with 16 bits used in LPCM quantization. DSD employs a technique called *noise shaping* to drastically increase the apparent signal-to-noise ratio. Noise shaping alters the spectral shape of the quantization error such that the amount of noise below 20 kHz is reduced at the expense of increasing the amount of noise above 20 kHz. As a result, the amount of noise above 20 kHz rises sharply. Moving the noise to these ultrahigh frequencies is not considered to be a problem, because this high-frequency noise is uncorrelated to the signal below 20 kHz, and the human ear is only sensitive to high-frequency content that is correlated (Pohlmann, 2011). Figure 5.9 shows a typical example of the frequency spectrum of a DSD coded audio signal. Due to the noise-shaping process, DSD achieves a dynamic range of 120 dB (equivalent to 20 bits) in the audio frequency band from 20 Hz to 20 kHz. At higher frequencies the dynamic range rapidly decreases, because of the increased noise level. Consequently, DSD has a usable bandwidth of about 50 kHz. Hence, DSD provides better than CD-quality sound and qualifies as HRA. Its performance is more or less comparable to 20-bit LPCM with a sampling rate of 96 kHz (Siau, 2015; Waldrep, 2017).

Figure 5.9
Typical example of the frequency spectrum of a DSD audio signal. The amplitude of the spectrum smoothly decreases but starts to rise again above 20 kHz due to the high-frequency noise that is the result of noise shaping.

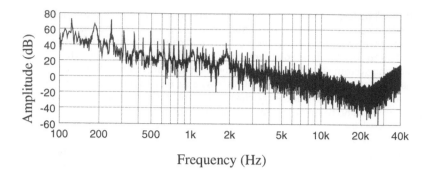

Sometimes DSD is used with higher sampling rates to achieve a larger bandwidth by pushing the high-frequency noise further away from the audio band. Commonly used higher sampling rates for DSD are double rate 5.6448 MHz and quad rate 11.2896 MHz. The standard sampling rate 2.8224 MHz equals 64 times the sampling rate of the CD: 44.1 kHz. Therefore, it is also referred to as DSD64. Similarly, the higher sampling rates are also known as: DSD128 and DSD256.

Which format is better: LPCM or DSD? Since DSD has a much larger quantization error than LPCM, it requires sophisticated digital processing such as noise shaping to reduce the noise in the audio frequency band. It can be shown that such processing of DSD signals always results in higher distortion and noise levels compared to LPCM (Lipshitz and Vanderkooy, 2001). Nevertheless, in controlled listening tests no significant differences can be heard between DSD and 24-bit LPCM with a sampling rate of 176.4 kHz. It appears that DSD and LPCM are perceptually indistinguishable from one another (Blech and Yang, 2004).

The name DSD implies a more direct path from recording to playback, but this is a myth. Despite the marketing claims that DSD requires less signal processing, in practice the opposite is true. Most professional digital audio recording equipment is based on LPCM. In fact, DSD releases of modern music that rely on multitrack techniques are almost always recorded and mixed using LPCM and only converted to the DSD format during the final mastering stage (Katz, 2015; Waldrep, 2017). This final conversion adds noise and distortion. The culprit is the DSD noise-shaping process that is required to achieve an acceptable dynamic range in the 20 Hz to 20 kHz frequency band. Because of the high-frequency noise, DSD cannot match the technical performance of 20-bit LPCM sampled at 96 kHz.

> **Recommendation 5.8** *DSD recordings often sell for a higher price than equivalent LPCM ones. There is absolutely no compelling reason to pay extra for the DSD version if an LPCM version with at least 20 bits and 96 kHz sampling is available.*

5.2.4 Lossless Audio Compression

Digital audio can take up quite a bit of bandwidth and storage space. Similar to digital video, data compression is desirable. A typical 5.1-channel LPCM audio track for DVD-Video has a bit depth of 24 and a sampling rate of 48 kHz. Such a track will fill up a single-layer DVD in about 90 minutes, without leaving any room for the video data (see box 5.4).

Lossless data compression techniques reduce the amount of data by identifying redundancies and recoding these redundancies in a more efficient way using less bits. With these techniques there is no loss of data. The original data can be reconstructed perfectly. Lossless audio compression is analogous to computer utilities such as Zip that pack the data more efficiently without changing the content. There is absolutely no difference in sound quality

---

BOX 5.4 AUDIO DATA RATE

LPCM audio for DVD-Video consists of 5.1 channels each having a bit depth of 24 and a sampling rate of 48 kHz. The number of bits per second that such a data stream produces equals $6 \times 24 \times 48000 = 6912000$ bit/s. Since 8 bits equal a byte, this is equivalent to 0.864 MB/s. A single layer DVD holds about 4.7 GB of data, which at this rate fills up in about 90 minutes ($4700 / 0.864 \approx 5440$ seconds).

---

between uncompressed digital audio and losslessly compressed digital audio. This is in sharp contrast with lossy compression techniques that reduce the amount of data by discarding parts of it based on properties of human perception. With lossy compression it is not possible to retrieve the original data. Depending on how much data is discarded, the sound quality will deteriorate. Both lossless and lossy compression techniques are being used to efficiently store and transmit digital audio. Let's first take a look at the commonly used lossless techniques. Since lossless data compression techniques do not discard any data, these techniques generally yield better sound quality than lossy compression techniques.

Recommendation 5.9 *For the best sound quality use either uncompressed or losslessly compressed digital audio. When given a choice, always prefer lossless audio compression over lossy compression.*

While general compression algorithms like Zip reduce the size of an audio file by about 10–20%, lossless audio codecs take advantage of the particular structure of the audio data and can achieve a reduction of 50% or 2:1. Examples of some commonly used lossless audio codecs are: 1) Free Lossless Audio Codec (FLAC) from Xiph (xiph.org/flac); 2) Apple Lossless Audio Codec (ALAC) from Apple (macosforge.github.io/alac); 3) Monkey's audio (monkeysaudio.com); and 4) Meridian Lossless Packing (MLP), a proprietary audio codec (www.meridian-audio.com) that is used for DVD-Audio, Dolby TrueHD, and DTS-HD Master Audio (Gerzon et al., 2004).

Dolby TrueHD (www.dolby.com) is a lossless audio codec based on MLP that is widely used on both Blu-ray discs and UHD Blu-ray discs. It achieves a compression ratio of 2:1 to almost 4:1, depending on the content of the audio signals. Dolby TrueHD is based on a hierarchical data structure that makes it possible to downmix any number of audio channels to a two-channel mix. Dolby TrueHD has also been expanded to include Dolby Atmos compatibility. A special Dolby Atmos substream can be included that represents a losslessly encoded object-based mix. When played on a Dolby Atmos sound system, the Dolby Atmos substream will be decoded and rendered to fit your specific loudspeaker configuration. A Dolby Atmos enabled soundtrack is backward compatible. If you play such a track on a system that has no Dolby Atmos decoding capability, you will experience a standard 5.1 or 7.1-channel surround mix.

DTS-HD Master Audio (www.dts.com) is a lossless audio codec that is very similar to Dolby TrueHD. It is also based on MLP and used on both Blu-ray discs and UHD Blu-ray discs. Like Dolby TrueHD, DTS-HD Master Audio can consist of several hierarchical substreams for different surround channel configurations. It has also been extended to include DTS:X, a lossless object-based surround mix that is the DTS equivalent of Dolby Atmos. DTS:X soundtracks are backward compatible and can be played on a standard 5.1- or 7.1-channel surround sound system.

Recommendation 5.10 *Blu-ray discs usually contain multiple audio tracks. For the best audio quality select the LPCM track (sometimes referred to as 'uncompressed') or an audio track with lossless compression: either DTS-HD Master Audio or Dolby TrueHD.*

5.2.5 Perceptual Coding

Perceptual coding is a lossy data compression technique in which properties of human perception are used to discard as much data as possible without perceptually degrading the data. While lossless compression can achieve a data reduction of 2:1, higher compression ratios are often desirable for audio and video streaming, and also for storing multichannel audio on DVDs. With perceptual coding only a fraction of the data is needed to convey perceptually important information in an audio signal. Perceptual coding is much more efficient than lossless compression; data reductions of more than 6:1 are feasible.

The goal of perceptual coding is to significantly reduce the data rate. The data rate depends on the number of channels, the bit depth, and the sampling rate. Perceptual coding reduces the data rate by selectively decreasing the bit depth based on the content of the audio signal. Thus, certain parts of the audio signal are coded using less bits. As a result, the quantization noise in these parts will increase. A psychoacoustic model of the human hearing system is used to keep this rise in quantization noise inaudible. The validity of the psychoacoustic model is crucial to the success of perceptual coding. The combination of this model with a mechanism for bit allocation forms the heart of a perceptual codec. It is its most proprietary part, and many companies keep it secret (Pohlmann, 2011).

Perceptual coding takes advantage of masking properties of human hearing. Masking is the process by which the threshold of audibility for one sound is raised by the presence of another louder sound (Moore, 2013). Masking is a common everyday experience. For example the music from your sound system may mask the sound of a nearby refrigerator, or the music from your car's sound system may mask the sound of the car's engine. A sound is most easily masked by a sound that has a similar frequency content. Loud sounds at certain frequencies may mask softer sounds at neighboring frequencies. The louder sound, called the *masker*, locally reduces the frequency-resolving power of the basilar membrane of the human hearing system. This phenomenon is called *amplitude masking.* Masking can also occur when a loud sound occurs just before or after a softer sound. The closer the sounds are in time, the greater the masking effect. This is called *temporal masking* (Moore, 2013; Pohlmann, 2011).

Perceptual coding takes advantage of both amplitude and temporal masking by dividing the audio signal into blocks with a fixed number of samples and analyzing the frequency content of each block. The frequency spectrum of each block is in turn divided into frequency bands of equal or varying widths, depending on the psychoacoustical model that is being used. Next, the intensity of sound in each frequency band is analyzed to determine how much masking will occur in the neighboring frequency bands. The frequency bands that only contain sounds that are completely masked do not need to be coded and can be completely ignored. In the other bands, the number of bits used for quantization is chosen such that the level of quantization noise lies just below the threshold of audibility. Since masking from neighboring frequency bands can significantly raise this threshold, some bands can be coded with just a few bits. With multichannel audio the number of bits needed for coding can be further reduced by taking advantage of redundancies between the different channels (Pohlmann, 2011; Taylor et al., 2009).

MP3 is unmistakably the most widely known perceptual audio codec. Its official name is MPEG-1 Layer III, and it was developed by the Moving Picture Experts Group (MPEG), the group that also developed several lossy video codecs (see section 5.1.4). The MPEG-1 standard (ISO/IEC 11172, 1993) is the oldest standard and only supports mono and stereo. It consists of three different compression techniques called *layers.* Layer I is the least sophisticated technique that requires the highest data rate. Layer II is more efficient than Layer I, and Layer III is the most efficient. The MPEG-2 standard (ISO/IEC 13818-3, 1998) is a more modern codec that supports multichannel audio in two different ways: 1) MPEG-2 BC, which is backward compatible with the MPEG-1 standard and also comes in three layers, and 2) MPEG-2 AAC (Advanced Audio Coding), which is not backward compatible. MPEG-2 AAC deals with all the audio channels simultaneously, which makes it more efficient than MPEG-2 BC (Pohlmann, 2011; Taylor et al., 2006). The MPEG-4 standard (ISO/IEC 14496-3, 2009) encompassed the previous MPEG standards. It adds support for object-oriented audio, lossless audio coding with ALS (Audio Lossless), and low bit rate coding with AAC+, also known as HE-ACC (High Efficiency).

Multichannel audio is often perceptually coded using one of the Dolby Digital or DTS codecs. Dolby Digital (www.dolby.com), also known as AC-3, is widely used to provide 5.1 multichannel sound in consumer applications, such as DVD-Video and digital television. Dolby Digital Surround EX is an extension that provides 6.1 multichannel surround sound. The sixth channel is a single surround back channel that the decoder derives from the left and right surround channels (LS and RS) using the Dolby Prologic surround upmixer. In fact, Dolby Digital Surround EX only has 5.1 discrete channels; the extra surround channel is only created upon playback. Dolby Digital Plus, also known as Enhanced AC-3, provides 7.1 discrete channels.

DTS Digital Surround from Digital Theater Systems (www.dts.com) uses the Coherent Acoustics perceptual codec. It consists of a core bitstream that provides 5.1 multichannel audio and supports several extension bitstreams. The hierarchical structure of a core bitstream with extensions ensures backward compatibility with older DTS decoders. DTS Digital Surround ES is based on the XCH extension that provides an additional discrete surround back channel. DTS Digital Surround 96/24 is based on the X96 extension that provides support for sampling rates of 96 kHz. The number 24 in DTS Digital Surround 96/24 refers to the bit depth. DTS-HD High-Resolution Audio is a perceptual codec based on two extensions: the XXCH extension adds two discrete back surround channels to provide 7.1 multichannel surround sound, and the XBR extension increases the maximum bit rate to provide higher quality, but still perceptually coded sound.

Two other perceptual audio codecs worth mentioning are Vorbis from Xiph (vorbis.com) and Windows Media Audio (WMA) from Microsoft (www.microsoft.com). Vorbis is a lossy codec with a compression performance that is comparable to the MPEG-4 AAC codec. Vorbis is commonly referred to as Ogg Vorbis where Ogg refers to the file format from Xiph. The original WMA codec is a lossy codec that is comparable in performance to MPEG-1 Layer III (MP3). The WMA Pro codec is a more advanced codec that achieves better compression ratios and supports multichannel audio. There is also a lossless variant of the WMA codec simply called WMA lossless.

The audio quality of the perceptual codecs mentioned above strongly depends on their data rate. With a high enough data rate, perceptual codecs can achieve remarkable transparency, in the sense that their perceived sound quality is indistinguishable from the original audio data. Perceptual codecs are highly nonlinear, and as a result their perceived sound quality cannot be expressed using traditional audio specifications and measurements. The only way to evaluate their sound quality is to actually listen to their sound. The human ear is the final arbiter. The gold standard for evaluating the performance of perceptual codecs consists of performing controlled listening tests using a large number of expert listeners and objectively evaluating the results using appropriate statistical analysis (Pohlmann, 2011).

As the data rate is reduced the audio quality of a perceptual codec will start to suffer. Discarding too much data will take its toll. It will result in an overall loss of clarity, poor reproduction of transients, and a less pronounced separation between different sound sources. The sound may also become slightly harsh. Some people have taken the compression to extremes in their zeal to achieve the lowest possible data rate. Their horribly compressed audio files have given perceptual coding a bad reputation (Harley, 2015; Taylor et al., 2009).

In general, a higher bit rate means more information and hence a higher sound quality. However, when you are comparing different codecs, it is not possible to judge the sound quality based on data rate alone. Each codec has a different data rate at which it begins to exhibit audible problems. Some codecs are more efficient than others and are able to achieve the same sound quality at a lower data rate. Furthermore, comparison of different codecs is complicated because they exhibit different kinds of audible artifacts when their data rate gets lowered.

Recommendation 5.11 *When you are given a choice between high bit rate and low bit rate perceptually coded audio streams or files (that are encoded with the same audio codec), always choose the high bit rate ones.*

How high of a data rate should you use? Figure 5.10 shows the outcome of a controlled listening test in which the subjective audio quality of some commonly used two-channel codecs has been evaluated (Soulodre et al., 1998). The sound quality is expressed using a five-point grading scale; the higher the score the better the sound quality. At a bit rate of 128 kbit/s (64 kbit/s per audio channel) the MPEG-2 AAC outperforms the other codecs. The Dolby Digital codec requires a bit rate of 192 kbit/s to achieve the same sound quality. Today, most streaming services and file downloads use either 256 kbit/s or 320 kbit/s, which should be more than sufficient for any perceptual codec.

Recommendation 5.12 *Prefer MPEG-2 AAC over MPEG-2 Layer III (MP3). Use two-channel MPEG-2 AAC streams or files with a bit rate of at least 128 kbit/s, and preferably higher (256 kbit/s or 320 kbit/s).*

The perceptual coding of 5.1 multichannel sound is most often done using the Dolby Digital or DTS Digital Surround codecs. Figure 5.11 compares their performance with the multichannel MPEG-2 AAC codec. The results are based on extensive listening tests conducted by the European Broadcasting Union in eight different test laboratories in several countries

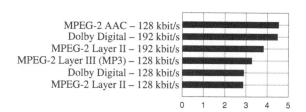

Figure 5.10
Subjective audio quality of some commonly used two-channel codecs for a two-channel reference signal with 16 bits and a sampling rate of 44.1 kHz.

Based on data from Soulodre et al. (1998).

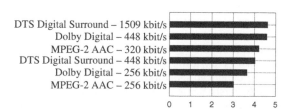

Figure 5.11
Subjective audio quality of some commonly used multichannel codecs for a 5.1-channel reference signal with 16 bits and a sampling rate of 48 kHz.

Based on data from EBU-Tech 3324 (2007).

(EBU-Tech 3324, 2007). The best sound quality is achieved using either DTS Digital Surround at 1509 kbit/s or Dolby Digital at 488 kbit/s. These data rates are the maximum allowed data rates for DVD-Video. However, not all DVD-Video releases use these maximum data rates: Dolby Digital is also used at a rate of 384 kbit/s, and DTS Digital Surround at 768 kbit/s. Obviously, the sound quality will not be as good as at their maximum rates. Although figure 5.11 shows that Dolby Digital and DTS Digital Surround have a comparable sound quality, other studies have found some small differences. It appears that Dolby Digital and DTS Digital Surround perform equally well for ambient surround sounds, but that DTS is slightly better at reproducing discrete surround mixes (Barbour, 2005).

Recommendation 5.13 *DVDs usually contain multiple perceptually coded audio tracks. For the best audio quality, select the multichannel DTS Digital Surround track. If this track is not available, choose the multichannel Dolby Digital track instead.*

Some advanced audio codecs combine perceptual and lossless coding. Two noteworthy examples are MQA and Auro-3D. MQA stands for Master Quality Authenticated (www.mqa.co.uk). It is a nearly lossless codec developed by Meridian that packs high-resolution audio into a 24-bit digital audio stream that is backward compatible with CD-quality LPCM digital audio. The top 16 bits of the MQA stream constitute a lossless LPCM audio stream that can be played on almost any device. The remaining 8 bits are cleverly used to store the additional data that the MQA decoder uses to perceptually reconstruct the high-resolution audio stream (Craven et al., 2013; Stuart and Craven, 2014). Auro-3D (www.auro-3d.com) is a nearly lossless audio codec that packs several overhead surround sound channels into an existing multichannel 24-bit digital audio stream. The top 20 bits of this stream constitute a regular LPCM audio stream with 5.1 or 7.1 channels. The remaining 4 bits are used by the Auro-3D Octopus codec to perceptually code the additional overhead channels (Van Daele and Van Baelen, 2011). Since MQA and Auro-3D audio consist of an LPCM stream with some hidden data, their data can be further compressed using any lossless audio codec such as FLAC, Dolby TrueHD, or DTS-HD Master Audio.

5.3 Digital Media

Digital audio and video can be obtained from a plethora of sources. The devices you use to play different media are called *source components*. They come in different varieties and can play different types of media. Figure 5.12 shows an audio-video system that consists of a

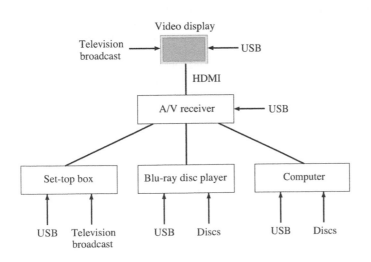

Figure 5.12
Audio-video system with three source components: a set-top box, a Blu-ray disc player, and a computer. The video display and the set-top box have a built-in tuner to receive television broadcasts. Both the Blu-ray disc player and the computer are able to play optical discs. All devices have a USB connector to play media files from an external storage device.

Figure 5.13
The audio-video system
of figure 5.12 with each
component connected to a
router to form a local home
network. The router is
connected to a modem that
connects the local network
to the Internet.

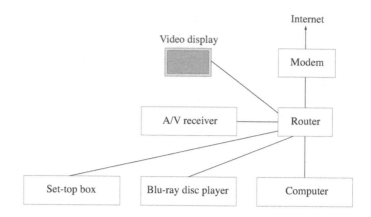

video display, an A/V receiver, and several source components. The Blu-ray disc player is used to play optical discs, such as CD, DVD, and Blu-ray discs; the cable set-top box is used to receive digital television and radio broadcasts; and the personal computer is used to play digital audio/video files and streams from the local network or the Internet. All devices also have a USB (Universal Serial Bus) connector in which you can plug a USB storage device and play the audio or video files stored on it.

When you connect all these devices to your home network the real fun starts. Not only can they share content (streams and files), you can also control the devices using apps on your smartphone or tablet. Many companies make nice remote control apps for mobile devices running iOS or Android. These apps let you use the touch screen of your mobile device to search, browse, and play the content using the different source components that are hooked up to your audio and video system. Figure 5.13 shows the same audio-video system as figure 5.12, but focuses on the network connectivity of the different devices and their integration in your local home network. The router is the grand central station of your network. It interconnects all devices on your local network and also connects your home network to the outside world. The connection to the outside world is usually established using a modem. The modem connects your local network to the infrastructure of your Internet service provider.

Different protocols are being used to share content on the network. They can be divided into two main groups: 1) standard file sharing protocols for computer networks and 2) specialized media sharing protocols. The standard file sharing protocols are not specifically tailored to audio and video sharing, they just provide access to the files on the network and make them available as if they are stored locally on the device. The most widely used file sharing protocol is SMB (Server Message Block), also known as CIFS (Common Internet File System). It is the default protocol used on Microsoft Windows computers and is therefore also known as Microsoft Windows Network (microsoft.com). DLNA is a specialized media sharing protocol from the Digital Living Network Alliance (dlna.org). It uses Universal Plug and Play (UPnP) as the underlying connectivity technology such that devices seamlessly discover each other's presence on the network. More and more devices from different manufacturers are becoming DLNA compliant. Besides DLNA, there are a number of other proprietary protocols in use. The most prominent ones are Apple AirPlay (apple.com) and Google Cast (google.com/cast).

As I already mentioned in section 2.1.2, most modern audio and video devices serve double duty as a source component. Because of this blending of functionalities, you have many options for playing content. The different types of digital source components can typically be categorized as follows:

•   *Set-top box:* A set-top box is a device that is used to receive and decode terrestrial, satellite, and cable television broadcasts. As such it is connected to an antenna, satellite dish, or cable outlet. Some set-top boxes double as a digital video recorder. They have an internal hard disk to store the recordings. Some set-top boxes also connect to the Internet and the home network and can be used as a digital media player.

- *Digital media player:* A digital media player device plays streaming audio and video from the Internet and the home network. Some players can also play media files that are shared on the home network or that are stored on an attached hard disk or other storage device. Some media players are limited to music, others do music, videos, and photos. Many devices have integrated Internet services for browsing websites, accessing social media sites, checking e-mail, and video conferencing. Some also include functionality to play computer games. Examples of popular media players are: Amazon Fire TV (amazon.com), Apple TV (apple.com), Google ChromeCast (google.com), and Roku (roku.com).
- *Optical disc player:* The main function of an optical disc player is to play optical discs like Blu-ray discs, DVDs, SACDs, and CDs. Many modern disc players have integrated digital media player functionality. They connect to the Internet and the home network to play streaming audio and video. Most modern disc players can also play media files from the home network or from an attached storage device.
- *Gaming console:* A gaming console is a dedicated device for playing computer games. All gaming consoles have integrated digital media player functionality, though the specific capabilities vary between brands. Some consoles also play Blu-ray discs and DVDs. Examples of popular gaming consoles are: Microsoft Xbox (xbox.com), Nintendo Wii (nintendo.com), and Sony PlayStation (playstation.com).
- *Computer:* Computers, desktops, or laptops can also be used as entertainment devices. With the right software they can play streams, files, and even optical discs. A computer that is used for music only is often called a *music server,* while a computer dedicated to playing both audio and video is often referred to as a *home theater PC,* or *HTPC* for short (Briere and Hurley, 2009; Tittel and Chin, 2006).
- *Television set:* A television set (TV) is of course used to watch digital television broadcasts. Therefore, a TV is always equipped with one or more tuners for terrestrial, satellite, or cable television broadcasts. Many modern TVs also have built-in media player capabilities. Most of them can play streams from the Internet and the local network and also files that are shared on the home network or stored on an attached storage device. A number of modern TVs is equipped with Android TV (android.com/tv), a smart TV platform developed by Google.
- *A/V receiver or controller:* Many modern A/V receivers and controllers have at least some built-in digital media player functionality. Most of them are able to connect to the Internet and the local network to access streams and media files.

As you can see there is quite some overlap in functionality between the different devices. Nevertheless, in most cases you still need several devices to be able to play all the content you want.

5.3.1 Digital Television

You can receive digital television in three ways: terrestrial, satellite, and cable. In *terrestrial systems* the digital television signals are transmitted by radio waves from a terrestrial (earth-based) television station to an antenna attached to your television or receiver. In *satellite systems* the television signals are also transmitted by radio waves, but the ground station sends these signals to a geostationary telecommunication satellite that relays the signal back to the earth. You use a satellite dish receiver to pick up the signal. In *cable systems* the digital television signals are delivered to your home using an underground cable. The advantage of cable systems, compared to terrestrial and satellite systems, is that their transmission is unaffected by atmospheric conditions. Changing weather and solar flares can disrupt the reception of radio waves resulting in garbled or blocky digital video or in some cases no video at all.

Today, most digital television broadcasts are in HD video (720p, 1080i, 1080p), some even in UHD (4K) and they often come with surround sound. But, be aware that not all digital television is broadcast in HD, many channels still have a standard resolution of 480i or 576i. The video codecs used are either MPEG-2 or MPEG-4 AVC. For UHD broadcasts the codec is often MPEG-H HEVC. The bit rates used for digital video vary between 5 and 40 Mbit/s. Some content providers use low video bit rates to be able to transmit more television channels without increasing their occupied bandwidth. Digital sound is transmitted using MPEG-2 layer II for stereo and MPEG-2 AAC or Dolby Digital for surround sound. Again the bit rates vary.

5.3.2 Internet Streaming

The Internet is a great resource for all kinds of content. Streaming audio and video content using the Internet has become very popular. Streaming is a delivery method in which you receive the content continuously. As soon as you hit play, you can start to watch or listen. The next part of the content is delivered to you just in time as you continue watching and listening. Streaming is an alternative to downloading content, where you have to wait until you have obtained the entire file before you can start watching or listening.

Streaming media over the Internet can be difficult, because the Internet is a packet-switched network. Packet-switching divides the content to be transmitted into several small packets of data. Each packet is tagged with the IP address of the receiving party and sent to a router. The router reads the address and sends the packet on to some other router on the Internet. Together the routers determine the best possible path to the receiving party. Hence, packets may travel different routes to spread the load across the network, to reduce travel times, or to circumvent temporary disruptions. Packet-switching increases the reliability and efficiency of the Internet, but packets can be delivered out of order, multiple times, or not at all. The packets that arrive at the destination address first need to be reassembled in the correct order to make any sense of it. When certain packets go missing, or do not arrive in time, dropouts may occur in your audio or video stream. The stream may temporarily freeze and continue only when the right packets have arrived. To make the streaming experience as seamless as possible, most media players buffer part of the data. After you push play, the player first fills up its buffer and only then starts playing the content. This creates a small time delay between the packets arriving and the content playing. This delay provides some margin to deal with temporary dropouts. As long as the right packets arrive before all the data in the buffer has been used up, you will not notice a temporary dropout.

You need a sufficiently fast and reliable Internet connection for the streaming of digital video. The higher the resolution, the faster your connection needs to be. Table 5.8 lists the recommended minimum Internet connection speeds for different video resolutions. A video stream might still work with a slower speed, but it will play with a reduced bit rate, and thus a lower quality. Providers often use a technique called *adaptive bit rate streaming,* which detects your connection speed in real time and adjusts the bit rate accordingly. To experience high-quality streaming video, you need a fast Internet connection. You can easily test the speed of your connection using a speed test website, like speedtest.net.

Recommendation 5.14 *Make sure your Internet connection speed does not limit the quality of the streaming video that you receive. A minimum speed of 15 Mbit/s would be a good starting point for HD and UHD video.*

In addition to your Internet speed, also check if your Internet provider imposes a monthly data limit or bandwidth cap on your connection. Some do, and if you receive lots of streaming content you might quickly reach such a limit. For example, one hour of streaming HD video at 8 Mbit/s already results in a total of 3.6 GB of data. This means that you reach a typical bandwidth cap of 250 GB within about 70 hours of streaming.

Usually your Internet connection speed is the limiting factor in playing streaming video. The distribution through your home network should not be the limiting factor, even if you use wireless connections (Wi-Fi) instead of more reliable wired connections (Ethernet). All modern wireless connections handle streaming video with ease. However, multiple active devices on the same network can drastically reduce the actual speed that can be achieved. Radio interference from other electronic devices and boundaries such as walls and ceilings can reduce the achievable speed as well.

Streaming video is encoded with different MPEG codecs. The MPEG-H HEVC codec is attractive for both UHD and HD content, because of its highly efficient compression. The video bit rates that are being used depend on the resolution of the content (see table 5.8). The streaming bit rates are on the low side if you compare them to the typical bit rates used on Blu-ray and UHD Blu-ray discs (see table 5.5). Therefore, if you are after the ultimate performance in digital video, streaming is currently not the way to go; you will still be spinning Blu-ray discs for a while. The digital audio that accompanies streaming video is mostly encoded with MPEG-4 AAC or Dolby Digital with a total of 5.1 channels, but sometimes in two-channel stereo only. Again, if you want ultimate performance, Blu-ray discs are still the way to go, because of their lossless audio tracks in Dolby TrueHD or DTS-HD Master Audio.

Music streaming requires much less bandwidth than video streaming, because of the much lower bit rates. Most music streaming services only provide lossy encoded audio. Popular codecs for music streaming are MPEG-2 Layer III (MP3), MPEG-2 AAC, Vorbis, and WMA. Table 5.9 shows the bit rates that are typically used. Low bit rates are meant for mobile applications where Internet speed over cellular networks might be severely limited. Internet speed will not be an issue when listening to streaming music at home using your own broadband Internet connection. Only a few services provide CD-quality lossless audio (44.1 kHz sampling with 16 bits). The bit rate for streaming CD quality is still reasonable: uncompressed it amounts to 1411 kbit/s (see box 5.5), but when losslessly compressed using FLAC it can be reduced to about 700 kbit/s without any loss of fidelity.

Table 5.8 Recommended minimum Internet connection speed for streaming video.

Content	Resolution	Speed
SD	480i, 480p, 576i, 576p	3 Mbit/s
HD	720p	6 Mbit/s
HD	1080i, 1080p	8 Mbit/s
UHD	4K	15 Mbit/s

Table 5.9 Typical quality of music streaming services.

Content	Bit Rate	Codec
Lossy low quality	96 kbit/s	MPEG, Vorbis, WMA
Lossy medium quality	128 or 192 kbit/s	MPEG, Vorbis, WMA
Lossy high quality	256 or 320 kbit/s	MPEG, Vorbis, WMA
Lossless CD quality	1411 kbit/s	FLAC

BOX 5.5 BIT RATE OF CD-QUALITY AUDIO

CD-quality audio is sampled at 44.1 kHz and has a bit depth of 16. With two audio channels (stereo) the bit rate becomes $2 \times 16 \times 44100 = 1411200$ or equivalently 1411 kbit/s.

Recommendation 5.15 *To obtain the best audio quality with music streaming services make sure the service delivers lossless CD-quality streaming (44.1 kHz sampling with 16 bits).*

5.3.3 Media Files

Media files containing digital audio and video have become a popular alternative to disc-based media, especially for audio-only formats. It is much more convenient to play digital audio files than old-fashioned CDs. You can store your entire music collection on a large hard disk and use a computer or other media player to search through it and play your favorite tracks. To make things even more interesting, you can control most media players using an app on your smartphone or tablet. No need to change discs after every track and no more searching among stacks of CDs to find that favorite one. On the video side, disc-based content still rules supreme. The media files for HD and UHD feature-length movies can be huge. Using optical discs instead of hard disks still makes a lot of sense. The inconvenience of having to change discs is less of an issue, because most people watch an entire episode or movie that lasts from one to three hours. Quite a contrast to listening to music where you may want to change tracks every few minutes.

Media files are containers for digital audio and video data. Some media files can contain only audio, while others can hold both audio and video. But there is much more that can be stored inside a media file. Media files also contain audio/video synchronization data to make sure that the audio and video are in sync. Media files can contain subtitles, chapter information, multiple audio tracks, and multiple angles of view. In addition, they often store different types of metadata such as the title of the song or movie, artist, genre, and copyright information. And, music media files sometimes contain still pictures for album art.

Digital media files can be stored on a multitude of storage media: USB sticks, memory cards, CD-ROMs, hard disks, and solid-state drives (SSD). The hard disk remains the device of choice when you want to store a large number of files like your entire music collection. It is a good idea to have at least one backup drive with a copy of all your media files. Hard drives can and do fail. There is a saying that there are two kinds of people in the world: those who have had a hard drive failure and those who will. The purpose of a backup is to have a copy of all your files in case a disaster occurs. Hazards to the integrity and availability of your files come in many different forms: device failure, computer malware, theft, fire, water damage, human error, etc.

Different types of media files have different file formats and data formats. The file format describes what kind of container is being used. The data format describes the way in which the audio and video data is encoded. It is important to distinguish between the two, because the file format does not always tell you how the audio and video data is encoded. For example, a media file with the MP4 file format may contain MPEG-4 AVC video and AAC audio, or it may instead contain MPEG-2 video and Dolby Digital audio. Table 5.10 lists some common media file formats that are used to store both audio and video data. Most of them can contain different types of audio and video data formats. Some common combinations are listed in table 5.11.

Not all media files are compatible with all media players. Some devices or computer software programs are not able to play certain types of files. The most obvious reason for incompatibility is that the player does not recognize the file format. But, even when it does, compatibility may be an issue. If the player does not support a certain audio or video codec, it might be able to read the file, but not be able to decode the audio or video. This may result in a moving picture without sound, sound without a picture, or nothing at all.

Table 5.10 Common media file formats that are used to store both audio and video data.

Name	Proprietor	File Extensions
3GP	3GPP	.3gp
Advanced Systems Format	Microsoft	.asf .wmv .wma
Audio Video Interleave	Microsoft	.avi
BDAV MPEG-2 Transport Stream	Blu-ray Disc Association	.m2ts .mts
DivX Media Format	DivX	.divx
Flash Video	Adobe Systems	.flv .f4v
Matroska	Matroska	.mkv .mk3d .mka
MPEG-4 Part 14	ISO/IEC	.mp4 .m4p .m4v .m4a
MPEG Program Stream	ISO/IEC	.mpg .mpeg .m2p .ps
MPEG Transport Stream	ISO/IEC	.ts .tsv .tsa
Ogg	Xiph	.ogg .ogm
Quicktime	Apple	.mov .qt
Video Object	DVD Forum	.vob

Table 5.11 Some examples of file format and data format combinations. These are just examples; many other codec combinations are possible.

File Format	Video Codec	Audio Codec
.avi .divx	MPEG-4 Part 2	MPEG-2 Layer III
.avi .divx	MPEG-4 Part 2	Dolby Digital
.mpg	MPEG-2	MPEG-2 Layer II
.mp4 .mkv .mov .vob	MPEG-2	Dolby Digital
.mp4 .mkv .mov .f4v	MPEG-4 AVC	MPEG-2 AAC
.mp4 .mkv .mov .m2ts	MPEG-4 AVC	Dolby Digital
.m2ts	MPEG-4 AVC	Dolby TrueHD
.m2ts	SMPTE VC-1	Dolby TrueHD

Recommendation 5.16 *When you buy a media player (computer, smart TV, A/V receiver), make sure it can play all the different types of media files that you require. Check compatibility for both the file formats and the data formats (codecs).*

Certain media file formats are specifically meant for storing digital audio. These digital audio files do not contain video data. The difference between file format and data format is less important for these files, because there is more or less a one-to-one relationship. Table 5.12 shows the most commonly used file types. These files are often referred to by their acronym.

Audio files can be uncompressed, losslessly compressed, and perceptually coded. For the highest playback quality you are advised to either choose uncompressed or losslessly compressed files and leave the perceptually coded files alone. Perceptually coded files use lossy compression techniques that make it impossible to retrieve the original audio data. Depending on the bit rate that is being used a perceptually coded file might sound just alright or really awful (see section 5.2.5). Perceptually coded files are still great for less critical casual listening, for example while you are on the move and traveling in a car or

Table 5.12 Common digital audio file formats.

Acronym	File Extensions	File Format	Audio Codec
AAC	.aac .m4a	MPEG-4 Part 14	MPEG-2 AAC
AIFF	.aiff .aif	Audio Interchange File Format	LPCM
ALAC	.m4a	MPEG-4 Part 14	ALAC
APE	.ape	Monkey's Audio	Monkey's Audio
DSD	.dsf .dff	DSD file format	DSD
FLAC	.flac	FLAC file format	FLAC
MP3	.mp3	MP3 file format	MPEG-2 Layer III
OGG	.ogg	Ogg file format	Vorbis
WAV	.wav	Resource Interchange File Format	LPCM
WMA	.wma	Advanced Systems Format	WMA

airplane. The most commonly used perceptually coded audio files are AAC, MP3, OGG, and WMA. While WMA does have a lossless variant, most WMA files use lossy compression. Uncompressed DSD audio is always stored in a DSD file, while uncompressed LPCM audio is either stored in a WAV file or an AIFF file. The WAV file format has one notable limitation: it does not support the storage of metadata such as title, artist, or genre in the file. The AIFF file format can store such metadata and is therefore preferred over WAV. Since the audio data is not compressed, WAV and AIFF files have the largest file size compared to the other formats. The ALAC and FLAC file formats losslessly compress the LPCM data, which results in a file size that is approximately 50% smaller. Because the compression is lossless, the original audio data can be retrieved. Therefore, ALAC and FLAC files sound exactly similar to WAV and AIFF files. Nevertheless, some audiophiles ridiculously claim to hear a difference. Note that the file extension used for ALAC files equals '.m4a', which is also used for AAC encoded files. To distinguish the two you need to look at the metadata stored in the file. FLAC is more widely used than ALAC, but Apple's devices and iTunes only natively support ALAC.

Recommendation 5.17 *For the best audio quality use lossless LPCM files like AIFF, ALAC, or FLAC with at least 16 bits and a sampling rate of at least 44.1 kHz.*

Music files can be bought online from many websites. The audio quality of the files varies considerably. Some sites only offer perceptually coded files with bit rates of 256 kbit/s or 320 kbit/s, while other sites mainly sell HRA audio files with increased sampling rates and bit depths. You should at least aim for CD-quality music (44.1 kHz / 16 bit) and nothing less. A good place to start your search for high-quality lossless audio files is the list of online retailers hosted at findHDmusic.com. The HRA files are worth considering, especially for more recent recordings. Older material might not contain any high-frequency content above 20 kHz and might also have a reduced dynamic range, due to the performance limitations of the recording and mixing equipment used at that time. However, older material that has been remastered might still be worth checking out in the HRA format. Carefully remastered tracks might sound better, regardless of the presence of any ultrahigh-frequency content. Often it is difficult to determine the origins of an HRA file, because of the limited information provided by music retailers.

Besides the uncertainty of the source of the recording, there is also the possibility that the HRA file does not utilize the whole frequency range or bit depth that is advertised. In the past some online retailers have just upconverted CD-quality files to a higher sampling rate and bit depth and sold them at a higher price. Such upconversion doesn't improve the sound quality at all, because the original file didn't utilize the higher dynamic range and didn't include any ultrahigh frequencies. You can use special software on your computer to try to identify fake HRA files (Goodwin, 2012). MusicScope is a convenient program to analyze digital audio files. Alternatively you could use the free Audacity software and the Bitter plug-in from Stillwell Audio. If you don't want to analyze the files yourself, you can try the Lossless Audio Checker program (Lacroix et al., 2015), which is limited in functionality but attempts to automatically detect upsampling and bit depth usage. See appendix B for more details on these computer programs.

Another interesting aspect to look at is the actual dynamic range used in the recording. A digital audio file, even when it is an HRA one, will never sound good if it has a heavily compressed dynamic range. As I have mentioned before, recordings aimed at the mass consumer market are often heavily compressed to make them sound louder and stand out in casual listening conditions. Such compression destroys the dynamics in the recording and makes the overall presentation flat and uninvolving. The compression of the dynamic range makes the recording louder in comparison to other recordings. Music producers believed that louder recordings had a commercial advantage and this led to the so-called *loudness war* (Vickers, 2010).

A well-established measure for the dynamic range of a recording is *loudness range* (LRA). It is a measure for the *macrodynamics* in the audio signal, that is, the loudness differences between different sections of a musical piece (EBU-Tech 3342, 2016). LRA is a suitable measure for the evaluation of musical dynamics in the classical sense, such as *pianissimo* and *fortissimo*. LRA is measured in *loudness units* (LU); 1 LU corresponds to 1 dB. A high value of LRA indicates that there are some very quiet parts and some very loud parts in a musical piece. LRA is a well-defined statistical measure that is not easily swayed by periods of silence or short duration peaks in the program (see box 5.6 for further explanation). Music with a high LRA usually sounds more natural and more exciting (Katz, 2015). The MusicScope software (see appendix B) computes, among other statistics, the LRA of a digital media file. The Roon media player (see appendix B) also uses LRA to display the dynamic range of a media file.

Another measure for the dynamic range of a recording is the DR value, which is essentially the ratio of the peak values to a time-averaged value of the amplitude of the audio signal (see box 5.7). DR is a measure of the *microdynamics* in the audio signal, that is, the saliency of the peaks (Deruty and Tardieu, 2014). DR can be computed using the TT Dynamic Range Meter (see appendix B). The higher the DR value, the higher the dynamic range used in the recording.

BOX 5.6 LOUDNESS RANGE (LRA)

The Loudness Range (LRA) is essentially the difference between the highest and lowest loudness in a particular program. The loudness is computed using the standardized Program Loudness as described in ITU-R BS.1770 (2015). Since the lower edge of the loudness range should correspond to the weakest musical signal and not to the noise floor, a relative gating-threshold is used to discard the weakest sounds. The change of loudness in time is computed using a sliding window with a length of 3 seconds. The distribution of the computed loudness levels is then quantified using a percentile range: the upper edge of the loudness range is the loudness that 95% of the program sits below, and the lower edge is the loudness that 10% of the program sits above. The use of these percentiles prevents a single silent or loud part dominating the computation of the loudness range (EBU-Tech 3342, 2016).

BOX 5.7 DR VALUE

The TT Dynamic Range Meter computes a DR value that describes the compression of the inner dynamics of the recording. It is the average cumulative difference between peak and rms values over the duration of a song or album, taking into account only the highest 20% of the levels. The DR value is expressed in dB and always rounded to the nearest whole number (pleasurizemusic.com, dr.loudness-war.info).

Table 5.13 Suggested LRA and DR values for different styles of music. Based on data from Katz (2015) and pleasurizemusic.com.

Type of Recording	LRA	DR
Pop, rock, blues	4–13	7–14
Classical, jazz, country, folk	8–23	9–20

While LRA and DR roughly portray the same information about dynamic range, their numerical values are not directly comparable. A good recording with a natural dynamic variation should have high LRA and DR values. The amount of natural dynamic variation differs among music styles. For example, disco music typically has a much lower dynamic range than jazz music. Table 5.13 lists some suggested LRA and DR values for different styles of music.

It would be useful to know the dynamic range of a particular recording before you buy the corresponding digital audio file. Well, you may be lucky to find it in the community-based Dynamic Range Database at dr.loudness-war.info. You could use this database to decide whether it is worthwhile buying an HRA replacement for a certain CD-quality media file that you already own. If the DR value is on the low side, an upgrade to HRA may not be worthwhile.

5.3.4 Computer Audio

Playing music is best done using a computer. Even traditional audiophiles have embraced computers as their main means for playing digital audio (AES, 2012). The reason that computers are the way to go is twofold. First, the computer gives you instant access to your entire music collection. Music management software on the computer makes it easy for you to organize your files, to quickly find a particular piece of music, and to create playlists with multiple tracks on the fly or in advance for uninterrupted listening. Second, a computer is a highly customizable device and you can take advantage of that by configuring it such that it delivers the best possible audio quality. Though it requires some patience and some technical experience, it is well worth the effort.

Any modern computer can be turned into a high-quality music server or player. Both desktops and laptops qualify for the job. However, it is best to have your computer dedicated to playing music and not use it for other purposes. Popular choices for dedicated music servers are small form-factor computers such as the Apple Mac mini (apple.com/mac-mini) and the Intel NUC (intel.com/nuc). If you have the required expertise you could decide to custom build your own music server. This is actually less complicated than it sounds; it merely boils down to selecting components, such as a motherboard, processor, and memory, and assembling these parts in an attractive computer case (Tittel and Chin, 2006).

It is important that the music server doesn't create too much noise. Almost all computers have a fan that provides cooling for the electronics inside. The sound of a rotating fan contributes to the background noise in the room (see section 4.3.2) and can be a significant distraction from the music. Therefore, if the computer is in the same room as the loudspeakers, make sure it is quiet. Larger fans are preferable, because they move more air and can rotate slower while providing the same cooling power. Slow rotation means less fan noise. Another culprit can be the hard drive inside the computer or the external drive attached to it. The mechanical movements of their internal parts can also create distracting sounds. Always take into account how much noise a computer device generates when you want to use it for your music server setup (Harley, 2015; Tittel and Chin, 2006).

All modern computers are equipped with a sound card that can output analog audio, but you do not want to use it. Built-in sound cards have low-quality digital-to-analog converters (DACs) that will not provide the best possible audio quality. Furthermore, the inside of a computer is a hostile environment for delicate line-level analog audio signals. The rapid switching of digital signals in the computer generates high-frequency noise that can couple into the analog audio signals, resulting in unwanted electromagnetic interference (EMI). You are much better off letting an external device handle the digital-to-analog conversion. Your computer should send the digital audio signals to an A/V controller or receiver that uses high-end DACs for the conversion. Alternatively, you could use a dedicated external DAC to handle the conversion. This option is popular among audiophiles who have a conventional two-channel stereo playback system. It is less useful in a multichannel system, because the A/V controller or receiver has to convert the analog signals from the external DAC back to digital to be able to perform bass management and other signal processing. Since this extra conversion step can slightly degrade sound quality, it is better to feed the digital signal directly to the A/V controller or receiver. While HDMI is the most commonly used digital connection method for digital audio and video, it is less frequently used for audio-only digital connections. Common ways to connect a music server to an A/V controller, receiver, or external DAC are S/PDIF and USB Audio, which I will describe in more detail in section 5.4.

> **Recommendation 5.18** *Do not use the built-in sound card of your computer to convert digital audio to analog audio. For the best audio quality send the digital audio from the computer to an A/V controller, A/V receiver, or dedicated external DAC, and let them handle the conversion.*

To get the best sound quality from your music server, you need to properly configure your music player software, the operating system, and the sound driver. I will give you some general guidelines. I intentionally refrain from giving very specific instructions for certain programs and operating systems, because such detailed instruction would differ among players and systems and would quickly become obsolete when new versions of the software come out. There are plenty of websites that offer more detailed guidelines complete with screenshots of the specific programs. A good place to start is: www.thewelltempered-computer.com.

You want your music server to send the digital audio unaltered to the DAC in your A/V controller or receiver. More specifically, you want to avoid having your music player or operating system distort or corrupt the digital audio data. When your server is able to send the data unaltered to the DAC, it is called *bit perfect* or *bit transparent*.

There are many reasons why a music server may not be bit perfect. One culprit is the sound mixer of the operating system. This mixer allows the computer to play multiple audio streams at once, for example the output of your music player and the warning sounds of your operating system. Since the mixer includes volume controls for the different streams, it alters the original data. The mixer also converts all streams to the same bit depth and sampling rate, otherwise it cannot mix them together. Depending on your actual operating system, driver, and software used, it is sometimes possible to bypass the mixer completely and give your music player software exclusive access to the sound driver. If you are not able to bypass the mixer, you should do two things: First, turn off all operating system sounds. Second, mute or turn down the volume of all the other sources except your music player software.

BOX 5.8 DIGITAL VOLUME CONTROL

A digital volume control can only alter the volume in discrete steps. The smallest step is of course related to the number of bits being used. To lower the volume the most significant bit is set to zero and the other bits are shifted one position to the right, for example:

```
1111   1111   1111   1111
0111   1111   1111   1111
0011   1111   1111   1111
```

The least significant bits that fall off on the right side are simply discarded. In the example, the effective bit depth is reduced from 16 to 15 bits and again to 14 bits. In the process, low-level detail is being lost and the quantization error increases.

The digital volume controls found in your operating system mixer and in your music player software can also prevent you from achieving bit perfect playback. To avoid data corruption, you must always set them to their maximum volume (100% or 0 dB). Note that in the mixer you should set both the master volume and the volume of your music player software to the maximum setting. All the other volume sliders in the mixer should of course stay at their minimum value. You should always adjust your playback volume on your A/V controller or receiver and never use the digital volume controls on your computer. The reason is that a digital volume control discards bits. To substantially lower the volume, the least significant bits are discarded (see box 5.8) and you end up with a reduced bit depth, which means that you lose low-level detail and effectively increase the noise.

To achieve bit perfect playback, it is also important to switch off any sound enhancements that alter the original audio data. Examples are equalizers, surround effects, bass boost, and other DSP (digital signal processing) plug-ins. Sound enhancements can be added by your sound driver, operating system, or music player software. So, make sure you go through their settings and disable all enhancements.

Recommendation 5.19 *To get the best sound quality from your computer, take the following steps:*

- *If possible, give your music player exclusive access to the sound driver.*
- *Set the digital volume control of your music player to its maximum level.*
- *Set the master volume of the operating system's sound mixer to its maximum level.*
- *Set the volume of your music player in the mixer to its maximum level, and turn down or mute all the other sources.*
- *Disable all operating system sounds.*
- *Disable all sound enhancements in the sound driver, operating system, and music player.*

Two other key settings are the bit depth and sampling rate of your sound driver. You should always set the sound driver to the highest bit depth that the DACs in your A/V controller or receiver can handle. Usually, this is 24 bits. With this setting, your sound driver will convert all digital audio to 24 bits, regardless of the bit depth of the digital audio files. So, when you play 16-bit CD-quality files, the sound driver will output them in 24 bits. This conversion to 24 bits doesn't change the sound quality of those files. They will not sound better nor will they sound worse, the reason being that the conversion from 16 to 24 bits only consists of adding new bits with a value of zero (see box 5.9). Since no information is added or discarded, the sound quality remains the same.

BOX 5.9 BIT DEPTH CONVERSION

Reducing the bit depth of an existing digital audio file will possibly also reduce the sound quality. For example, going from 24 to 16 bits, you need to discard 8 bits. This reduction in low-level detail may be perceptible, depending on the content of the original file. By contrast, increasing the bit depth of an existing digital audio file will not change the sound quality. To convert 16 bits to 24 bits all that is done is adding 8 bits that have a zero value. For example,

```
1111   1111   1111   1111
```

simply becomes

```
1111   1111   1111   1111   0000   0000
```

Since no information is added or discarded, the 24-bit file will not sound any better or worse than the 16-bit original.

Recommendation 5.20 *Always set the audio bit depth of your computer to the maximum value that your DACs can handle (usually 24 bits).*

The choice of the sampling rate in your sound driver requires some more thought. Contrary to what you might expect, it is not always a good idea to set it to the highest sampling rate that your DACs can handle. First of all, increasing the sampling rate of a digital audio file does not increase its sound quality. If the higher frequency information is not present in the original file, upsampling will not magically add it. Upsampling merely converts the existing information to a higher sampling rate. Upsampling is done by a sampling rate converter. Such a converter uses sophisticated digital signal processing to change the sampling rate. If the new sampling rate is an integer multiple of the original one, the sampling rate converter only has to compute the missing samples. For example to convert a sampling rate of 48 kHz into 96 kHz, the sampling rate converter needs to compute one additional sample between each two of the original samples. This is illustrated in figure 5.14. However, if the conversion is not an integer multiple things become a bit more complicated and the sampling rate converter has to recompute every sample. A typical example is the conversion from 44.1 kHz to 96 kHz. Figure 5.15 shows that the new samples do not align in time with the original samples. In this case, the digital audio is clearly altered and playback of the upsampled audio is not bit perfect. In fact, sampling rate conversion, whether it involves integer or noninteger multiples, is never bit perfect, it always introduces errors in the original audio. How bad these errors are very much depends on the particular computations used by the sample converter to estimate the new samples. Some converters sound terrible while others perform very well with no noticeable degradation of sound quality. The differences in their sound can be attributed to the various implementations of the upsampling filters used (Katz, 2015).

To avoid the pitfalls of sampling rate conversion and to achieve bit perfect playback, your computer should output the digital audio in its native sampling rate. However, the sound driver can only output one sampling rate at a time. This means that when you switch between playing audio files that have different sampling rates, the sound driver should also switch sampling rate. In the unlikely case that all your audio files have the same sampling rate you could just set your sound driver once to that particular sampling rate and never again worry about it. But it is more likely that sampling rates differ among files. Since it is too cumbersome to manually reconfigure your sound driver each time you play a file with a different sampling rate, many music players can communicate with your sound driver and automate the switching of sampling rates. Some popular players that can automatically switch are: Audirvana

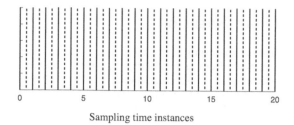

Sampling time instances

Figure 5.14
Comparison of the sampling time instances in a 48 kHz sampled signal and a 96 kHz sampled signal. The *solid lines* represent the time instances of the 48 kHz sampled signal. The 96 kHz sampled signal uses the exact same sampling time instances with one additional sample, indicated by the *broken lines* between each two samples of the 48 kHz sampled signal.

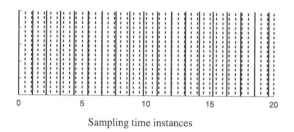

Sampling time instances

Figure 5.15
Comparison of the sampling time instances in a 44.1 kHz sampled signal and a 96 kHz sampled signal. The *solid lines* represent the time instances of the 44.1 kHz sampled signal. The *broken lines* represent the time instances of the 96 kHz sampled signal. Unlike the example of figure 5.14 the sampling instances do not line up nicely.

Plus, JRiver MediaCenter Foobar2000, MediaMonkey, Roon, and Signalyst HQPlayer (see appendix B). Apple's iTunes, which is widely used, doesn't support automatic switching of sampling rates. To get around this limitation, you can use a special program that runs alongside iTunes that takes over the actual playing of the files. In this way you can still use the iTunes interface while at the same time get bit perfect playback. Some examples of such programs are Amarra, Bitperfect, and Pure Music (see appendix B).

So, is automatic sampling rate switching always better than sampling rate conversion in the computer? Well, in practice it depends on the DACs in your A/V controller or receiver, because these DACs often also perform sampling rate conversion. Many DACs upsample their digital input signals to the highest sampling rate that they support, usually 96 kHz or 192 kHz, before they convert the digital signal to an analog one. So despite the fact that your computer is bit perfect and sends the digital audio out in its native sampling rate, your digital audio is still subject to sampling rate conversion before it is being converted to analog. This raises the questions whether your DAC or your computer does a better job with respect to sampling rate conversion. In theory, the computer might do a better job because it has more computational power than a DAC and can therefore use a more sophisticated sampling rate conversion algorithm. In practice, the quality of the sampling rate converters used in different operating systems vary. Technical measurements show that Windows 10 does a very poor job, while Linux (with the appropriate driver) and macOS do a much better job (Archimago's Musings, 2015). Unfortunately, there is no simple answer to the question whether sampling rate conversion or sampling rate switching gives you the best possible sound quality from your computer. Therefore, I recommend you listen and compare the two different methods for your own particular setup.

Recommendation 5.21 *Set the audio sampling rate of your computer either to the native sampling rate of the files that you play (using software that automatically switches the sampling rate of the sound driver) or set it to the maximum sampling rate that your DACs can handle (96 kHz or 192 kHz). Listen and compare the two methods and choose the one that sounds best.*

Another setting in your media player that requires your attention is *volume leveling* or *loudness normalization*. This feature automatically adjusts the playback volume of each individual media file such that upon playback they appear to have the same loudness. It works by computing the loudness of the file and subsequently adjusting its level to a certain predefined loudness standard (see box 5.10). Many people like to enable loudness normalization for convenience, as it avoids manual volume changes when playing multiple tracks of different origin. When you enable loudness normalization your playback chain is no longer bit perfect. Nonetheless, it can be a good idea to enable it anyway. Many modern popular recordings are mastered way too loud and the peaks in their audio signals are close to the maximum digital level. If two subsequent samples are close to the maximum digital level, the peak of the reconstructed analog signal between these two samples can easily exceed the maximum digital level. This is called *intersample clipping* and results in distortion of the analog waveform during the digital-to-analog conversion. Lowering the digital playback volume using loudness normalization can avoid such clipping. If you use loudness normalization it is important to set your bit depth to 24 bits. This will give you enough precision not to worry about any audible degradation of the digital audio signal. Recall from section 5.2.2 that the best DACs only have a precision of 21 bits. So with 24 bits you have 3 bits to perform loudness normalization without audibly changing the precision of the digital-to-analog conversion. These 3 bits provide 18 dB of possible volume correction. Loudness normalization seldom needs to apply a correction of more than 18 dB.

5.3.5 Digital Audio Extraction

Playing CDs is not as convenient as playing media files. A music server can organize your media files and make them easier to find and play than the old-fashioned CD. You can control your music server from a smartphone or tablet and never have to walk to the player to change CDs. Unsurprisingly many people like this convenience and consequently, the vast majority of CDs sold today are only played once: to transfer the content to a music server. Not only are media files convenient, they also have the potential to sound better than the original CD. This holds especially true for older CDs that may have accumulated some scratches or other minor abrasions. Let me explain why. Audio data on a CD is stored as a continuous track. The optical system in your CD player is supposed to read that track continuously. It has only one pass to get the sound right and has to perform error correction on the fly. If it is unable to correct certain errors, the reproduced sound may have audible clicks and pops or at least differs in

---

BOX 5.10 LOUDNESS NORMALIZATION

A standardized way to normalize the loudness of different audio programs is to first measure the Program Loudness using the method described in ITU-R BS.1770 (2015) and then adjust the level of the audio signal such that its loudness equals an agreed-upon reference value. The choice of this reference value differs among implementations. The European Broadcast Union recommends a reference loudness of −23 LUFS (loudness units below full scale). Their recommendation is widely used and is known as EBU R 128 (2014). The Sound Check implementation used in iTunes aims at a reference loudness of −16.5 LUFS (Katz, 2015).

some way from the data stored on the disc. Computer data, like a media file, is not stored as a continuous track, but in sectors with unique addresses. These sectors are read incrementally and are reread when errors occur. Unless some major disaster occurs, the media file that you play is always an exact bit-for-bit accurate copy of the original data. By playing media files instead of CDs, you eliminate the uncertainty of optical reading problems and the sound that you hear will be bit-for-bit accurate (Harley, 2015; Pohlmann, 2011).

Many digital audio extraction (DAE) computer programs exist that convert your CDs into media files, but these programs are not all created equal. It is essential to use a program that has a so-called *secure-ripping mode.* Secure ripping aims to recover the audio data from a CD without any errors. It reads the data on the disc multiple times to try to obtain consistent results. If it detects an error, it attempts to correct it by rereading the erroneous part of the audio track. Secure ripping ensures that the media files you create are bit-for-bit accurate copies of the data on the CD. Secure ripping software is often able to recover perfectly accurate audio even from damaged CDs, but in such a case the ripping can easily take several hours. However, sometimes a perfect rip may just not be possible. The ripping program will then report on the errors it encountered so that you get an idea of the quality of the rip. Most programs will also apply an automatic glitch correction at the erroneous samples as a last attempt to obtain an audio file without any pops and clicks. The error report will include a list of suspicious positions allowing you to listen to the erroneous parts and decide whether the created audio file is still acceptable.

Secure ripping software has to work in unison with your computer's CD-ROM drive. It has to take into account the limitations and capabilities of this drive. Some drives use caching, which means the drive holds a certain amount of data in its buffer so that it is readily available when requested more than once. This generally increases the drive's performance but can cause problems for secure ripping: when the ripping software asks the drive to reread an erroneous section of the disc, instead of actually rereading the data, the drive returns the previously cached data from the buffer. To circumvent this problem, the ripping software can request more data than the buffer can hold, forcing the drive to flush its cache. Circumventing the cache in this way will increase the time needed for ripping the CD. Therefore, it is better to have a drive without cache, or a drive in which the software can instruct the drive to neglect the cache (Force Unit Access or FUA for short). Almost all modern drives support a feature called Accurate Stream, which allows the drive to precisely locate an area on the disc. This feature greatly reduces the time the drive needs to position itself for successive rereads. Without Accurate Stream secure rips are still possible, but they will not be as fast.

Some secure ripping programs support AccurateRip (www.accuraterip.com) as an additional means to verify the accuracy of ripped audio data. AccurateRip is an online database that contains checksums of ripped CD tracks. It allows you to compare the checksum of your ripped track with the checksums of other people. If they are the same, it is likely that your rip is truly error free. Secure ripping programs that support AccurateRip are dBpoweramp, Exact Audio Copy (EAC), Rip, and X Lossless Decoder (XLD); for more details see appendix B.

> **Recommendation 5.22** *Convert your CDs into digital media files using DAE software that supports secure ripping and AccurateRip.*

When ripping CDs it is important to make sure the media files contain the right metadata. You need metadata tags like artist, album title, song title, year, and genre to be able to organize your music library in a logical and consistent way. Most ripping programs automatically download the appropriate tags from an online database. However, it remains a good idea to carefully review the automatically added tags, because sometimes they are incorrect, contain small errors, or follow a different convention than the other files in your music library (Harley, 2015).

5.3.6 Optical Discs

The CD, DVD, and Blu-ray disc (BD) form three different families of optical discs. They are all being used to store digital audio and video. Data is stored in the form of microscopic pits. The pits form a spiral track from the inside edge of the disc to its outer edge. The pits on the disc are read by spinning the disc underneath a laser beam. The pits change the intensity of the beam, and this change can be detected by an optical sensor. There is no physical contact between the sensor and the disc or between the laser and the disc. Therefore, playing the disc causes virtually no wear. This is a big advantage that ensures long media life. The smaller the pits on the disc, the more data can be stored. The CD is the oldest format and has the largest pits and hence the smallest data capacity. The DVD and BD have smaller pits and can hold much more data. The physical characteristics of the different discs are listed in table 5.14. To increase the capacity, DVD and BD can have a second or even a third layer (Taylor et al., 2009).

CD, DVD, and BD come in different physical formats and in different application formats. The physical format depends on the physical construction of the disc. It determines how the data is stored on the disc. Besides the mass produced read-only discs, there exists a multitude of different recordable discs that have a different physical makeup. Table 5.15 lists the most common physical formats. The CD-DA format in this table is the widely used audio CD; CD-DA simply stands for Compact Disc Digital Audio. The CD-ROM is the read-only format used to store computer files. DVD-ROM and BD-ROM are the read-only formats widely used to release feature films, documentaries, and concerts. Several incompatible recordable formats exist: BD-RE, DVD-RAM, DVD-RW, and DVD+RW are re-recordable; BD-R, DVD-R, and DVD+R are only recordable once.

The application format determines what kind of data the optical disc contains. Table 5.16 lists some commonly used application formats. Today, the most common application formats are CD-DA, DVD-Video, and HDMV. The audio-only formats SACD and DVD-Audio have not taken off in the marketplace and are now difficult to find. Nowadays, high-resolution audio is more commonly distributed as downloadable files instead of optical discs.

Table 5.14 Physical characteristics of some optical discs (BDA, 2015a; Pohlmann, 2011; Taylor et al., 2009). The BDXL discs are used for UHD Blu-ray discs.

	CD	DVD	BD	BDXL
Pit length (μm)	0.833	0.400	0.149	0.112
Track spacing (μm)	1.60	0.74	0.32	0.32
Correctable error (mm)	2.5	6.0	7.0	7.0
Capacity single layer (GB)	0.7	4.7	25	33
Capacity dual layer (GB)	–	8.5	50	66
Capacity triple layer (GB)	–	–	–	100

Table 5.15 Examples of physical media formats for optical discs. The 'R' stands for recordable; 'RW' and 'RE' stand for rewritable (Taylor et al., 2006, 2009).

Disc Family	Physical Formats
CD	CD-DA, CD-ROM, CD-R, CD-RW
DVD	DVD-ROM, DVD-RAM, DVD-R, DVD+R, DVD-RW, DVD+RW
BD	BD-ROM, BD-R, BD-RE

Table 5.16 Examples of application formats for optical discs (Taylor et al., 2006, 2009).

Disc Family	Application Formats
CD	CD-DA (Compact Disc Digital Audio)
	VCD (Video CD)
	SVCD (Super Video CD)
DVD	DVD-Video
	DVD-Audio
	SACD (Super Audio Compact Disc)
	DVD-VR (Video Recording)
	DVD-AR (Audio Recording)
	DVD-SR (Stream Recording)
	AVCHD (HD video on recordable DVD)
	AVCREC (BDAV on recordable DVD)
BD/BDXL	HDMV (HD and UHD Movie)
	BD-J (Blu-ray Disc Java)
	BDAV (Audio Visual for recordable media)

While music is best played using digital files, it is still worthwhile using discs as your primary source of high-quality video. Blu-ray discs continue to offer the best video quality, because their bit rate is higher than the bit rate of streams and most downloaded files. Not all source components can play optical discs. You basically have three choices: 1) a dedicated disc player, 2) a computer with an optical drive, and 3) a gaming console with an optical drive. Players differ in the physical and application formats they can play. Most of the players are compatible with CD, DVD-Video, and Blu-ray, but not all players can play DVD-Audio and SACD. Compatibility remains an issue with recordable discs (Taylor et al., 2009). Playing DVD and Blu-ray discs on a computer requires special software that can read the copy-protected and region-encoded content. Most players can also play media files stored on a data disc (CD-ROM, DVD-ROM, BD-ROM). They are able to play photos, music, and movies depending on the particular file formats used on such a disc. The most commonly used media file formats are the ones that I described in section 5.3.3.

Recommendation 5.23 *When you buy a disc player, make sure it can play the different types of optical discs that you require. Check compatibility for the physical formats, application formats, and media file formats.*

For the best video quality I recommend a dedicated player, either a Blu-ray disc player or a UHD Blu-ray disc player. A good-quality Blu-ray player is equipped with dedicated video hardware that can easily outperform a computer or gaming console with respect to image quality. The dedicated hardware of the Blu-ray player takes care of some essential video processing steps such as scaling, chroma upsampling, and deinterlacing. I provide a more in-depth look at these video processing steps in section 6.3. While the video quality is very important, the audio quality of your Blu-ray disc player should not be of much concern, the reason being that the player just sends the digital audio signal to your A/V controller or receiver that handles the digital-to-analog conversion. In such a setup, the quality of the DACs in the player is irrelevant. They are not being used.

The decoding of surround sound can either be done in the player or in the A/V controller (or receiver). You can make your choice in the setup menu of the player. If you set the audio output of the player to *bitstream* audio, the digital audio from the disc is sent directly to the A/V controller or receiver without the player decoding it. This is the most common setting. However, if you have an older A/V controller or receiver that is not capable of decoding the latest surround sound formats, you could let the player handle the decoding instead. In this case, you set the audio output of the player to LPCM audio. With this setting, the player is going to decode the compressed audio stream and convert it into multichannel LPCM digital audio, which is then sent to your A/V controller or receiver.

Another audio setup item in the player's menu that deserves your attention is *secondary audio*. Blu-ray discs support a secondary audio stream that is sometimes used for audio commentary or for sound effects in the disc's menu. To be able to hear the secondary audio, your player has to mix it with the primary audio stream and output the end result as one merged stream. Before the streams can be mixed they must be decoded to LPCM audio (Briere and Hurley, 2009; Taylor et al., 2009). The process of mixing and decoding these streams has the potential to slightly degrade sound quality. Therefore, to obtain the best sound quality from your player it is best to turn off the secondary audio feature in the player's setup menu. You may consider turning it on temporary, for example, to listen to the audio commentaries, but don't leave it on by default for general watching.

Recommendation 5.24 *Disable the secondary audio stream in the setup menu of your Blu-ray disc player to achieve the best possible sound quality.*

Optical discs are tolerant to a certain amount of surface contamination and scratches. One reason is that the laser reading the disc is not focused on the outer surface, but on the data layer that is buried inside the disc away from the surface. Small scratches and contaminations appear out of focus and hence do not interfere with the laser's ability to read the data of the disc. Another, more important reason is that redundant data is added to the original audio and video data such that certain reading errors can be detected and corrected (Pohlmann, 2011; Taylor et al., 2009).

If you encounter problems during playback, check the disk for surface contaminations. Often a simple cleaning of the disc solves your problem. Gently wipe the disc with a clean soft lint-free cloth. Always wipe the disc radially, from the inside towards the outside. Never wipe the disc in a circular motion. The reason is that the data is arranged in a spiral from the inner edge to the outer edge. If you wipe in a circular motion you might scratch the disc along its track causing consecutive data loss that results in unrecoverable errors. If on the other hand you wipe radially, any scratches that might occur are perpendicular to the data track and cause intermittent errors that are much easier to correct by the player (Taylor et al., 2009).

5.4 Digital Connections

Your source components all need to be connected to your audio system and to your video display. Today's standard is to use HDMI (High-Definition Multimedia Interface) for these connections. One popular option is to connect all your source components to your A/V controller or receiver, and use the controller or receiver to send the digital video on to the display. In such a setup, your controller or receiver acts as the grand central station of your entire system. You use it to switch between different source components.

To get the best possible quality from your source components, I recommend using only digital connections to transfer the audio and video to the A/V controller, A/V receiver, and display. Stay away from analog connections whenever you can. A modern display is inherently a digital device. If you send it an analog video signal, it will first convert it back to a digital signal to be able to process it. Since almost all source components are also digital devices, what happens is that the source component first converts its digital source material to an analog video signal, which in turn is converted back to digital by the display. These conversions are not only unnecessary, they also degrade the quality of the digital video.

A similar case can be made for digital audio. A/V controllers and A/V receivers do have analog audio inputs, but internally they often convert the analog audio signals back to digital in order to perform bass management, equalization, and other types of digital audio processing (see section 7.5). Only after such processing do they convert the digital signals to analog signals suitable for amplification. Thus, using analog audio connections for your source components results in extra conversion steps that can possibly degrade the audio quality. Some A/V controllers and A/V receivers have a special *direct mode* in which the analog audio input signals are fed directly to the analog electronics without any digitization in between. However, this mode also bypasses bass management, equalization, and all other audio processing. When you use direct mode, all digital audio processing has to be done by the source component before it converts its digital source material to the analog domain. The main drawback of the direct mode approach is that you need to configure the audio processing separately in each of your source components instead of setting it up once in your controller or receiver. Another drawback is that the audio processing capabilities of most source components are rather limited compared to the capabilities of controllers and receivers. The direct mode approach is only worth considering if the built-in digital-to-analog converters (DACs) of your source component are of a much higher quality than the ones in your controller or receiver. However, except for some specialty products aimed at the audiophile market, the converters built into source components are not of the highest quality. It makes much more sense to invest in a high-end A/V controller or A/V receiver that is equipped with excellent DACs and let this controller or receiver handle the digital-to-analog conversion for all your source components.

Beside the ubiquitous HDMI connections digital audio can also be transferred in alternative ways. The older S/PDIF (Sony/Philips Digital Interface) is still a popular option for audio-only source components. An S/PDIF connection comes in two variants: one using a single coaxial cable and another using an optical Toslink cable. Computers and computer-related devices can also transfer digital audio using a USB connection. HDMI, coaxial S/PDIF, optical S/PDIF, and USB are good choices for transferring digital audio from one device to another. I will elaborate a bit more on their properties in the upcoming sections so you can decide which one to use for each of your source components.

Wireless Bluetooth technology can also be used to transfer digital audio from one device to another. Many lifestyle products such as smartphones, wireless loudspeakers, soundbars, and automotive systems have built-in Bluetooth technology. Some A/V controllers and A/V receivers have also been equipped with this technology. However, due to the limited bandwidth of the Bluetooth connection the digital audio needs to be compressed using perceptual codecs, and some loss of sound quality is unavoidable. The best audio quality over Bluetooth is obtained with the aptX and aptX HD codecs (aptx.com), which both support digital audio up to 48 kHz sampling. The aptX codec is limited to a bit depth of 16, while aptX HD can handle a bit depth of 24. While aptX HD is marketed as providing better than CD-quality audio, bear in mind that it utilizes lossy compression techniques.

5.4.1 *HDMI*

HDMI (High-Definition Multimedia Interface) is a proprietary interface for transmitting digital audio and video between different devices (hdmi.org). It comes in different versions. Table 5.17 summarizes the video capabilities of the different HDMI versions. Higher versions have higher bandwidth capabilities that allow increased video resolutions and frame rates. Higher versions also support more multichannel audio formats and have additional features. Table 5.18 lists the most important features along with the minimum HDMI version that supports these features.

All HDMI versions support HD video (1920 × 1080). Stereoscopic 3D video is supported from version 1.4. Although the UHD video resolution (3840 × 2160) is supported from version 1.4, it doesn't support the Rec. 2020 color space used for UHD material. You need at least HDMI version 2.0 to be able to fully enjoy UHD video content. However, if you also want HDR

Table 5.17 Video capabilities of different HDMI versions (hdmi.org).

	1.0/1.1/1.2	1.3	1.4	2.0	2.1
Maximum bandwidth (Gbit/s)	4.9	10.2	10.2	18	48
Maximum resolution	1920×1200	2560×1600	4K	4K	10K
Refresh at max. resolution (Hz)	60	60	30	60	120
Video bit depth	8	12	16	16	16

Table 5.18 Supported features for different HDMI versions. The HDMI version listed is the minimum version required for the listed feature. All higher versions also support the feature (hdmi.org).

Feature	Version
Deep Color (12 bits per channel)	1.3
Stereoscopic 3D / 1080p / 24 Hz	1.4
Stereoscopic 3D / 1080p / 60 Hz	1.4b
Rec. 2020 Color Space	2.0
HDR Dolby Vision	2.0
HDR10	2.0a
HDR HLG	2.0b
LPCM 7.1 (192 kHz / 24 bit)	1.0
Dolby Digital/DTS Digital Surround	1.0
DSD (SACD)	1.2
Dolby TrueHD/DTS-HD Master Audio	1.3
Dolby Atmos/DTS:X	1.3
Audio Return Channel (ARC)	1.4
Enhanced Audio Return Channel (eARC)	2.1

capabilities, you need at least HDMI 2.0a, which supports HDR10, the mandatory format for HDR content on UHD Blu-ray discs. In addition, your video devices need to support HDCP version 2.2 (Fleischmann, 2017). HDCP stands for High-bandwidth Digital Content Protection (digital-cp.com). It is a digital copy protection technology that encrypts the digital audio and video data that is transmitted through the HDMI interface (or any other compliant interface). Only authorized devices that support HDCP are able to decrypt the audio and video data. HDCP ensures a secure end-to-end connection between HDCP compliant devices. Note that HDCP comes in different versions. While the 1.x versions are all backward compatible, the 2.x versions are not. Hence, a device that only supports a particular subversion of the HDCP 1.x will not be able to receive content from a device that uses HDCP 2.x.

Recommendation 5.25 *Make sure your display supports both HDMI 2.0a and HDCP 2.2 if you want to be able to enjoy UHD content.*

All HDMI versions support 7.1 multichannel surround sound in LPCM, Dolby Digital, or DTS Digital Surround format. The modern lossless surround formats, Dolby TrueHD and DTS-HD Master Audio, require at least HDMI version 1.3. The object-based surround sound formats, Dolby Atmos and DTS:X, also require at least version 1.3.

Many older A/V controllers and receivers do not support the latest HDMI versions. They will not be able to deal with HDR video content. If you own such a controller or receiver, you can still enjoy HDR content by connecting the source component directly to the display and use an alternative means to send the audio signal to the controller or receiver. Many modern UHD

Figure 5.16
Connecting a Blu-ray disc
player that has two HDMI
outputs. One HDMI output
is used to send the digital
video directly to the display,
while the other HDMI
output is used to send the
digital audio to the A/V
controller or receiver.

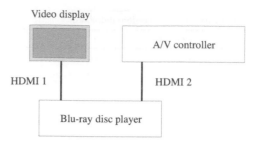

Blu-ray players have two HDMI outputs. These players allow you to use one HDMI output to send the video signal directly to the display and use the other HDMI output to send the audio signal to the controller or receiver (see figure 5.16). This setup allows for the best possible video quality, because the digital video is sent directly to the display and is not processed by the A/V controller or receiver. Many controllers and receivers are known to slightly degrade the video quality. Hence, it is better to leave them out of the loop. I explain the possible degradations that controllers and receivers can introduce in section 6.3.3 on output color spaces.

The Audio Return Channel (ARC), supported by HDMI version 1.4 and higher, can be used to send an audio signal from the display back to the A/V controller or receiver. This is a useful feature because the display itself is also a source component: it receives television broadcasts and streaming content from the Internet, and can also play media files from the local network or an attached storage device. You don't want to hear the sound of this content through the tiny built-in loudspeakers of the display. You want this sound to emerge from your high-quality front and surround loudspeakers, and that is why the sound of your display needs to be sent to your A/V controller or receiver. The ARC allows you to send the audio from the display back to the controller or receiver using the same HDMI cable that already connects the controller or receiver to the display. The ARC feature of HDMI obviates the need to run a separate cable. Not all HDMI ports on a display support ARC functionality. Typically, only one or a few are ARC compliant. On most displays the HDMI ports that support ARC are clearly labeled with the letters 'ARC'. To use them, you often also need to enable ARC and CEC (Consumer Electronics Control) in the display's setup menu. The ARC is limited in the audio formats it can pass. It supports multichannel LPCM, Dolby Digital, and DTS Digital Surround. The ARC doesn't support Dolby TrueHD, DTS-HD Master Audio, Dolby Atmos, or DTS:X. These modern surround sound formats require the enhanced ARC (eARC) feature that comes with HDMI version 2.1. If your components don't support ARC or eARC, you need to run a separate digital audio cable from your display back to your A/V controller or receiver to enjoy your display's sound through your A/V system. The cable used for this purpose is often an S/PDIF coaxial or optical Toslink cable. Figure 5.17 compares this alternative connection with the ARC enabled HDMI connection.

Die-hard audiophiles worry about *jitter* in the HDMI digital audio interface and believe that the type of interface and the cables being used can make a profound difference in sound quality. Jitter is the variation in the time instances of the samples in a digital audio signal (Dunn, 2000; Pohlmann, 2011). Jitter influences the digital-to-analog conversion process, because accurate reconstruction of the analog audio signal requires that the digital samples occur at the right times. Jitter causes the right samples to occur at the wrong times and as a result the analog waveform gets distorted. An example of jitter induced distortion is shown in figure 5.18.

The HDMI interface is notorious for its relatively high amount of jitter with timing errors as high as 4 ns. However, this is nothing to worry about, because modern well-designed DACs are practically immune to jitter. Such DACs use a buffer memory to absorb any timing variations. This buffer is filled with data as it pours in at irregular time intervals. After a slight delay, the data in the buffer is read out at an accurately controlled regular rate. In this way,

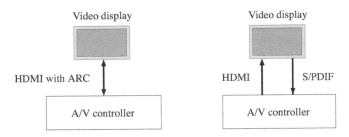

Figure 5.17
Two different ways to send the digital audio from the display to the A/V controller or receiver. *On the left,* only one HDMI cable is used. This cable carries the audio and video from the controller to the display, and it also carries the audio from the display to the controller using the ARC feature of HDMI version 1.4 and higher. *On the right,* two cables are used. One HDMI cable that carries the audio and video from the controller to the display, and one S/PDIF cable that carries the audio from the display to the controller.

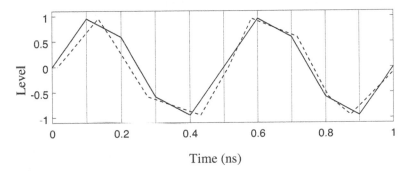

Figure 5.18
Example of waveform distortion due to jitter. The *solid line* represents the original audio signal, which is sampled at the times indicated by the *vertical dotted lines.* In the presence of jitter, the samples occur at the wrong times, some a bit earlier and some a bit later. This results in a distorted waveform that is represented by the *broken line.*

any jitter present in the incoming signal is removed (Lesso, 2006; Pohlmann, 2011). Different manufacturers use different designs and achieve different maximum levels of jitter. Modern DACs have a maximum jitter amplitude that is typically less than 2 ns (Nishimura and Koizumi, 2004). The best designs can have an overall maximum jitter level as low as 50 ps (0.05 ns) or even less (Dawson, 2011). Controlled listening tests have been performed to determine the audibility of jitter. It appears that 10 ns is the lowest level of periodic jitter that is audible (Benjamin and Gannon, 1998). Purely random jitter is only audible if it has an amplitude larger than 250 ns (Ashihara et al., 2005). Thus, the audibility thresholds for jitter are much larger than the maximum jitter typically produced by modern audio equipment. Furthermore, the low-level jitter is almost always masked by sounds in the recording or in the listening environment. This leads to the conclusion that jitter isn't audible when you use modern audio equipment under typical listening conditions (Nishimura and Koizumi, 2010).

You should choose your HDMI cables based on the maximum bandwidth that they can sustain. The higher HDMI versions support higher resolutions and frame rates, and this requires more bandwidth (see table 5.17). The different types of HDMI cables are:

- *Standard HDMI Cable:* a cable that has been tested for a bandwidth of 4.9 Gbit/s. It only supports lower resolution digital video up to 1080i and 720p.

- *High Speed HDMI Cable:* a cable that has been tested for a bandwidth of 10.2 Gbit/s. The cable of choice for HD video. It supports 1080p, stereoscopic 3D, and 4K at 30 fps.
- *Premium High Speed HDMI Cable:* a cable that has been tested for a bandwidth of 18 Gbit/s. The cable of choice for UHD video. It supports 4K at 60 fps.
- *48G HDMI cable:* a cable that has been tested for a bandwidth of 48 Gbit/s. The cable of choice for the higher resolutions and frame rates provided by HDMI 2.1.

The longer an HDMI cable, the more difficult it becomes to sustain the required bandwidth. For a length of 5 meters (16 feet) or less, there is not much to worry about. However, at longer cable lengths problems may occur, even with cables that are rated at higher bandwidths. A poor cable results in an intermittent picture or a picture with white sparkles. The cable either works or it doesn't. As long as it works, there is no difference in picture quality between the cheapest and superduper high-end cables. If you encounter connectivity problems with cables longer than 10 meters (33 feet), you may consider using an *active HDMI cable.* An active HDMI cable has built-in electronics that boost the HDMI signal to sustain it over longer lengths. Such a cable requires an external power source.

Recommendation 5.26 *For reliable transmission of HD video over HDMI, use an HDMI cable that supports at least 10.2 Gbit/s (high speed). For UHD video, use an HDMI cable that supports at least 18 Gbit/s (premium high speed).*

5.4.2 S/PDIF

The Sony/Philips Digital Interface or S/PDIF is a digital audio interface for consumer applications (IEC 60958, 2016). It is an older interconnection standard that is still widely used. It supports two-channel LPCM audio with a sampling rate up to 192 kHz and 24 bits. It also supports 7.1-channel Dolby Digital and DTS Digital Surround. It doesn't support newer surround sound formats such as Dolby TrueHD, DTS-HD Master Audio, Dolby Atmos, and DTS:X, because of its limited bandwidth. Neither does it support DVD-Audio (MLP) or SACD (DSD) due to the lack of a proper copy protection mechanism such as HDCP.

The S/PDIF interface comes in two variants: an electrical one and an optical one. The electrical connection uses a single coaxial cable with RCA connectors that carries an electrical signal with a 0.5 V peak-to-peak amplitude. To ensure proper transmission of digital audio, the cable should have a shield coverage at least 90%, an impedance of 75 Ω, and a low capacitance of about 44 pF/m or 13 pF/ft (Lampen and Ballou, 2015; Pohlmann, 2011). The optical connection uses a plastic or glass fiber to transmit the digital audio. It is commonly known by its Toshiba brand name Toslink. It is used in many mass-market audio products and is considered to be inferior with respect to its coaxial relative. The coaxial S/PDIF interface has a potential bandwidth of 500 MHz that is much higher than the 6 MHz of the optical interface. The coaxial interface is therefore less prone to cable induced jitter (Harley, 2015). However, with modern well-designed DACs, this appears to be no longer an issue. The fact that different S/PDIF cables result in different amounts of jitter that can lead to audible differences is largely a thing of the past (Harley, 2015; Pohlmann, 2011). Jitter in the S/PDIF interface is typically less than 600 ps (0.6 ns), which is well below the audibility threshold of 10 ns (Benjamin and Gannon, 1998).

5.4.3 USB Audio

On computers the USB interface (usb.org) can be used to transfer digital audio. It offers a useful alternative to the S/PDIF and HDMI interfaces. A number of modern A/V controllers and receivers are equipped with a USB audio input so you can directly connect a computer using a standard USB cable. Note that this USB audio input differs from the typical USB port that is used to connect a USB stick or hard disk to play media files. The USB audio input has a type B

Figure 5.19
Two distinct USB receptacles. *On the left*, the USB type A receptacle. *On the right*, the USB type B receptacle.

receptacle, while the other USB ports use type A receptacles. These two distinct receptacles are shown in figure 5.19. If your A/V controller or receiver doesn't have a USB audio input, but does have an S/PDIF input, you can still take advantage of the USB audio feature by using one of the many available USB to S/PDIF converters.

The use of the USB audio feature on a computer or portable device requires that the device supports a version of the USB *audio device class* (ADC). There are currently three versions: ADC 1.0, ADC 2.0, and ADC 3.0 that are backward compatible (usb.org). These device classes ensure the interoperability between different computer devices and audio equipment. Note that these classes differ from the USB 1.0, USB 2.0, and USB 3.0 designations that specify the version of the USB communication protocol that is being used. ADC 1.0 supports digital audio with 24 bits and a sampling rate up to 96 kHz. Higher sampling rates require the use of ADC 2.0 or ADC 3.0. All major computer operating systems, macOS, Linux, and Windows, natively support USB ADC 1.0. If you want to use ADC 2.0 or ADC 3.0 you may need to install an additional hardware driver for your operating system. Most Apple devices natively support USB ADC 2.0.

USB audio has different transfer modes that differ in the way they clock the digital audio signal. It is crucial to select the right one if you care about audio quality. Remember that the accuracy of the clock determines the amount of jitter and thus the accuracy of the digital-to-analog conversion. Yes, I just told you not to worry about jitter, but with computers it doesn't hurt to pay some attention to it, the reason being that many computers produce a very high amount of jitter. It is very difficult to maintain an accurate and stable audio clock inside a computer, because it continually performs loads of other tasks aside from the digital-to-analog conversion. In addition, the electronics inside a computer are tightly packed together, creating lots of electromagnetic interference (EMI) from power lines and other digital circuits. Thus, the audio clock gets easily destabilized and contaminated resulting in irregular timing.

USB audio offers three different transfer modes for transmitting digital audio from a computer device to your A/V controller or receiver (thewelltemperedcomputer.com):

- *Synchronous mode:* The computer controls the transfer rate. It derives the clock signal from the USB bus. Since the computer is always multitasking, this results in a highly irregular clock signal and thus lots of jitter.
- *Adaptive mode:* The computer controls the transfer rate, but the DAC in the A/V controller or receiver adapts its timing to the rate at which the data is pouring in. This mode is far less sensitive to jitter in the clock signal generated by the computer.
- *Asynchronous mode:* The DAC in the A/V controller or receiver controls the transfer rate. A buffer in the DAC is filled with the data coming from the computer and the DAC uses its own accurate clock signal to read the data from this buffer. Since the clock signal used by the DAC is independent of the clock used in the computer, jitter on the computer's clock doesn't influence the operation of the DAC.

Avoid the use of the synchronous transfer mode, as it can result in very high amounts of jitter. It is much better to use the adaptive or asynchronous modes. With a high-quality A/V controller or receiver, the asynchronous mode should give you the best results. The clock used to drive the DACs in high-quality audio equipment is much more stable and accurate than the clock used in a typical computer.

Recommendation 5.27 *Preferably use the asynchronous transfer mode with a USB audio interface. If this mode is not supported, use adaptive mode. Never use synchronous mode.*

You can use any decent USB cable for transmitting digital audio. It is best to keep the cable length below 5 meters. There is no need to buy an expensive audiophile-quality USB audio cable. If you use adaptive or asynchronous transfer mode with a high-quality A/V controller or receiver, the construction of the USB cable will have no influence on the final sound quality at all.

6
VIDEO EQUIPMENT

The quality of your video display matters a lot. It has a huge impact on the quality of the video images that you are reproducing: beautiful colors, deep blacks, and lots of detail and contrast. Video-wise, the display is the most important piece of equipment in your room. Besides its image quality, its size matters as well. The larger the display, the more intense the experience.

To get the most out of your video display you need to properly calibrate it. Most displays have several picture modes that provide a completely different look in terms of colors and contrast. These picture modes have fancy names like sports, daylight, game, and movie. The manufacturer has invented these modes to stand out from the competition and attract your attention in the showroom. Mostly, these modes have nothing to do with the faithful reproduction of the original material. Recall from Chapter 3 that our goal is to reproduce the video images as the artist has created and intended them. Nothing less and nothing more. Instead of the fancy settings from the manufacturer, we want our display to be accurate. You can achieve an accurate reproduction of video by performing some simple calibration steps that I will outline in this chapter.

Video quality not only depends on a properly calibrated display, it is also influenced by the video processing employed by the display and the source components. Video processing alters the original video signals, but if used correctly it can actually improve the quality of the

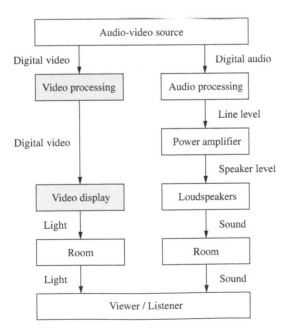

video. This is especially true with low-resolution (SD) or low-quality video. Examples of video processing that can be used to improve video are scaling to a higher resolution, converting from interlaced to progressive video, increasing sharpness, and reducing motion blur. However, too much processing can actually harm high-quality material. Therefore, it is important to properly configure the video processing capabilities of both your display and your source components. I will show you how to do exactly that in this chapter.

6.1 Video Displays

Large high-quality displays can be prohibitively expensive. Therefore, trade-offs have to be made. You basically have two choices: 1) a front-projection system or 2) a flat-panel display. A front-projection system is similar to the display used at your local movie theater. It projects its image onto a screen. These screens can be made quite large: more than 2 meters (80 inch) diagonal screen size. This is much larger than the typical flat-panel display. Flat-panel displays with a diagonal of more than 2 meters do exist, but these very large panels are also very expensive. Currently, such a large flat-panel display costs about ten times more than a front-projection system at the same screen size. Reasonably priced flat-panel displays are always smaller than 2 meters in diagonal, often much smaller: a popular size is 1.4 meter (55 inch). Although providing large screen sizes at moderate costs, a front-projection system also has its drawbacks. It provides a much dimmer picture than a flat-panel display, because its luminance output is much lower than the luminance provided by a flat-panel display. Therefore, a front-projection system has to be watched in a darkened room, otherwise the picture will be washed out by the ambient light. A flat-panel display, on the other hand, has been designed to be watched in a room with a certain amount of ambient light. You don't need to darken the room completely to be able to get a high-quality picture with lots of contrast and deep blacks. This is a major advantage of a flat-panel display.

Certain flat-panel displays and home theater projectors can be used to display stereoscopic 3D images. All practical 3D-capable displays require you to wear special glasses. These glasses work together with the display to ensure that your left and right eye receive two different images. These images slightly differ in perspective, and this difference tricks your eyes in seeing 3D depth. As of 2017 stereoscopic 3D is essentially dead in home theater (Fleischmann, 2017). The major manufacturers of flat-panel display have ceased producing 3D capable displays, and only a limited number of stereoscopic 3D projector models remain.

6.1.1 Performance

Your video display should show the video images as they were intended by the director and producers without altering or enhancing them (EBU-Tech 3320, 2017; ITU, 2009b). To be able to do that your display needs certain performance capabilities. Display manufacturers all tend to invent fancy marketing terms to highlight their products' capabilities. The terms used differ among manufacturers, which makes it difficult to compare models. It is always a good idea to consult a number of technical reviews online or in magazines to get a better and often objective view on the performance of certain models. The best reviews include technical measurements of some key performance metrics like peak luminance output, contrast, and color accuracy. Some good websites to check out are given in table 6.1. In addition, don't forget to consult the user's manual before you buy a new display. This manual can be a great help in clarifying certain points about the capabilities of the display.

Resolution

Your display should at least be able to show all the pixels of the source material. Thus, for UHD it must have 3840 × 2160 pixels (4K), and for HD it must have 1920 × 1080 pixels (2K). The vast majority of modern flat-panel displays have a 4K resolution. With projectors this is not (yet) the case. Many projectors are limited to 1920 × 1080 pixels; only a few models offer true 4K resolution. Some manufacturers use pixel shifting to simulate a 4K resolution

Table 6.1 Some websites that offer detailed technical reviews of video displays.

Website	Displays
cnet.com	Flat panels and projectors
flatpanelshd.com	Flat panels
hdtvtest.co.uk	Flat panels and projectors
hometheaterhifi.com	Flat panels and projectors
hometheaterreview.com	Projectors
projectorcentral.com	Projectors
projectorreviews.com	Projectors
rtings.com	Flat panels
soundandvision.com	Flat panels and projectors

with a 1920 × 1080 panel. This works as follows: the projector takes a 4K input signal and first displays only half the pixels, then shifts the panel a fraction of the pixel size to display the second half of the pixels. This all happens in a few milliseconds such that human eye blends the two halves together into one picture.

Recommendation 6.1 *Your display should be capable of displaying 3840 × 2160 pixels (UHD, 4K), or at least 1920 × 1080 pixels (HD, 1080p).*

It is important to distinguish between the *native resolution* of the display and its compatible *input resolutions*. The native resolution equals the actual number of pixels displayed on the screen, while the input resolution is merely the resolution of the video material that the display accepts. For example, a display with a native resolution of 1080 lines may accept a 4K input signal, but it cannot show the true 4K resolution, because it doesn't have enough pixels. Most displays can take multiple input resolutions and convert them to the native resolution of the display. The idea is that video with a resolution that is lower than the native resolution is upscaled to fill the entire display. After such upscaling it is common to perform some advanced sharpening to improve a blurry look. I explain a bit more about such video processing in section 6.3.

The higher the native resolution, the less likely that you see the pixel structure. With HD and UHD resolution displays, you have to sit very close to be able to see the individual pixels. At normal seating distances this should not be a problem, and the individual pixels will blend together in a smooth continuous image (see also section 4.2.2). A few years ago the pixel structure was sometimes visible on lower resolution projectors. The term coined for this visible artifact is the *screen-door effect*. With modern displays this effect is largely a thing from the past.

Contrast

Contrast is one of the most important aspects of image quality. High-contrast pictures look lifelike and portray a certain depth. Contrast is the ratio between the highest and lowest luminance that a display can deliver. Recall from section 3.2.1 that the peak luminance of a flat-panel display is much higher than the peak luminance of a home theater projector. Modern flat-panel displays have absolutely no trouble reaching the desired peak luminance of 100 cd/m² for SDR content (see table 3.2). Most displays marketed as HDR-capable can go higher than 400 cd/m², but not all HDR displays can achieve the recommended target of 1000 cd/m². Currently, no projector can go high enough to present HDR material in a way that a flat-panel display can. A projected image from a home theater projector is always much dimmer; the target for the peak luminance of a good projector is only 38–58 cd/m². Given such a limited luminance output, contrast in a projected image is largely determined by the lowest luminance that is used for the black parts of a picture. Ambient light and stray light from other lighter parts of the projected image conspire to make the darkest black gray. This ruins the achievable contrast. With a projection system it is quite a challenge to achieve sufficient contrast (see section 4.3.1). In a rather dark room, there are some differences in achievable contrast that depend on the particular technology used in the projector. Some projectors are able to produce blacker blacks than others. I will come back to these differences in section 6.1.3 on projector technology.

Recommendation 6.2 *Your flat-panel display should be capable of achieving a peak luminance of 100 cd/m² for SDR content and 1000 cd/m² for HDR content. When you use a projector as your display, it should at least be capable of achieving a peak luminance of 38–58 cd/m².*

Flat-panel displays can produce contrasty images even in the presence of some ambient light. They have a big advantage over projection-based systems, because they can output a higher peak luminance. Still, the black levels in a display are relevant to achieve good contrast. Again, the particular technology employed in the display determines how dark the darkest black on the screen can be. I will discuss the differences when explaining the different flat-panel display technologies in section 6.1.2.

Given a certain peak luminance, the minimal requirements for black level can easily be derived from the recommended contrast values in table 3.2. You simply divide the peak luminance by the desired contrast value. This yields the maximum allowable black level. For example with a peak luminance of 100 cd/m² and a recommended sequential contrast ratio of 2000:1, the maximum allowable black level becomes 0.05 cd/m². This is the recommended black level for displaying both SDR (EBU-Tech 3320, 2017) and HDR video (VESA, 2017), although for HDR also the more difficult to achieve 0.005 cd/m² is recommended (EBU-Tech 3320, 2017; ITU-R BT.2100, 2017). The lower your black level, the more contrast you get. Note that manufacturers often state ridiculously high-contrast ratios. This is a marketing gimmick. They have come up with contrast measurement methods that have nothing to do with the conditions in which you typically watch the display. You are better off making your own contrast measurements (as I explain in sections 4.3.1 and 6.2.3) or rely on measurements provided in a technical review from one of the websites listed in table 6.1.

Recommendation 6.3 *To achieve a high contrast ratio, the blacks on your display should have the lowest possible luminance. Aim for 0.05 cd/m² or lower.*

Color

Accurate color reproduction is another quality aspect to look for in a high-quality display. Displays differ in the number of colors that they can reproduce. There are two important aspects to be aware of. First, the bit depth of the colors, and second, the size of the color gamut. Modern displays, especially all HDR-capable ones, use 10-bit colors. This should be more than sufficient, because HD video material only uses 8 bits (ITU-R BT.709, 2015) and UHD Blu-rays are limited to 10 bits (BDA, 2015b), although the UHD video specifications also allow for 12 bits (ITU-R BT.2020, 2015).

All flat-panel displays and projectors can display the colors of the Rec. 709 color gamut (ITU-R BT.709, 2015). This gamut includes all the colors that can occur in SD and HD video, including all DVDs and Blu-ray discs (see also table 3.6 and figure 3.12). The UHD video specification allows for a much wider color gamut, the Rec. 2020 (ITU-R BT.2020, 2015). Currently, no display can reproduce all the colors in this very wide gamut. For the time being, we lack the technology to reproduce all these colors. Most modern HDR-capable displays can display 90% or more of the DCI-P3 color gamut (VESA, 2017). This gamut is considered to be an intermediate step between the Rec. 709 and Rec. 2020 gamuts. Wide color gamut material released on UHD Blu-ray is almost always within the DCI-P3 color gamut, because this is the standard gamut of the movie industry. While the use of Rec. 2020 is possible according to the UHD Blu-ray specifications, it is not likely to be used. Of course, this may change in the future as display technology advances.

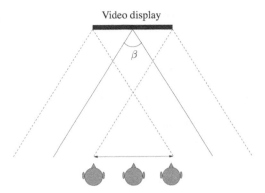

Figure 6.1
The viewing angle β of a display is usually defined as a cone that covers both sides of the display. Viewers who watch the display at a larger angle than β will see a deteriorated picture that lacks contrast and suffers from color shifts. The viewing angle is defined for the middle of the screen, but also the two sides of the screen must be viewed from an angle smaller than β. Therefore, all viewers must be within the area indicated by the *double-sided arrow* between the two *dashed lines*.

> **Recommendation 6.4** *Your display should at least be capable of displaying all the colors in the Rec. 709 color gamut with 8-bit precision, and for UHD and HDR all the colors within the DCI-P3 gamut with 10-bit precision.*

Viewing Angle

The image quality of a video display varies with the angle from which you view the screen. This angle is called the *viewing angle* and is usually defined as a viewing cone that covers both sides of the display, as illustrated in figure 6.1. Note that although the name is similar, this angle differs from the viewing angle α discussed in section 4.2.2. For a flat-panel display the amount that the image changes with viewing angle depends on the specific type of panel that is used. For a projection system it solely depends on the reflective properties of the projection screen, and not on the projector. You get the best picture quality when you are straight in front of the display. If you sit farther towards the side, the luminance drops, contrast starts to fade, and colors get distorted. At extreme angles colors can even start to change so much that they seem inverted. A large enough viewing angle is important, as most displays are viewed with several audience members seated next to each other.

The viewing angle specified by a display manufacturer is the maximum angle at which the image quality is still acceptable. Unfortunately, there is no standardized definition of what is deemed to be acceptable; it varies among manufacturers. Some specify the viewing angle as the angle at which the luminance has halved, while others use a certain drop in the contrast ratio. With modern flat-panel displays viewing angles can almost approximate a full 180° cone in the horizontal plane. In the vertical plane, the viewing angle may still be restrictive. You should not rely on the specifications of the manufacturers to get an idea of the acceptable viewing angle; it is best to actually have a look at the display yourself. In general, if you are seated within a viewing cone of 70° you should not encounter any problems.

6.1.2 Flat-Panel Displays

Flat-panel displays are the television sets of today. There are currently two different display technologies being used for flat panels: LCD (Liquid Crystal Display) and OLED (Organic Light Emitting Diode). Flat-panel displays have completely replaced the older Cathode Ray Tube (CRT) and rear-projection TVs. Plasma panels are also a thing of the past.

A flat-panel display consists of a matrix of pixels in which each pixel can be activated by a small switch. When the switch is opened it supplies a voltage to the pixel, which causes an optical effect. The switches used in a flat-panel display are thin-film transistors (TFTs). Both LCD and OLED displays use a matrix of TFTs to activate the pixels. The pixels themselves do differ. The LCD pixels work like a light valve: in order for them to work, they need to be backlighted. Only when the LCD pixel is turned on does it allow the backlight to pass through, otherwise it blocks the light. By contrast, a pixel in an OLED display is a light source by itself (Lee et al., 2008).

Flat-panel displays no longer need to be completely flat. Since 2014 they can also be slightly curved. These curved screens are sometimes claimed to be superior to their flat-panel counterparts. Well, it turns out that this claim is not true. A curved screen does nothing to improve picture quality. Proponents of curved displays say that these screens are more immersive. However, the curve is only very shallow, so at a typical viewing distance (see section 4.2.2) the field of view is almost the same as that of a perfectly flat display. The angular difference in the field of view is too small to have any noticeable effect on your immersive experience. As a matter of fact, a curved display may actually slightly degrade the picture quality. It introduces a small geometric distortion: a slight bowed effect on the top and the bottom. And, if you are not sitting directly in front of the screen, but next to the prime seat, you will also see a slight foreshortening of the image. Finally, curved screens reflect room light and windows in a strange distorted fashion. That being said, it is of course always better to avoid reflections off any display, flat or curved, as much as possible by closing the curtains, dimming the lights, and darkening the room.

LCD

In a Liquid Crystal Display (LCD), each pixel consists of a liquid crystal that acts as a light valve. It can pass or block light, based on the *polarization* of the light. To understand polarization, you need to know that light is an electromagnetic wave that oscillates perpendicular to the direction of travel. Unpolarized light oscillates randomly in all these perpendicular directions. By contrast, polarized light oscillates only in certain restricted directions. The liquid crystal is backlighted with polarized light. Varying the voltage applied to the pixel changes the alignment of its liquid crystals and allows polarized light to pass. The amount of light that passes is controlled by the voltage. To produce colors, each pixel is divided into three subpixels that have a red, green, and blue color filter. These subpixels are small enough for the human eye to see them as one (color) pixel. By varying the amount of light that passes through the red, green, and blue color filters, the display is able to produce all necessary colors (Brawn, 2015; Lee et al., 2008). Recall from section 2.2.2 that because of the trichromacy theory of color vision, a large range of colors can be created by mixing only the three primary colors red, green, and blue.

When a still image is displayed for a prolonged period of time, say several hours in a row, *image retention* may occur on an LCD. The displayed image may remain visible as a residual image even when the display is showing other video material. Image retention occurs, because over time electrically charged particles collect at the electrodes of the liquid crystal and cause a parasite electrical field that keeps the crystal from realigning. In most cases, image retention is only temporary and the residual image will disappear after the display has been shut down for several hours. Do not display fixed images, like a menu screen, for prolonged periods of time, and use the screensavers built into your display or source component to avoid image retention.

Modern LCDs use Light Emitting Diodes (LEDs) as their backlight. Older displays used Cold Cathode Fluorescent Lamps (CCFLs), which resemble a thin fluorescent tube. The use of LEDs provides numerous advantages over CCFLs. They have fast turn-on and -off times, high brightness, and a long lifetime (more than 50 000 hours). They are less harmful to the environment, because contrary to CCFLs they contain no mercury, and they are more energy efficient (Brawn, 2015; Lee et al., 2008). LCD televisions that use LEDs as their backlight are often referred to as *LED TVs*. This is a bit of a misnomer, because they are in fact LCD TVs with a LED backlight. They are not to be confused with an OLED display, which forms its image in a completely different way than the LCD does.

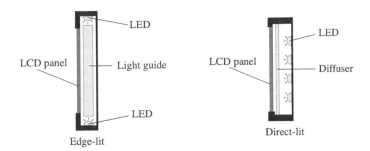

Figure 6.2
Two different backlight constructions for an LCD display. *On the left*, edge-lit panel where the LEDs are placed in a row along the edges and their light is distributed through a light guide behind the panel. *On the right*, direct-lit panel where the LEDs are directly behind the panel.

The backlight of an LCD can be constructed in two different ways: 1) *edge-lit*; or 2) *direct-lit*, also known as *full-array* (Bodrogi and Khanh, 2012; Fleischmann, 2017; Lee et al., 2008). In an edge-lit design, the LEDs are placed in a row at one or more edges of the screen. A *light guide plate* is placed behind the LCD panel to diffuse the light and make it more uniform across the entire screen surface. Figure 6.2 shows a side view of an edge-lit display. By contrast, in a direct-lit design a full array with several rows of LEDs is placed behind the LCD panel. Again, a diffusing plate is used to distribute the light across the screen. This construction is also illustrated in figure 6.2. The main advantage of an edge-lit design is that the cabinet can be made slimmer and lighter. Another advantage is that it is more energy efficient than a direct-lit display, because it requires fewer LEDs. However, an edge-lit design may suffer from screen uniformity problems; certain parts of the screen may look brighter than others, especially the edges and the corners. It is much easier to achieve a uniform light distribution across the entire screen with a direct-lit design, because the LEDs are more evenly distributed over the entire screen surface. But there is another advantage of a direct-lit design that is even more important: it can have better contrast.

The contrast performance of an LCD can be drastically improved by *locally dimming* some of the LEDs of its backlight (Bodrogi and Khanh, 2012; De Greef and Hulze, 2007). The idea is that the dark parts of the image are made even darker by dimming the LEDs that illuminate these parts of the screen. The reason that local dimming is important is because liquid crystals always leak some light, even when they are turned off. This light leakage raises black level and reduces contrast. Since contrast is one of the most important image-quality aspects, local dimming can make a huge impact on the perceived image quality. However, local dimming has to be done in the right way or it will generate unwanted *blooming* artifacts. Blooming is a term used to describe the appearance of soft light halos around small bright objects. What happens is that the light of the LED that illuminates the bright object of the image spills into the darker area surrounding the object. Blooming can be completely avoided by using a direct-lit display that has one LED for every pixel in the display. In this way, local dimming can be done on a pixel-by-pixel basis, and the light can be controlled very precisely. However, the cost of the display would become prohibitive. Therefore, manufacturers divide the screen up into a number of zones that consist of blocks of several pixels and use one LED for each zone. Obviously, the more zones a display has, the more precisely the light can be controlled and the less likely you are to be bothered by blooming artifacts. Unfortunately, manufacturers almost never specify how many zones they use in a particular model. Sometimes you can find the number in a technical review. Obviously, the best local dimming results are obtained with a direct-lit display having many independently controlled zones. Edge-lit displays also use local dimming, but the amount of control is limited to bands across the screen.

Recommendation 6.5 *To achieve high contrast with minimal blooming on a flat-panel display choose a direct-lit full-array LCD that features local dimming with many independently controlled zones.*

Local dimming should not be confused with a technique called *dynamic contrast*. With local dimming the individual LEDs in the backlight are controlled based on the dark and light areas in the picture, while with dynamic contrast all the LEDs are dimmed at once based on the average brightness of the picture. So, a predominantly dark picture is made even darker by dimming the entire screen, including the smaller brighter areas. Dynamic contrast improves sequential contrast but does nothing to improve simultaneous contrast (see section 3.2.1). While local dimming significantly improves the image quality, dynamic contrast can mess up the tonal gradations such that they no longer follow the ideal gamma curve (see figure 3.9). Therefore, if your display provides a dynamic contrast feature it is best to switch it off.

The type of LEDs being used in the display influences the achievable color gamut; *RGB LEDs* and *quantum dots* are currently being used in top-tier displays to be able to increase their color gamuts to the DCI-P3 gamut. An RGB LED combines red, green, and blue light such that the combined light appears white. A quantum dot is a microscopic particle that glows a certain color when hit with the light from an LED backlight. Both RGB LEDs and quantum dots can be used to specifically create the right kind of red, green, and blue light that is needed for the three primary colors that the pixels are made of. Which specific technique is used in your display doesn't matter, as long as the display can show more than 90% of the colors of the DCI-P3 gamut. The best way to find out is to have a look at the measurements accompanying a good technical review of the display (see table 6.1).

OLED

In an OLED display every pixel is made up of several small Organic Light Emitting Diodes (OLEDs). These OLEDs are light sources by themselves and are used to produce colors. As with an LCD, each pixel consists of red, green, and blue subpixels that are small enough to blend together. In an OLED display a fourth completely white subpixel is added to increase the luminance output of the display. So, every pixel consists of four self-illuminating subpixels. Hence, there is no need for a backlight (Brawn, 2015).

OLED displays are capable of excellent picture quality. They have a wide viewing angle and good screen uniformity. Their biggest advantage is that they have excellent contrast ratios. Since every pixel is a light source that can be completely turned off, there is no light leakage as with backlit LCDs. By turning the OLED pixel off, it can reach pure black and the contrast ratio becomes infinite. In practice, the contrast ratio will be large, but not infinite due to the ambient light in the room and light from the display that is reflected back to the screen. In terms of contrast, OLED displays outperform the best LCDs with full-array local dimming. Currently, OLED displays offer the best possible picture quality. However, they tend to be much more expensive than their LCD competitors due to the high manufacturing costs. Another slight disadvantage of OLED displays is that they cannot put out as much light as an LCD can. The peak luminance of an OLED display is much lower, but some top of the line models can reach the minimum 600 cd/m^2 needed for rendering HDR content (see table 3.2).

Recommendation 6.6 *To achieve high contrast without any blooming on a flat-panel display choose an OLED model.*

6.1.3 Front Projectors

A projector is still the way to go if you want a very large screen. A projector allows you to reproduce a real theater experience at home, often at a surprisingly low cost compared to the largest flat-panel displays. But, you really must watch your projection system in a darkened environment, otherwise the blacks on the screen will appear washed out and contrast will be way too low.

A projection system consists of two parts: a projector and a projection screen. Both matter quality-wise. Never project your image onto an arbitrary surface, like a white-painted wall. Always use a dedicated projection screen to obtain a high-quality image.

As its name suggests a projector projects the video image onto a screen. It uses a lamp to generate the light, one or more panels with a matrix of pixels to generate an image, and an optical lens to focus the projected image on the screen. The image quality of a projector depends to a large extent on its three main parts: the lamp, the panels, and the lens. Most projectors use three different panels, one for each primary color. This requires two additional parts: a color separator to split the light from the lamp into a red, green, and blue component, and a color combiner to construct one color image from the light coming from the three separate panels. Figure 6.3 shows this basic setup with three *transmissive panels*. Transmissive panels act as light valves. Their pixels can either pass the light or block it. Another option, used by many projectors, is to use *reflective panels*. The pixels in a reflective panel either reflect the light or don't. The use of three reflective panels is illustrated in figure 6.4. Transmissive panels are based on LCD technology. Reflective panels are either based on Liquid-Crystal-on-Silicon (LCoS) or on Digital Light Processing (DLP) (Brennesholtz and Stupp, 2008).

As with flat-panel displays, the panels used in a projector consist of multiple rows and columns of pixels. Each pixel can be independently activated by a TFT that sits next to the pixel and acts as a switch. The pixels modulate the light coming from the projector's lamp. The lamp always produces its maximum luminance. The individual pixels attenuate this maximum light such that an image can be formed. How well the panels can block the light is one of many factors that determine how dark the blacks in the projected image can be. Different types of panels produce different black levels. Therefore, the type of panel used in the projector determines how much contrast can be achieved with a certain type of projector.

Besides the panels, the lens used in the projector also affects image quality. High-quality projectors use a lens that consists of multiple coated glass and plastic elements that work together to reduce geometric image distortion, preserve contrast, and avoid *chromatic aberrations*. Chromatic aberrations occur if a lens fails to focus all the different colors at the same convergence point. These aberrations show up as colored fringes along lines and shapes, especially in the corners of the image.

In projectors with three panels, it is crucial that the panels are properly aligned. Any misalignment shows up as color fringes along lines and shapes in the image, because the red, green, and blue colors do not overlap properly. Misalignment is easy to spot by displaying a test pattern that consists of a grid of white lines on a black background. Such a pattern can be found on the AVS HD 709, DVE HD Basics, and HD Benchmark video test discs (see appendix

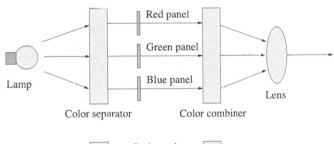

Figure 6.3
Schematic diagram of a projector that uses three transmissive panels.

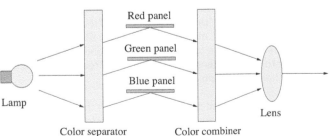

Figure 6.4
Schematic diagram of a projector that uses three reflective panels.

Table 6.2 Test patterns for aligning the panels in a three-panel projector.

AVS HD 709	Misc. Patterns—Convergence	Crosshatch with Circles
		Horizontal Convergence
		Vertical Convergence
		Mixed Convergence
DVE HD Basics	Basic Patterns	Geometry
HD Benchmark	Advanced Video—Setup	Geometry

A and table 6.2). Typically, misalignment is more pronounced near the edges of the screen due to a slight optical distortion in most projectors. Misalignment problems are not to be confused with chromatic aberrations. Both create color fringes, but it is easy to distinguish between them by looking at two opposite corners of the screen. If the fringes in the other corner are in the opposite direction, it is a chromatic aberration problem. If they are in the same direction, it is a panel misalignment. Many projectors have a panel alignment feature in their setup menu. You can use this feature to fine-tune panel alignment, in case you experience any color convergence problems. However, sometimes perfect alignment is simply not possible, because of slight differences in optical path lengths between the panels and the lens. Any magnification or distortion difference between the three optical paths will make it impossible to achieve perfect alignment everywhere on the screen (Brennesholtz and Stupp, 2008).

Recommendation 6.7 *If you have a three-panel projector, check if the panels are correctly aligned using a test pattern that consists of a grid of white lines on a black background. Use the projector's setup menu to adjust the panels if necessary.*

LCD

Liquid Crystal Display (LCD) projectors (www.3lcd.com) use panels in which each pixel is made of liquid crystals. There are three separate panels, one for red, one for green, and one for blue. Dichroic mirrors are used to direct the light to the panels and also to filter out all unwanted color and only pass the red, green, and blue that is appropriate for each panel. After passing through the panel the three colored light beams are combined using a prism and leave the projector lens as a full-color image (Brawn, 2015; Brennesholtz and Stupp, 2008).

LCD projectors use transmissive panels: their pixels act as light valves. Similar to flat-panel LCDs, the liquid crystals cannot completely block out the light. There is always some residual light leakage. This is the main disadvantage of LCD projectors. They cannot produce blacks very well and are thus limited in the amount of contrast they can provide. On the plus side, LCD projectors are typically more affordable than LCoS and DLP projectors.

LCoS

Liquid-Crystal-on-Silicon (LCoS) projectors also use panels in which each pixel is made of liquid crystals. Unlike an LCD projector, the LCoS type uses reflective panels. The liquid crystals are placed upon a reflective metallic mirror. The light from the projector lamp passes twice through the liquid crystal layer: once to reach the mirror, and once to reflect off the mirror and leave the projector through the lens. Again, there are three separate panels, one for each primary color (Brennesholtz and Stupp, 2008).

LCoS panels can be made smaller and of a higher resolution than the traditional LCD panels. The LCoS pixels can be smaller, because they have a fill-factor of about 90% instead of the 65% in an LCD panel. So, almost no space is lost between the pixels and the projected image appears to be seamless (Lee et al., 2008). While on older LCD projectors you could be plagued by the screen-door effect, there is absolutely no need to worry about it if you use an LCoS projector.

LCoS projectors are better at reproducing black levels than LCD projectors, mainly because the light passes twice through the liquid crystals. As a result LCoS projectors deliver better contrast. However, LCoS projectors tend to be more expensive than their LCD competitors. There are only a few manufacturers of LCoS-projection systems. The two major ones are JVC (jvc.com) and Sony (sony.com). JVC uses the term D-ILA (Direct Drive Image Light Amplifier) to refer to its LCoS projectors, while Sony calls them SXRD (Silicon X-tal Reflective Display).

> **Recommendation 6.8** *Choose an LCoS projector over an LCD projector to get better black levels and increased contrast.*

DLP

Digital Light Processing (DLP) projectors use reflective panels in which each pixel is a tiny mirror that can be flipped between two different positions. These panels are fabricated from crystalline silicon and are known as DMDs, which stands for Digital Micromirror Devices or Deformable Mirror Devices (Brennesholtz and Stupp, 2008; Hornbeck, 1983). Each micromirror on the panel can be individually controlled. In one position the micromirror reflects the incoming light towards the projector's lens, while in the other position it reflects the light away from this lens. The brightness of the reflected light that reaches the lens is varied by flipping the mirror at different high rates. As long as the flipping rate is high enough, the human eye cannot distinguish the individual flashes of light, and the different rates create the impression of varying amounts of light (Brawn, 2015; Brennesholtz and Stupp, 2008).

DLP projectors can either use one or three DMD panels. The projectors that employ three panels are rather heavy and expensive. They are mainly used in large venues, such as movie theaters. Most DLP projectors intended for home theater use have only one DMD panel. Single-panel projectors are much cheaper and have another big advantage: they do not suffer from panel misalignment problems (Brennesholtz and Stupp, 2008).

How does a single-panel DLP projector generate a full color image? The basic idea is to display the three primary colors in rapid succession. First, the panel is illuminated by red light and the pixels on the panel are activated according to the red components in the image. Then the panel is illuminated by green light and the pixels change to represent the green component. And finally, the same thing happens for the blue component of the image. As long as the three colored images are presented in rapid succession, the human eye fuses them into one full-colored image.

In most single-panel DLP projectors a color wheel is used to generate the red, green, and blue light. A color wheel is a translucent disc that sits between the projector's lamp and the DMD panel. It has multiple color segments, often two red, two green, and two blue ones. These six segments correspond to two cycles of red, green, and blue colored light. The color wheel rotates permanently and its rotation speed is synchronized with the flipping of the micromirrors on the DMD panel (Brennesholtz and Stupp, 2008).

DLP projectors are much brighter than LCD and LCoS projectors, because the light does not have to travel through a liquid crystal layer; it reflects directly off the micromirrors. DLP projectors can also deliver deep blacks, because in the off position the micromirrors do not reflect any light towards the projector's lens and consequently there is no light leakage. As a result DLP projectors offer the best possible contrast ratios, compared to LCD and LCoS projectors (Brawn, 2015; Texas Instruments, 2013).

Recommendation 6.9 *Choose a DLP projector over an LCoS or LCD projector to get better black levels and increased contrast.*

However, a small number of people might experience an unwanted visual artifact when viewing a DLP projector that uses a color wheel. This artifact is called the *rainbow effect* or *color breakup*. People who are sensitive to this effect will see flashes of color when they look across the display from one side to the other. These flashes appear most often when bright objects on a mostly dark background move across the screen. As the eye tracks the moving object one color slightly leads the eye, another is centered at the point of the eye's gaze and the third slightly lags the eye (Brennesholtz and Stupp, 2008). Some people perceive the rainbow artifact frequently, while others may never see it at all. Extremely high frame rates (more than 1500 Hz) are needed to totally eliminate the rainbow effect. At a frame rate of 180 Hz, most people no longer notice the rainbow effect (Brennesholtz and Stupp, 2008). The appearance of the rainbow effect can be further reduced by using a color wheel that in addition to the red, green, and blue primary colors, also has segments that correspond to the secondary colors yellow, cyan, and magenta. As an added benefit, the use of such a color wheel also boosts the overall color intensity. Projectors that use such color wheels are marketed as having 'BrilliantColor' technology (Brawn, 2015; Texas Instruments, 2013). Another way to get rid of the rainbow effect is to replace the lamp and the color wheel by three separate LED light sources (one for red, one for green, and for blue) and flash these LEDs sequentially in time. Using LEDs has the additional benefits of lower energy consumption, longer lifetime, and the potential for a larger color gamut. Finally, also note that three-panel DLP projectors do not suffer from the rainbow effect, because they do not utilize a color wheel.

Recommendation 6.10 *To avoid the rainbow effect with a DLP projector, choose a single-panel model that has either a BrilliantColor color wheel, or uses three separate LED light sources for the primary colors. Or alternatively, use a DLP projector that has three separate panels, one for each primary color.*

Projector Lamps

Projectors require a light source that is capable of providing an extremely high brightness. Most projectors use a *high-intensity discharge* (HID) lamp, either an *ultrahigh-performance* (UHP) lamp or a *Xenon lamp*. The UHP lamp dominates the home theater projection market. UHP lamps are available in versions from 50 W up to about 900 W. They have a long lifetime that can exceed 10 000 hours. However, they do not provide pure white light. Their color spectrum contains lots of blue and green, but only very little red. It can be a challenge to achieve accurate color reproduction with a UHP lamp. It requires the absorption of the dominant blue and green light, and this will typically reduce the overall intensity of the light. The Xenon lamp is mostly used in very high-power digital cinema applications, and much less in home theater projectors. The lamps typically use 7000 W electrical power or more. They are very expensive and typically have a shorter lifetime than UHP lamps. Nevertheless, their big advantage is that they can provide purer white light. Their color spectrum is nearly flat across the visible range (Bodrogi and Khanh, 2012; Brennesholtz and Stupp, 2008; Derra et al., 2005).

Projector lamps have a very low efficiency. Only part of their electrical power is converted into light, and the remainder is dissipated as heat. Consequently, projectors can get very hot. They are equipped with a mechanical ventilation system that blows air through the projector's casing and prevents the lamp from getting too hot and failing. The fans used to cool the lamp can make a lot of noise. Fan noise can be a nuisance and distract you from the movie, especially during the quieter scenes. In section 4.3.2, I recommended that the total amount of ambient noise in your room should stay below 35 dBA. It is therefore a good idea to select a projector that does not produce more than 30 dBA. Most projectors that have been designed specifically for home theater use satisfy this requirement.

Recommendation 6.11 *Choose a projector that does not produce more than 30 dBA fan noise.*

In practice, a UHP lamp does not last the full 10 000 hours, because its intensity is reduced considerably after about 2500–4000 hours (Bodrogi and Khanh, 2012). So lamp life can be an issue when you have a projector. You might need to replace the lamp multiple times, before your projector is at the end of its life. Lamp life is often specified in the number of hours until the light output is reduced to 50% or the time until the lamps fails, whichever comes first (Brennesholtz and Stupp, 2008). It is often a good idea to replace the lamp after about 2000 hours, because after that it may no longer deliver sufficient light output. Besides becoming less bright with age, the lamp also shifts color. If you want your projector to deliver accurate colors over time, you need to calibrate it regularly. I explain how to calibrate your display, both flat panels and projectors, in section 6.2.

Recently, projectors that use LEDs or solid-state lasers as their light source instead of an HID lamp have appeared on the consumer market. LEDs and lasers have a much longer life span, more than 50 000 hours, which is similar to the life span of a flat-panel display. They are capable of excellent color rendition, and they do not shift color as they age.

6.1.4 Projection Screens

The projection screen reflects the light from the projector towards the viewers. To achieve good image quality it is critical that you to choose a high-quality screen. Never just project your image on any surface that is available. Always use a projection screen. A high-quality screen has a neutral color such that all the colors of light are reflected equally. A high-quality screen also reflects the light uniformly across its entire surface without any locally brighter or darker areas. And finally, a high-quality screen doesn't have any visible grain structure when viewed from an appropriate viewing distance.

The *gain* of the screen is a measure of how reflective the screen is. A screen having a gain of 1.0 reflects the light equally in every direction. A screen with a gain higher than 1.0 reflects the light back to the viewer over a narrow angle. The light that would be reflected into the other directions is all concentrated in this narrow angle, and as a result the screen looks brighter. The higher the gain, the narrower the angle of reflection and the brighter the screen. Too high a gain results in a very narrow angle of reflection. A very narrow angle of reflection is not desirable for two reasons. First, the viewers who are not centered in front of the screen, but seated further towards the sides will see a much dimmer picture than the person sitting in the prime seat directly in front of the screen. In other words, a high gain screen has a limited viewing angle. Second, the person sitting in the prime seat may experience *hotspotting:* some parts of the screen, especially the center, look brighter than other parts (Brennesholtz and Stupp, 2008; Rushing, 2004).

For home theater use, a screen with a gain of 1.3 or lower is recommended (Harley, 2002). Besides having the advantage of slightly increasing the brightness of the projected image, such a screen would also reject some of the ambient light coming from the sides. This can potentially increase contrast. To further improve on the rejection of unwanted ambient light, screens can also be made angular selective in their reflection. For example, the screen may be constructed to reflect the light coming from the projector, but reject as much as possible the ambient light coming from other directions. These *ambient light rejecting* (ALR) screens improve both black level and contrast, but they only work when the projector and the audience are within certain specified angular ranges (Brennesholtz and Stupp, 2008). Another option to improve black level is to use a gray screen. Such a screen reflects less ambient light. However, it also reflects less light from the projector. Gray screens have a gain lower than 1.0, because they attenuate all light. They require quite powerful projectors, and as such it is often wiser to stay with a traditional white screen.

The maximum brightness of the image on the projection screen depends on three factors: 1) the gain of the screen, 2) the size of the screen, and 3) the maximum light output of the projector. Obviously, increasing the gain of the screen or the light output of the projector also increases maximum image brightness. On the other hand, increasing the screen size decreases the maximum image brightness. With a larger screen, the light is spread out over a larger surface area and hence the image gets dimmer. A larger screen requires a more powerful projector to be able to reach the target screen luminance of 38–58 cd/m² (see table 3.2). The light output of a projector is specified in *ANSI lumen* (lm). The lumen is a measure of *luminous flux*. Home theater projectors typically have a light output in the range of 500–2500 lm (Brennesholtz and Stupp, 2008). Figure 6.5 shows the amount of lumen that you need to reach the target luminance on screens of different sizes. With a high gain screen you need less lumen from your projector to reach the same maximum image brightness. Figure 6.5 shows the amount of lumen that you need for both a 1.0 and 1.3 gain screen. If you prefer to perform your own calculations, you can use the formula given in box 6.1.

Recommendation 6.12 *Make sure that your projector provides enough lumen output for the size of your projection screen. The peak luminance on the screen should be at least 38–58 cd/m².*

Unfortunately, there is no mandatory standard for specifying the lumen output of a projector. This makes it difficult to compare the lumen specification for projectors made by different manufacturers. There are two complicating factors. First, the color temperature (see figure 3.13) at which the lumen output is measured has a huge influence on the outcome. Most projectors have trouble reproducing red. Out of the box, their white point is bluish with a color temperature approaching 9300 K. For accurate color reproduction, we need the color temperature to be 6500 K. Calibrating the projector (as I explain in section 6.2) to a color temperature of 6500 K effectively removes some of the blue and green light from its output and as a result its rated lumen output will be lower. The difference between the calibrated lumen output and the one specified by the manufacturer may differ about 20% (Brawn, 2015) and sometimes even 50% (projectorreviews.com). It is best not to rely on the lumen specification provided by the manufacturer. Instead check out a good technical review of the projector in which the lumen output after calibration to 6500 K has been measured. Some websites to look for such reviews are given in table 6.1. The second factor complicating the lumen specification is the

---

BOX 6.1 PROJECTOR LUMEN OUTPUT

The screen luminance L_v (cd/m²) obtained from a projector with a lumen output Φ_v (lm) is given by (Brennesholtz and Stupp, 2008).

$$L_v = \frac{\Phi_v}{\pi A} R$$

where A is the surface area in m² and R is the gain of the screen. Expressing A in terms of the diagonal d (m) of a 16:9 screen, and rewriting the equation yields:

$$\Phi_v = \frac{\pi L_v}{R} \cdot \frac{d^2}{\frac{9}{16} + \frac{16}{9}} \approx 1.34 \cdot \frac{L_v d^2}{R}$$

You can use this formula to compute the light output Φ_v required from the projector to produce a maximum luminance of L_v on a 16:9 projection screen with diagonal d and gain R.

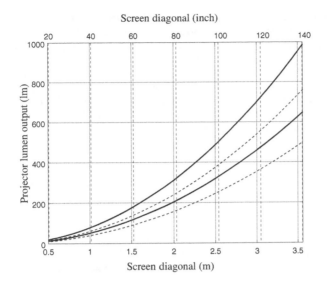

Figure 6.5
Required projector lumen output as a function of the screen diagonal (in meter or inch). To achieve a desired luminance of 38–58 cd/m² the projector lumen output should be between the *solid lines* for a screen with gain 1.0 and between the two *broken lines* for a screen with gain 1.3.

aging process of the projector's lamp. With time the lamp will get less bright. The largest drop in brightness often occurs after the first 100 hours of use. The remaining brightness will then stay at about 75% of the initial value for quite some time. You might want to take that into account to avoid the screen luminance dropping below 38 cd/m² after only a few hours of use. Eventually, after 1000 hours or more, you may need to replace the lamp, because it becomes too dim. But, you sure don't want that to happen after only 100 hours of use. So, take into account some safety margin when determining the required lumen output of your projector.

Recommendation 6.13 *Do not rely on the specifications of lumen output provided by the projector's manufacturer. The lumen output after calibration to a white point of 6500 K is often much lower (20–50%).*

Your projection screen should be perfectly flat and remain flat. If it isn't, you will experience geometric distortions that will look especially awkward when you watch a scene in which the camera pans. The easiest way to have a perfectly flat screen is to install a fixed flat screen in which the screen material is stretched and fastened to the edges of an aluminum frame. However, not everyone wants a permanently installed fixed screen. A retractable screen can be an attractive option, because you can get it out of the way when it is not in use. Such a screen can be pulled down like a window shade, and rolled back up when you no longer need it. However, it is more difficult to keep such a retractable screen perfectly flat. If you want to use such a screen, look for a tensioned one. A *tensioned screen* flares out at the bottom and is held flat by a special tensioning system integrated in the border on the sides. These screens are also known as *tab-tensioned screens*. Retractable screens can be fitted with a motor such that you can lower and raise your screen with a remote control or control it from your home automation system.

Recommendation 6.14 *Use a projection screen that remains perfectly flat. Use either a fixed flat screen or a tab-tensioned retractable screen.*

The projection screen in a cinema is acoustically transparent. It is perforated to allow the sound from the loudspeakers behind it to pass through. It is not recommended using such a screen in your home theater. The term acoustically transparent is a bit misleading. These perforated screens do alter the sound and attenuate some of the high frequencies (Harley, 2002). To achieve the best possible sound quality do not install your loudspeakers behind the screen, instead place them to the sides and below the screen (see figure 4.17).

Most projection screens have a 16:9 aspect ratio. However, a lot of movies are in the Cinema-Scope aspect ratio (2.35:1 or 2.4:1). When projected on a 16:9 screen, you end up with black bars on the top and bottom of the screen (see section 5.1.1). It is possible to install a much wider screen and use the optical zoom feature of your projector to keep the image constant while switching between the 16:9 and the CinemaScope aspect ratios. Some projectors are equipped with a motorized zoom lens and a lens memory system for easy switching. These projectors can often also get rid of the black bars while in CinemaScope mode. In this way you are able to get a professional cinema look in your home theater room when watching CinemaScope material (CinemaSource, 2001b).

6.2 Display Calibration

The already good-looking image on your high-quality display can look even better if you properly calibrate your display. The goal of calibration is to make sure that your display reproduces the video material accurately with the correct brightness, contrast, and color. A calibrated display shows the material in exactly the way that the creator (director or producer) intended it to be shown. All displays, flat panels and projectors alike, require calibration in order to produce their best possible picture. When carefully calibrated a modestly priced display can easily outperform a much more expensive one that hasn't been calibrated.

Some displays have a picture mode that is already quite accurate, but most don't. The reason is that manufacturers try to distinguish themselves from the competition and want their display to have a unique look, preferably a look that attracts the most buyers. In order to grab your attention they deliberately tweak their displays to look bright, contrasty, and colorful, at the expense of being accurate (Harley, 2002; Taylor et al., 2006). After you calibrate such a display it will have a very different look. You will notice that it will look darker and that the colors are less vivid. If you are not used to a calibrated display, this look may be a bit disconcerting at first (Briere and Hurley, 2009). Please allow yourself some time to get used to this new look. Keep in mind that a calibrated display is no longer screaming for your attention, but has been adjusted to reproduce the content more accurately. It has been toned down, such that the content that you are watching grabs your attention instead of the display itself. A properly calibrated display will show much more detail and nuances and will look similar to what you are used to seeing at the movies.

Basic calibration of your display is easy to achieve. All you need is a Blu-ray test disc or a collection of digital video files with the appropriate test patterns. You look at these patterns on your display and adjust the display's controls according to the instruction that comes with the patterns. The AVS HD 709, DVE HD Basics, and HD Benchmark video test patterns, listed in appendix A, can all be used for basic calibration. If you do not already possess a video test disc, I recommend buying the HD Benchmark disc, because it includes advanced test patterns that I will use in the upcoming section 6.3 on video processing.

If you want to go further than the basic adjustments, you need special calibration software and a measurement device in addition to the test patterns. This allows you to take accurate measurements of the tonal values and colors on your display and use these measurements to fine-tune your display's settings.

A full calibration of your display consists of the following steps, each of which I will describe in more detail further on:

1. *White point:* set the color temperature to 6500 K.
2. *Black and white level:* maximize the dynamic range without losing detail in the darkest and brightest parts of the picture.
3. *Luminance output:* set the maximum brightness of the display.
4. *Color and tint:* establish correct color decoding with the right amount of saturation and hue.
5. *Grayscale tracking:* set gamma and avoid color casts along the entire grayscale.
6. *Color gamut:* adjust the primary colors that span the color gamut.

Steps 1, 2, and 4 only require video test patterns, while the other steps require the use of a measurement device and calibration software. If you do not want to invest in a measurement device, I urge you to at least perform the basic calibrations that require only test patterns. You could then try to fill in the gaps by looking at the recommended picture settings in a good technical review of your particular display model (see table 6.1). Reviewers often calibrate the display under review before they assess its picture quality. Some also list the settings of the display's controls after calibration. These settings are a good place to start and are certainly better than nothing if you do not have your own measurement device. Bear in mind though that every display and also its viewing conditions are a bit different, and therefore it is always better to perform your own calibration.

It is difficult to accurately judge colors and different tonalities by eye. That is why you need a measurement device, either a *colorimeter* or a *spectroradiometer*. A spectroradiometer is a professional device designed to measure the light energy as a function of its wavelength. It produces a spectral energy density curve. A colorimeter is a device with three separate photodetectors that measure luminance. Each photodetector in a colorimeter has a color filter that makes it respond to a specific primary color. A colorimeter produces three luminance values, one for each primary color. It is a small device that is much less expensive than a spectroradiometer. It is less accurate than a spectroradiometer, but accurate enough for calibrating your display (Brennesholtz and Stupp, 2008). Popular colorimeters for display calibration are the i1 Display Pro from X-rite (xrite.com) and the Spyder5 from Datacolor (datacolor.com). The accuracy of these meters can decrease over time as their color filters age. You can slow down this aging process by storing your colorimeter in a cool, dry, and dark place when it is not in use.

Colorimeters often come bundled with their own calibration software. This software is aimed at calibrating computer displays. It only works if you connect the computer to your display. It then shows a number of color and grayscale patches on the display, takes some measurements with the attached colorimeter, and automatically changes the settings of the computer's graphics card. This means that the display is now only calibrated for the computer on which you have run the calibration software. If you connect another source, for example a Blu-ray disc player, the display is still not calibrated. Therefore, you need special video calibration software that only relies on adjusting the controls of the display itself. Such software works as follows: instead of using the computer to show color and grayscale patches on the display, you play a Blu-ray test disc. You only use the computer to take measurements of the test patterns on the display. There is no need to connect the computer to the display. Based on the measurements, you then manually adjust the controls of the display. This is a trial and error process. You use the measured values to make a modification to one of the display's controls and you repeat the measurements to see whether this adjustment got you closer to the target values. You keep on repeating these steps until you get it right. Popular software programs for calibrating your display in this way are CalMan, ChromaPure, HCFR, and LightSpace (see appendix B).

HDR displays have different memories for the display controls: one for SDR and one for each type of HDR content (HDR10, Dolby Vision, HLG). Therefore, you need to perform several different calibrations: one for SDR and one for each HDR mode. Some displays can be manually switched from SDR to HDR, but mostly you need to feed it the appropriate SDR and HDR test patterns. The AVS HD 709 and HD Benchmark video test discs include SDR test patterns with the appropriate color and grayscale patches for the video calibration programs CalMan, ChromaPure, and HCFR. To calibrate for HDR10 or HLG content you need the corresponding DVS UHD test patterns (see appendix A). Currently calibration for Dolby Vision can only be done using the built-in test pattern generators of video calibration software such as CalMan. It also requires a so-called *Dolby Vision Golden Reference* file, which is a unique calibration target for each display model. This golden reference takes into account the limitations in luminance range and color gamut of the particular display and ensures that after calibration it performs as closely as possible to the desired target values (Schulte and Barsotti, 2016).

Displays often have individual memories for each different input. After you have calibrated your display using a Blu-ray test disc, you have altered only the controls associated with the input used by the Blu-ray player. You still need to set the controls for the other inputs. Most of the time this is simply a matter of manually copying the settings to the other inputs. However, sometimes this doesn't work and a certain device will look distinguishably different, for example much darker. The reason is probably that this device outputs a different black level or color space than the Blu-ray player. To correct this problem, search through the menus of the devices and set them to use the same color spaces and output the same black levels (more on this in section 6.2.2 and 6.3.3).

Recommendation 6.15 *Make sure that the picture controls are set to the optimal values found during the calibration process for each individual input of the display.*

The different display controls can influence each other. It is therefore a good idea to repeat the steps of the calibration process multiple times until no further adjustments are needed. After completing all the steps of the calibration process for the first time, at least repeat steps 2, 3, and 4. After you have completed the entire process perform a visual verification using some known source material that contains human skin tones. I personally like to use the demo material provided on the DVE HD Basics test disc (see appendix A).

Recommendation 6.16 *Display controls influence each other. Therefore, go back and forth between the different calibration steps until no further adjustments are needed.*

Almost every display has several picture mode presets. These presets provide different looks and have names like sports, game, or cinema. These picture modes do not necessarily give you the most accurate picture. Therefore, the first step, before you start the calibration process, is to determine which picture mode gives you the most neutral picture. That mode provides a good starting point for making further adjustments to the display's picture controls. Typically, the best mode to start off with is called 'movie' or 'cinema'. Other modes that you might want to check out are 'standard', 'natural', 'custom', or 'user'. Stay away from picture modes labeled 'vivid', 'dynamic', 'daylight', 'game', or 'sports'. These modes create a punchier image and add enhancements. They might look attractive at first, but they are not accurate and actually worsen the picture quality.

It is crucial to perform the calibration under the same lighting conditions that you generally use to watch the display. The amount of ambient light in the room has a huge impact on the perception of the darker tones in the picture. It is usually best to view the display is a darkened room (see section 4.3.1). If you watch movies in a darkened room, but other programs like sports with the lights on, it is worthwhile to calibrate the display separately for these two distinct lighting conditions. Most displays can store several user presets of their picture control settings. You could create one setting for daytime viewing and another for nighttime viewing. Some displays are equipped with an ambient light sensor and adapt the picture settings automatically based on the light level in the room. It is better not to use this feature, because it destroys your carefully calibrated settings. If your display has an ambient light sensor, turn it off in the setup menu. In fact, this holds true for all other automatic picture adjustments that change the picture on the fly, like dynamic contrast, contrast enhancement, color enhancement, live color, auto black level, and shadow detail. The only automatic feature that you want to enable on LCD displays is local dimming (see section 6.1.2). All the others should be turned off.

Recommendation 6.17 *Perform your display calibration under the same lighting conditions that you generally use for watching the display.*

Recommendation 6.18 *Turn off all automatic picture-adjustment features of your display, such as ambient light sensor, dynamic contrast, contrast enhancement, color enhancement, live color, auto black level, and shadow detail.*

Before you start the calibration process allow your display to warm up for about 10 minutes, preferably while displaying some video material on it. Most displays require a short amount of time to stabilize and reach their peak luminance and most accurate color reproduction (Bodrogi and Khanh, 2012). Displays can also change their characteristics over time. As your display ages, it will become less bright and its color may shift. This is especially true for projectors that use a UHP lamp. Therefore, you should recalibrate your projector at regular intervals, for example after every 200 hours of use.

Recommendation 6.19 *Allow your display to stabilize before you start calibration. Your display needs to warm up for approximately 10 minutes.*
Recommendation 6.20 *Calibrate your projector at regular intervals, for example after every 200 hours of use.*

6.2.1 White Point

The white point determines the color of white and gray. The white and gray tones on your display should look neutral. If they don't, all colors will have a certain color cast. Displays often have a setting for color temperature that determines the white point (see figure 3.13). Most displays have different presets for color temperature. Choose the preset that sets the color temperature to 6500 K. Unfortunately, no universal standard for labeling this preset exists. It may be labeled '6500K', or 'D65', but also less descriptive names are used such as 'warm', 'neutral', 'normal', or 'cinema'. You should definitely not choose the one labeled 'cool'. That preset will set the color temperature too high at 9300 K. This will make the white and gray tones look slightly bluish. Nevertheless, a lot of displays have 9300 K as their default setting, one reason being that at this setting the display has a higher luminance output. There is also a cultural preference in Asia for this color temperature (Poynton, 2012). However, video standards for HD video (ITU-R BT.709, 2015) and UHD video (ITU-R BT.2020, 2015) require the color temperature to be set at 6500 K for the most accurate reproduction of colors.

Recommendation 6.21 *Choose a color temperature preset that sets the display to 6500 K.*

Sometimes finding the right preset for 6500 K involves some trial and error, because of the obscure labeling employed by display manufacturers. It can help to display a grayscale pattern with different steps of gray, such as the ones listed in table 6.3. Try out the different color temperature presets of your display and see how the color of the grayscale changes. Choose the preset that makes the grayscale look neutral without a bluish cast to it.

Table 6.3 Grayscale patterns for visually evaluating color temperature.

AVS HD 709	Misc. Patterns—Additional	Grayscale Ramp Grayscale Steps
DVE HD Basics	Basic Patterns	Reverse Gray Ramps with Steps
DVS UHD	Miscellaneous Setup Patterns	Grayscale Ramps
HD Benchmark	Video Calibration	Color Temp (Crossed Steps White)

A more precise setting of the color temperature requires the use of a colorimeter. I describe how to use such a meter for this purpose in section 6.2.5 on grayscale tracking.

6.2.2 Black and White Level

Black level is the lowest luminance level in the video signal that should appear as pure black on the display. Similarly, white level is the highest luminance level in the video signal that should appear as pure white on the display. Remember from section 2.2.1 that the display is showing relative luminance levels. Therefore, the settings of the black and white level determine the dynamic range of the display, that is, they determine how the video signal is mapped to the darkest and brightest tones that the display is able to show. With the right setting of black level, the darkest tone in the video signal corresponds to the minimum luminance output of the display. And, with the right setting of white level, the brightest tone in an SDR video signal corresponds to the maximum luminance output of the display (usually 100 cd/m²).

It is important to understand that the setting of black and white level only influences the mapping of the video signal to the capabilities of the display. These settings do not directly influence the amount of luminance that the display outputs at its darkest and brightest settings. As I explained in section 6.1 the minimum luminance output of the display is determined by the technology that is being used in the display. Some types of displays, such as OLED flat panels and DLP projectors can go darker than other types that are based on LCD technology. The maximum luminance output of the display is determined by the power of the backlight, LEDs, or lamps used in the display. Often this maximum output can be independently controlled from white level. I will show you how in the upcoming section 6.2.3.

The control on the display for adjusting black level is often labeled 'brightness' by the manufacturer. The control for adjusting white level is often labeled 'contrast' or 'picture'. To complicate matters further, some LCD displays have a control labeled 'brightness' that doesn't change black level, but instead changes the luminance of the backlight (Poynton, 2012). The control for black level ('brightness') moves the entire video range up and down. The control for white level ('contrast') will make the range larger or smaller. This is illustrated in figure 6.6. In SDR video each *RGB* component is represented with 8 bits (see section 3.2.1). When all three *RGB* components have the same value, the color that appears is gray. Therefore, there are 255 levels of gray in an HD video signal. A value of 0 corresponds to black and a value of 255 to white. However, the video standard for HD video (ITU-R BT.709, 2015) places reference black (0%) at value 16 and reference white (100%) at value 235. This means that the video signal can go slightly below black (between 0 and 16) and slightly above white (between 235 and 255). This choice was made to avoid distortion of the video signal when it is being processed. You could think of reference black and reference white as safety margins that help to preserve the quality of the video signal. In HDR video the values for black and white are a bit different. Reference black is at value 64 for 10-bit digital video and at 256 for 12-bit digital video. Reference white is at value 940 (instead of 1024) for 10-bit digital video and at 3760 (instead of 4096) for 12-bit digital video (ITU-R BT.2100, 2017).

You should set black level ('brightness') as low as possible without losing any detail in the dark parts of the picture that are above reference black (value 16 for 8-bit SDR and value 64 for 10-bit HDR). The parts of the picture below reference black should not be visible and be as dark as reference black. Setting black level too high will result in the blacks appearing gray. This will wash out the picture and reduce the achievable contrast. Setting black level too low will result in loss of detail in the dark parts of the image. Figure 6.7 illustrates the optimal setting of black level.

Test patterns for setting the correct black level are listed in table 6.4. These patterns consist of a number of dark gray bars on a black background. The black background is at reference black, some of the dark bars are below reference black, and some are above reference black.

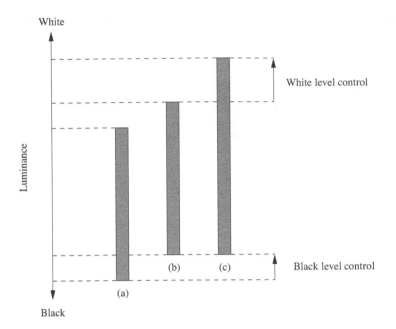

Figure 6.6
The effect of the black and white level controls on the mapping of the video signal to the luminance range of the display: *(a)* initial situation; *(b)* an increase of the black level control ('brightness') has moved the entire video range up; *(c)* an increase of the white level control ('contrast') has expanded the range towards higher luminance values while keeping black level fixed.

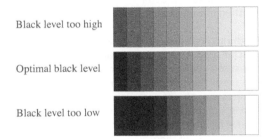

Figure 6.7
Examples of different black level ('brightness') settings. The *middle part* shows a grayscale from 0% (black) to 100% (white) in 10% steps. The *upper part* shows how this grayscale is displayed when black level is set too high; the darkest parts of the image will look gray instead of black. The *lower part* shows that setting black level too low will result in loss of detail in the dark parts.

Table 6.4 Test patterns for setting black level.

AVS HD 709	Basic Settings	Black Clipping
		APL Clipping
	Misc. Patterns—Additional	Dynamic Brightness
DVE HD Basics	Basic Patterns	PLUGE with Gray Scale
	Advanced Patterns	Window 100%–5% with PLUGE
DVS UHD	Basic Setup Patterns	Black Clipping
	Miscellaneous Setup Patterns	Dynamic Contrast
HD Benchmark	Video Calibration	Brightness (PLUGE 0%)
	Advanced Video—Setup	PLUGE 15%

You adjust the black level control of the display such that the bars below reference black disappear and the bars above reference black are just barely visible. Sit in the prime seat while adjusting black level and avoid looking at the screen from an extreme angle, because that will influence the appearance of the black bars. Also make sure that the ambient light level in the room matches your typical viewing conditions, because the optimal setting of black level changes with different amounts of ambient light in the room.

Recommendation 6.22 *Use a video test pattern to calibrate the black level of your display. Set the black level of the display as low as possible without losing detail in the dark parts (above reference black).*

Some Blu-ray players and displays may not show the dark gray tones that are below reference black. In that case, the bars that are below reference black will not show up in the test pattern, even when you crank up the control for black level to a very high value. Without these below-black bars it becomes a bit more difficult to determine the optimal setting of the black level control. In some devices you can make the bars appear by changing the black level settings of the HDMI interface: set the HDMI black level to 'Enhanced' or 'Extended'.

Some displays automatically vary black level depending on the average brightness of the picture (dynamic contrast, auto black level, shadow detail). You should turn this feature off if the display allows you to. If not, setting black level becomes a compromise. There are two different patterns for setting black level: one with a low average picture level (dark background) and one with a medium average picture level (light background). Use both test patterns to determine the setting of black level and use the highest setting of the two to avoid any loss of detail in the dark parts of the picture.

Now let's turn our attention to the white level control. For SDR video you should set white level ('contrast') as high as possible without losing any detail in the bright parts of the picture that are below reference white (value 235). Setting white level too low will make the brightest parts appear gray instead of white. It will also limit the maximum achievable contrast. Setting white level too high will result in loss of detail in the brightest parts of the image. Figure 6.8 illustrates the optimal setting of white level.

Test patterns for setting the correct white level are listed in table 6.5. The test patterns on the AVS HD 709 and HD Benchmark discs are more accurate than the one on the DVE HD Basics disc. These patterns consist of a number of nearly white bars on a pure white background. You should at least be able to see all the bars below reference white (value 235). It is even better if you can see all the bars (up to the value of 254). Raise the white level control until some of the bars disappear and blend in with the white background, then lower white level a bit until the bars are just distinguishable from the white background. The top three bars may be difficult to see, so don't worry too much about them.

Setting white level too high causes clipping. When the display is clipping, the video signal has reached its maximum brightness and everything in the video signal that is brighter will be fixed at this hard limit. That is why the white bars disappear in the test pattern if you raise white level too much. When the display clips the brightest parts of the video signal, the details in the highlights get lost. Often not all *RGB* color channels clip at the same white level. One of the channels may clip at a much lower setting of white level than the others do. When this happens the brightest parts of the picture shift in color. They are no longer white, but have a noticeable color cast. The channel that is clipped no longer keeps up with the other channels,

Figure 6.8
Examples of different white level settings. The *middle part* shows a grayscale from 0% (black) to 100% (white) in 10% steps. The *upper part* shows how this grayscale is displayed when you set white level too high; you will lose detail in the brightest parts of the image. The *lower part* shows that setting white level too low will result in the brightest parts looking gray instead of white.

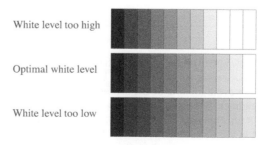

White level too high

Optimal white level

White level too low

Table 6.5 Test patterns for setting white level.

AVS HD 709	Basic Settings	White Clipping
	Misc. Patterns Additional	Grayscale Ramp
		Grayscale Steps
		Color Steps
		Color Clipping
DVE HD Basics	Basic Patterns	Reverse Gray Ramps with Steps
DVS UHD	Basic Setup Patterns	White Clipping
		Color Clipping
	Miscellaneous Setup Patterns	Grayscale Ramps
		Color Ramps
		HDR Clipping Patterns
HD Benchmark	Video Calibration	Contrast (Contrast Y)
	Advanced Video—Setup	Contrast R/G/B
		Crossed Steps White/ R/G/B
	Advanced Video—Evaluation	Clipping
		Dynamic Range RGB

and hence the brightest parts no longer contain an equal amount of red, green, and blue. If the bars and the background of the test pattern shift in color, this is a sure sign of clipping and you should reduce white level. Table 6.5 lists some additional clipping test patterns in which you can see the individual color channels and more easily judge whether they are clipped or not. Note that sometimes the clipping is the result of setting the color control of the display too high. If you experience clipping in one or two color channels, check if the color control is set to an appropriate value. I will explain in detail how to set the color control in section 6.2.4. Clipping can occur in both SDR and HDR video. You can use the DVS UHD test patterns from table 6.5 to check how your display handles clipping with HDR10 and HLG video.

> **Recommendation 6.23** *Use a video test pattern to calibrate the white level of your display. Set the white level of the display as high as possible without losing detail in the bright parts and without introducing color shifts.*

The black level ('brightness') and white level ('contrast') controls influence each other. It is therefore a good idea to readjust black level after you have set white level. You should go back and forth between these two controls several times until both appear to be set correctly. The black level control anchors the lower part of the entire luminance range. Changing it will slide the entire range up and down (see figure 6.6). Typically, the white level control increases the range, while keeping its lower value fixed. This is the proper way to implement the white level control. However, there are some displays in which the white level control increases the luminance range in both directions. With such a control, changing white level will also change black level. Consequently, this type of control is much more difficult to set correctly. If you encounter problems because of the interaction between the black and white level controls, I advise you do the initial adjustments using a grayscale pattern that shows both the upper and lower parts of the luminance range. Set the controls such that you avoid clipping at both ends of the range. After that fine-tune the controls using the regular test patterns.

After setting the black and white level controls you should check whether the entire luminance range is displayed smoothly without any banding or contouring. Recall from section 3.2.1 that in digital video the number of shades that make up the entire luminance range is always a finite number. When an insufficient number of shades is used, gradually changing areas of the picture will not be smooth, but will contain stripes and bands of different luminance steps. In some displays the white level control is a rather coarse control that at a high setting reduces the number of shades in the entire luminance range. To check whether your display has this problem it is worthwhile to look at a continuous grayscale ramp (see table 6.5). The ramp should look like a gradient. It should smoothly transition from dark to light without any banding or stripes. If it doesn't you should adjust white level control to see whether there is a setting close to the optimal setting that produces a smoother ramp. It is a good idea to reduce white level to obtain a smoother gradient, because banding is much more visible than a slight loss of highlight detail (Spears and Munsil, 2013).

Recommendation 6.24 *After setting the black and white level of your display, check if a grayscale ramp is displayed as a smooth gradient. If it is not, try to make the ramp smoother by reducing the white level.*

6.2.3 Luminance Output

The amount of light that your display produces should be in proportion to the amount of ambient light in the room. Although a very bright display offers a lot of contrast, a display that is way too bright is not pleasant to watch and induces eye strain. According to table 3.2, for SDR video the maximum luminance of your display should be around 100 cd/m² if it is a flat-panel display that is being watched in a dim room with some residual ambient light. It should be from 38 to 58 cd/m² if it is a projection system that is being watched in an almost dark room. These SDR guidelines apply to the maximum luminance of a 100% white screen, or a 100% white patch that covers not less than 50% of the screen. It doesn't apply to HDR video where small highlights are allowed to be much brighter (see section 3.2.1).

Preferably, you adjust the display's maximum luminance by directly changing the intensity of the light it produces, for example by reducing the intensity of the backlight in an LCD display. Alternatively, you could reduce the amount of light by turning down the white level control. However, only use the white level control as a final resort. Changing the white level control adjusts the scaling of the video signal and may result in losing bit precision (ITU, 2009b).

On an LCD flat-panel display you change the maximum luminance by adjusting the intensity of the backlight. The control for this is often aptly labeled 'backlight', but some manufacturers use the ambiguous label 'brightness'. If your display has a 'brightness' control, it could either be a backlight control or a black level control. Make sure to find out what it does, as these are two completely different controls. On an OLED display you change the amount of light simply by adjusting the light output of the OLEDs themselves. The control for this is called 'OLED light'. In a projection system the maximum luminance is already quite limited. Nonetheless, if the projected image is too bright, you have several options to dim it. First, you could switch the projector's lamp from 'full' to 'eco' mode. Second, you could close down the projector's iris. The iris is a small aperture of variable size that sits between the lamp and the lens and works as a light valve. Third, you could increase the size of your projection screen or install a screen with a lower gain (see section 6.1.4).

Recommendation 6.25 *Adjust the maximum luminance output of your display to 100–250 cd/m² for a flat-panel display and to 38–58 cd/m² for a projection system. Directly change the intensity of the light and try to avoid using the white level control for this purpose.*

The most accurate way to set the maximum luminance output of your display for SDR video content is by measuring it using a colorimeter and video calibration software such as CalMan, ChromaPure, HCFR, or LightSpace (see appendix B). Bring up a 100% white field or a large

Table 6.6 Test patterns for setting peak luminance of SDR video.

AVS HD 709	CalMan/ChromaPure/HCFR	Fields 100% White
		Windows 100% White
		APL 100% White
DVE HD Basics	Advanced Patterns	Window 100% with PLUGE
HD Benchmark	Advanced Video—Contrast Ratio	White 100%
	Equal Energy Windows	White 100%

100% white patch (100% white window) and place the colorimeter in front of your display. These SDR white fields and windows are included on the AVS HD 709, DVE HD Basics, and HD Benchmark video test discs (see table 6.6). The colorimeter should be facing the screen. On a flat-panel display you can press it lightly against the screen to avoid any ambient light influencing the measurement. With a projection system you position the meter about 10 cm (4 inch) in front of the screen at a slight upward angle such that the light reflected from the projection screen falls on the sensor of the colorimeter. Be careful not to aim the meter at its own shadow. To avoid the ambient light influencing the measurement, the room should be completely dark.

For HDR video you should set the peak luminance of the display to its maximum value. The HDR transfer function will take care of the rest. According to table 3.2, for HDR video the maximum luminance of your display should be 1000 cd/m². However, not all HDR displays are able to reach this peak luminance. To measure the HDR peak luminance of your display, you need to use an HDR test pattern with a 100% white patch that covers only 10% of the picture area, such as the ones included with the DVS UHD test patterns.

If you don't have a measurement device, you could try to set the maximum luminance of the display by hand. This is of course not as accurate, but it should get you in the right direction. The key to success is to ensure that the maximum amount of light is in proportion to the ambient light level. Therefore, you could use the test patterns from table 4.5 to judge if the amount of ambient light is right for your display. If not you have two options: 1) change the amount of ambient light or 2) change the maximum luminance output of the display. Setting the maximum luminance output of the display is largely a subjective matter, but if you are squinting every time a bright scene comes up, it is a sure sign that your display is too bright.

Finally, changing the maximum light output of your display influences the optimal settings for black level and white level. It is therefore recommended that you repeat the procedure in section 6.2.2 and recalibrate them.

6.2.4 Color and Tint

The color and tint controls of your display affect the color decoding process. Recall from section 5.1.4 that digital video is not stored as *RGB* components, but as a luma component Y' and two chroma components C_B and C_R. The luma component contains the luminance information of the picture and the two chroma components the color information. Displaying only the luma component Y' without the chroma components would result in a black and white picture.

The color control alters the relative amount of chroma with respect to luma (Poynton, 2012). Increasing the color control increases the saturation of the colors (see section 2.2.2); the picture becomes much more colorful. Decreasing the color control desaturates the colors and the picture becomes muted and washed out. On some displays the color control is labeled 'saturation'.

The tint control alters the balance between the two chroma components (Poynton, 2012). It changes the hue of the colors (see section 2.2.2). Therefore, on some displays this control is labeled 'hue'. If the tint control is not set correctly, all the colors of the display will be wrong.

The color and tint controls were introduced a couple of decades ago when electronics where not very stable. On old NTSC and PAL television sets the colors could change over time and these controls allowed the user to correct for these changes. Modern digital circuits are so stable that the color and tint control are no longer necessary. Moreover, PAL decoders are immune to hue errors and do not need a tint control (Poynton, 2012). Consequently, the tint control is no longer found on every display. By contrast, the color control is still very much present. Some manufacturers intentionally set the color control a bit too high to make the colors on their displays pop and stand out from the competition.

To get accurate colors, you need to check the settings of both the color and the tint controls. On a modern display, especially one that comes straight from the factory, these controls should not be set too far from their optimal settings. The tint control often needs no adjustment at all. The color control may need to be taken down a few notches, because its factory default may turn out to be a bit too high. But, keep in mind that it is very unlikely that the color control is set too low at the factory (Spears and Munsil, 2013).

Test patterns for setting the color and tint controls are listed in table 6.7. These patterns have differently colored bars, including the primary and secondary colors. The idea is to look at the bars through a colored filter, a red, green, or blue filter, and adjust the controls until certain bars have the same brightness. The most commonly used filter is a blue filter. Such a filter is included with the DVE HD Basics and HD Benchmark video test discs (DVE HD Basics includes all three color filters). Blue is used to set color and tint, because the eye is more sensitive to changes in blue than it is to changes in red or green. Looking through a blue filter, you adjust the color and tint controls while looking at a test pattern with several colored bars. Adjust the color control until the blue bar matches the gray bar, and adjust the tint control until the magenta bar matches the cyan bar. For SDR video you use 75% bars and for HDR video 50% bars. The 75% bars for SDR video are more accurate to use than 100% bars, which could be slightly clipped by the display. For HDR video 50% bars are needed as currently no display can produce the entire Rec. 2020 color gamut. The color and tint controls may influence each other, so it is a good idea to go back and forth between them until no other changes are necessary.

Recommendation 6.26 *Use a video test pattern and a blue filter to calibrate the color and tint controls of your display.*

If the color and tint controls are set correctly using the blue filter, the test patterns designed to be looked at through a red and green filter should also look correct (when viewed through their corresponding filters). Otherwise the color decoding doesn't adhere to the Rec. 709

Table 6.7 Test patterns for setting color and tint.

AVS HD 709	Basic Settings	Flashing Color Bars
	Misc. Patterns—Various	Flashing Primary Colors
		Flashing Color Decoder
DVE HD Basics	Basic Patterns	75% Bars with Gray Reference
DVS UHD	Basic Setup Patterns	Color Bars
	Miscellaneous Setup Patterns	Color Flashing Primary
HD Benchmark	Video Calibration	Color and Tint (Blue)
	Advanced Video—Setup	Color and Tint (Red)
		Color and Tint (Green)

standard for SDR video or the Rec. 2020 for HDR video. When you look through a red filter, you adjust the color control such that the brightness of a red bar matches the brightness of a gray bar, and you adjust the tint control such that a magenta bar matches a yellow bar. Similarly, with a green filter, you adjust the color control using a green and gray bar, and you adjust the tint control using a cyan and a yellow bar.

Sometimes the color and tint controls need to be adjusted quite a bit, or they seem to demand different settings when you look through different color filters. With SDR video the reason could be that the display wrongly maps the colors of the Rec. 709 to its own native color gamut. Many modern displays have the ability to show colors outside of the Rec. 709 color gamut. Such a display uses a *color management system (CMS)* to enlarge the standard color gamut. While this increases the color gamut, it is not an accurate representation of the colors as they are present in the SDR video signal. The larger color gamut should only be used for HDR video and never for SDR video (see figure 3.12). If possible, you should turn off the color mapping for SDR video in the setup menu of your display. The menu item that you are looking for can have several names; look for 'color space', 'color gamut', 'xvYCC', 'vivid color', 'live color', and 'color enhancement'. This item should be set to automatic, such that it is only activated for HDR video, or turned off completely if your display doesn't support HDR video.

6.2.5 Grayscale Tracking

Your display should accurately show all the different shades of gray between black and white. This involves two aspects. First, the shades of gray should be distributed such that for SDR video they follow a gamma curve (see figure 3.9) and such that for HDR video they follow the SMPTE ST 2048 transfer function (see figure 3.10), or the HLG transfer function (see figure 3.11). Second, each different shade of gray should look neutral and correspond to a white point with a color temperature of 6500 K.

Let's start with the setting for gamma and look at HDR and the color temperature aspect later on. Recall from section 3.2.1 that the optimal value of gamma depends on the amount of ambient light in the room (see table 3.5). Gamma should be such that the dark tones in the picture are as dark as possible without any loss of detail in the shadows. As most people will be watching their display in a dimly lit environment, the most common setting for gamma is 2.4 (EBU-Tech 3320, 2017; ITU-R BT.1886, 2011). Your room is considered to be dimly lit, if the ambient light doesn't exceed 15% of the maximum luminance output of your display (see section 4.3.1). Only if you tend to watch your display without dimming the lights and closing the curtains, a lower gamma of 2.2 might be appropriate. But note that, setting the gamma control too low for your ambient light level conditions results in a washed-out picture, because the darker gray tones are not dark enough. On the other hand, setting the gamma too high results in loss of detail in the darker parts of the image, because the darker gray tones are too dark to be clearly distinguishable. The different settings of gamma are illustrated in figure 6.9.

Many displays have a single dedicated control to adjust gamma. Some displays only have certain presets for gamma. More advanced displays have separate controls for adjusting the gamma of the *R, G,* and *B* components. And on some of these displays you even have a multipoint gamma control to make detailed adjustments to the shape of the gamma curve. It is difficult to set the gamma controls by eye. You could look at a test pattern with grayscale steps or a grayscale ramp to get a rough idea of the distribution of the different tones, but you have no absolute reference to base your gamma adjustments on. The more advanced multipoint *RGB* gamma controls are impossible to set by eye. To accurately set these controls you really need to perform measurements. These measurements are easy to perform using a colorimeter and a computer running video calibration software such as CalMan, ChromaPure, HCFR, or LightSpace (see appendix B).

Not all is lost, if you don't have access to this measurement equipment. In this case, the best thing you could do is to visually adjust the gamma of your display. Without any measurements

Figure 6.9
Examples of different
gamma settings. The
middle part shows a
grayscale from 0% (black)
to 100% (white) in 10%
steps. The *upper part*
shows how this grayscale
is displayed when you set
gamma too high; you will
lose detail in the dark parts
of the image. The *lower
part* shows that setting
gamma too low will result
in the dark parts looking
too bright.

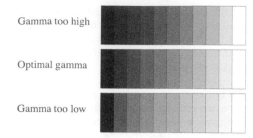

to guide you it is best to leave the advanced gamma controls alone. You should only adjust the overall gamma control or choose an appropriate gamma preset. One test pattern that you can use is the PLUGE pattern. This pattern has been mentioned in section 6.2.2 to set black level (see table 6.4). The PLUGE pattern has two dark gray bars that are above reference black. If you have set black level correctly, these two bars should be visible and not blend in with the dark background. The darkest black bar should be barely visible, and the other bar should be clearly visible. If the bars are much lighter than the background, then you should increase the gamma control. Another test pattern that you could use to visually adjust gamma is the 'Equal Energy Gamma' pattern on the HD Benchmark disc (see table 6.8). This pattern contains some fine-grained lines and checkerboards. The idea is that you view your display from a distance and slightly squint your eyes such that the lines and checkerboards appear as smooth gray patches. Your gamma control is set correctly if these patches match the brightness of the square in the middle of the test pattern. One warning though, this test pattern has its limitations. It only works if your display is unscaled and has a one-to-one pixel mapping with a resolution of 1920 × 1080 pixels (see section 6.3.1). In addition, this pattern doesn't work on LCoS projectors, because they are unable to display the fine lines at full amplitude (Spears and Munsil, 2013).

Regardless of whether you adjust gamma visually or use measurements, it is of paramount importance that your black level has been set correctly. Without a correct setting of black level, gamma will never be right (Poynton, 2012). As I explained in section 6.2.2 black level moves the entire video range up and down. Setting black level too low will crush all shadow detail and the adjustment of the gamma control cannot correct for this. As a result the gamma of your display will appear to be too high (compare figure 6.7 with figure 6.9). Check out box 6.2 for more details on the relationship between gamma, black level, and white level.

Table 6.8 Test patterns for visually setting gamma.

DVE HD Basics	Basic Patterns	PLUGE with Gray Scale
HD Benchmark	Video Calibration	Brightness (PLUGE 0%)
	Equal Energy Gamma	Display Gamma

BOX 6.2 ELECTRO-OPTICAL TRANSFER FUNCTION

The relative luminance L on your display depends on black level, white level, and gamma γ as follows (ITU-R BT.1886, 2011):

$$L = a \max(V + b, 0)^\gamma$$

where V is the video signal (normalized to $0 \leq V \leq 1$), a is a gain factor that corresponds to the setting of white level, and b corresponds to the setting of black level. The 'max' operator ensures that the relative luminance never drops below zero, even when b is set to a large negative value.

Table 6.9 Test patterns for measuring grayscale tracking.

AVS HD 709	CalMan/ChromaPure/HCFR	Grayscale
DVE HD Basics	Advanced Patterns	Window 100%–0% with PLUGE
DVS UHD	Advanced Setup Patterns	10% Window Patterns
	CalMan/ChromaPure/HCFR	Grayscale
HD Benchmark	Equal Energy Windows	White 0%–100%

The best way to set gamma is to measure it and use the measurement to make an informed change to the gamma controls of your display. The common way to measure gamma is to show different levels of gray on the display, one after another, and measure their luminance. Typically, the range from 0% (black) to 100% (white) is covered in steps of 10%. You could either use an entire gray screen (field) or a large patch (window). The AVS HD 709, DVE HD Basics, and HD Benchmark video test discs all have suitable SDR test patterns for performing these measurements (see table 6.9). If your display automatically adjusts its luminance according to picture content, your measurements of gamma will not be accurate. You should look in the setup menu of your display and turn this feature off. It is often called dynamic contrast. If you cannot turn it off, then you should only use the APL windows from the AVS HD 709 patterns, or the equal energy windows from the HD Benchmark disc. These windows have been specifically designed to defeat the dynamic contrast feature of your display. They gray windows have a surrounding background that changes such that each test pattern has the same overall average luminance regardless of the intensity of the gray window itself.

The measurements of the different gray tones are not only useful for determining gamma, but also for evaluating the second aspect of grayscale tracking, namely the color temperature of the different gray tones. All the gray test patterns from 0% (black) to 100% (white) should be a neutral gray without any color cast. Put differently, they should all correspond to the same white point with a color temperature of 6500 K. The relative amount of red, green, and blue should not change as the luminance of the gray test pattern changes. When it does, a color cast is introduced. For example, the darker gray tones may have a bit more red compared to the mid tones, or the light tones may have a bit more blue. In such cases the color of gray changes as the luminance changes, which is of course not desirable. An unbalance in color throughout the luminance range results in different gamma curves for the red (R), green (G), and blue (B) components. Therefore, to ensure proper grayscale tracking where the color temperature stays constant throughout the entire luminance range, the gamma curves for R, G, and B should be exactly the same.

Despite the fact that the controls for setting gamma differ among displays, the basic procedure for adjusting them is the same. You use video calibration software to measure the luminance Y, and the chromaticity coordinates x and y (see section 2.2.2) for each gray window from 0% to 100% (in steps of 10%). From the measurements of the luminance Y the calibration software will calculate a value for gamma. If this value is not equal to your target gamma (2.4), you need to adjust the gamma controls. Furthermore, to assess the color balance along the luminance range, the software will compare the x and y measurements of each gray window to the target white point with a color temperature of 6500 K. For each gray window the software computes the difference between the target white point and the measurement. These differences can be used to make informed adjustments to the individual controls for the gamma of the R, G, and B components. Figure 6.10 shows an example measurement made with HCFR. It shows the relative contributions of the individual R, G, and B components for the different luminance values of the gray windows. The top part of the figure shows the measurements before calibration of the gamma controls. The display in this example has a color temperature that is relatively constant along the entire luminance range, but the display is not neutral. There is clearly too much blue and also a bit too much red. The bottom part of the figure shows that after adjusting the gamma controls of the individual

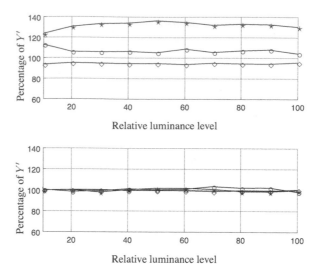

Figure 6.10
Grayscale calibration
example. The balance
of the *RGB* components
expressed as a percentage
of Y' for different relative
luminance levels that
range from 10% to 100%
in 10% steps. Red ○, green
◇, and blue ★. *Top,* before
calibration. *Bottom,* after
calibration.

Table 6.10 Grayscale calibration example. The reproduction error ΔE^*_{uv} before and after adjustment of the gamma controls of the display. The values shown are for the relative luminance range of 10–100% in 10% steps.

	10	20	30	40	50	60	70	80	90	100
Before	24.6	28.1	30.5	31.2	32.9	32.4	29.3	30.5	30.3	27.7
After	0.4	1.1	2.0	1.1	1.8	2.0	3.0	2.0	2.2	1.3

R, G, and *B* components, the color temperature has become close to neutral along the entire luminance range.

The graphical presentation in figure 6.10 is only one way to show the color differences. Usually the video calibration software also computes a numerical value that is called the *reproduction error* ΔE^*_{uv} to quantify the color errors. Table 6.10 shows these values for the example of figure 6.10. A reproduction error ΔE^*_{uv} below 4 is considered to be good as the corresponding color difference is almost indistinguishable by the human eye (EBU-Tech 3320, 2017).

It should be clear by now that you use the outcome of the grayscale tracking measurements to adjust both gamma and the color temperature of your display. Since the nature of the controls for setting gamma and color temperature differs among displays, there are several options. Find out which controls your particular display has and proceed according to the following instructions:

- *Single point gamma control:* This control consists of a single gamma slider. To adjust it, measure the grayscale tracking for the entire luminance scale from 0% to 100%. If the gamma value computed from these measurements is too low, increase the slider. If it is too high, decrease the slider.
- *Single point color temperature control:* This control consists of three sliders, one for each of the *RGB* components. These sliders are either labeled 'Color Temperature' or 'White Balance'. The sliders affect the entire luminance range from 0% to 100%. Balance the three *RGB* sliders such that a 50% gray window measures as a neutral gray with a color temperature of 6500 K.
- *Two-point RGB gamma control:* This control consists of two sliders for each of the three *RGB* components (six in total). One slider affects only the low end of the luminance range from 0% to 50%. This slider is called 'bias', 'cutoff', 'brightness', 'min', or 'low'. The other slider affects only the high end of the luminance range from 50% to 100%. It is

called 'gain', 'drive', 'contrast', 'plus', or 'high'. Balance the three *RGB* bias sliders such that a 30% gray window measures as a neutral gray with a color temperature of 6500 K. Then, balance the three *RGB* gain sliders such that a 70% gray window measures as a neutral gray with a color temperature of 6500 K. After these initial adjustments, measure the grayscale tracking for the entire luminance scale from 0% to 100% and readjust the bias and gain control if necessary.

- *Ten-point RGB gamma control:* This control consists of ten sliders for each of the three *RGB* components (30 in total). Each slider only affects one-tenth of the entire luminance range. There is one slider for each 10% increase in luminance. These sliders provide very fine-grained control to balance the three *RGB* components. To adjust them, measure the grayscale tracking for the entire luminance scale from 0% to 100% in steps of 10%.

If your display has a combination of these controls, then start with adjusting the single-point gamma and temperature controls. Then, if there is still room to improve, use the multipoint controls to fine-tune. After adjusting your display's controls, repeat the grayscale tracking measurements and evaluate the new results. Readjust the controls if necessary and repeat this procedure several times until your gamma and color temperature have reached their target values. You should aim for a gamma that is approximately equal to 2.4 and stays between 2.3 and 2.5 for the entire luminance range (EBU-Tech 3320, 2017). You should also aim for a color temperature of 6500 K, which means that the chromaticity coordinates of the white point are $x = 0.3127$ and $y = 0.3290$. This white point should stay constant along the entire luminance range with a reproduction error ΔE_{uv}^* smaller than 4 (EBU-Tech 3320, 2017).

Recommendation 6.27 *Calibrate your display for SDR video such that gamma is approximately equal to 2.4 and stays between 2.3 and 2.5 for the entire luminance range.*

Recommendation 6.28 *Calibrate your display such that the white point has chromaticity coordinates $x = 0.3127$ and $y = 0.3290$ (6500 K) and such that this white point is approximately constant across the entire luminance range with a reproduction error ΔE_{uv}^* smaller than 4.*

Calibrating the grayscale tracking is especially important for projection systems. In projection systems it is usually necessary to reduce the output of one or two of the *RGB* components to reach a neutral white point (Brennesholtz and Stupp, 2008). Often the blue and green components need to be reduced. If you still do not seem to have enough red at the end of the luminance range (90%–100%), you might need to reduce white level ('contrast') in order to be able to balance the three components. Since projector lamps shift color and lose power as they age, it is important to recalibrate your projector regularly, for example after every 200 hours of use.

Calibration of grayscale tracking for HDR video is a bit different from the procedure described above. The main difference is that the grayscale follows a particular HDR transfer function instead of the gamma curve. HDR10 and Dolby Vision video use the SMPTE ST 2084 (2014) transfer function (see figure 3.10), and HLG uses the HLG transfer function (see figure 3.11). Recall from section 3.2.1 that the SMPTE ST 2084 transfer function specifies absolute luminance levels, while the gamma curve and also the HLG transfer function specify relative luminance values. With SDR video you use the entire gamma curve and calibrate the display such that the maximum video output level corresponds to a luminance output of 100 cd/m². With HLG video you use the HLG transfer function and calibrate the display such that the maximum video output level corresponds to the maximum luminance output of the display. With HDR10 and Dolby Vision video you only use part of the SMPTE ST 2048 transfer function up to the maximum luminance capability of the display, the reason being that currently no display can produce the entire luminance range of the SMPTE ST 2084 transfer function. This luminance range goes from 0 to 10 000 cd/m². With today's technology, the best video displays can only reach about 1000 cd/m². Figure 6.11 shows an example of HDR grayscale tracking for a display that can produce a maximum luminance level of 700 cd/m². The display tracks the

Figure 6.11
Example of grayscale calibration for a video display with a maximum luminance output of 700 cd/m². The display tracks the SMPTE ST 2084 transfer function up to its maximum luminance and then clips all the highlights that are brighter. The *solid line* is the transfer function of the display, while the *broken line* is the SMPTE ST 2084 transfer function.

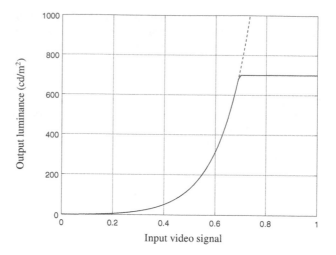

SMPTE ST 2084 transfer function up to its maximum luminance level and then clips all the highlights that are brighter. This is the best you can do given the capabilities of the display.

To calibrate grayscale tracking for HDR video, you first determine the maximum luminance output of your display with a 100% white patch that covers only 10% of the picture area (see also section 6.2.3). This white patch needs to be an HDR test pattern to make sure that your display is operating in HDR mode. Next, you measure grayscale tracking from 0% to 100% in steps of 10% using similar HDR patches that cover only 10% of the picture area. You adjust the *RGB* gamma controls such that each patch has a color temperature of 6500 K and such that the SMPTE ST 2084 or HLG transfer function is tracked.

Recommendation 6.29 *Calibrate your display for HDR10 and Dolby Vision video such that the grayscale tracks the SMPTE ST 2084 transfer function up to the maximum luminance capability of the display.*
Recommendation 6.30 *Calibrate your display for HLG video such that the grayscale tracks the HLG transfer function.*

6.2.6 Color Gamut

The rendition of colors on the display is not only governed by the color temperature setting and the color/tint controls. It also depends on the accuracy of the three primary colors, red, green, and blue, that span the color gamut. All colors that the display can render lie within the triangle spanned by these three primary colors in the CIE chromaticity diagram (see figure 2.13). Therefore, if the primary colors of the display are not accurate, none of the colors will be accurate.

You can use a colorimeter to measure the accuracy of the primary colors of your display. Table 6.11 lists the test patterns you can use to do this. These patterns consist of fields and windows with the primary and secondary colors at a relative luminance of 75% for SDR video and 50% for HDR video. Similar to the grayscale tracking measurements, you use a computer with calibration software to measure the luminance and the chromaticity coordinates of these patterns. Both the AVS HD 709 and HD Benchmark test discs contain SDR video, so the colors of their test patterns should match the Rec. 709 color gamut (ITU-R BT.709, 2015). Similarly, the DVS UHD test patterns contain HDR10 or HLG video, so the colors of these test patterns should match the Rec. 2020 color gamut (ITU-R BT.2020, 2015). However, currently no display can produce the entire Rec. 2020 color gamut and calibration with Rec. 2020 as a target is impossible. Therefore, color gamut calibration for HDR video is often done using the smaller DCI-P3 color space (see table 3.6).

The measurements of the primary and secondary colors for SDR video are done with patterns that have a luminance of 75% (instead of 100%), and those for HDR video with patterns that have a luminance of 50%. This reduction in luminance is deliberate to ensure none of the primary colors is clipped. Even a slight clipping of one of the colors can skew the measurements. The secondary colors are included in the measurements as an extra verification step for the accuracy of the color rendition. The secondary colors have a unique relation to the primary colors. Given the white point and the primary color, the secondary colors are fixed. If you draw a line from each of the primaries through the white point, the secondary colors lie exactly at the three points on the opposite side of the gamut triangle. This geometric relationship is illustrated in figure 6.12.

A display that has its primary and secondary colors in exactly the points shown in figure 6.12 is a perfectly balanced and calibrated display. This is very difficult to achieve in practice. Your display is considered to have a good color rendition if the reproduction error ΔE^{*}_{uv} for the primary and secondary colors is less than 7. And, if this reproduction error is lower than 4, the color rendition of your display is very accurate (EBU-Tech 3320, 2017).

Very few consumer displays have an adjustable CMS that can be used to fine-tune the primary colors. But, if your display does have them, you can use the measurement procedure above to guide you in adjusting them. The controls of an adjustable CMS differ among manufacturers and display models. Often you can adjust the hue, saturation, and luminance (also labeled 'brightness') independently for each primary color. This enables you to move the R, G, and B points that span the color gamut closer to the target values. The hue control rotates the R, G, or B point around the white point; the saturation control moves the R, G, or B point closer to or further away from the white point; and the luminance control adjusts the relative

Table 6.11 Test patterns for measuring primary and secondary colors.

AVS HD 709	CalMan/ChromaPure/HCFR	75% Color
DVS UHD	CalMan/ChromaPure/HCFR	Color
HD Benchmark	Equal Energy Gamut	75%

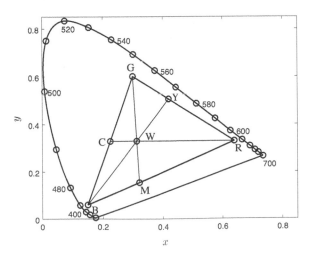

Figure 6.12
CIE 1931 chromaticity diagram with primary *(RGB)* colors, secondary *(CMY)* colors, and the white point *(W)* for the Rec. 709 gamut.

luminance of the *R*, *G*, or *B* point. Note that the color and tint controls can't be used to move the *R*, *G*, and *B* points independently of each other. The color control will either increase or decrease the entire gamut triangle. The tint control will rotate the entire triangle and will affect the hue of all the colors. These controls can't be used to correct for differences in accuracy of the three primary colors. If you adjust the color and tint control to correct one primary color, the other two primary colors may still be incorrect.

Recommendation 6.31 *If your display has an adjustable CMS, use it to calibrate your display such that the primary and secondary colors have a reproduction error ΔE^*_{uv} smaller than 7.*

Due to limitations in display technology and the limited CMS controls offered on most displays, it is often not possible to get all the secondary and primary colors exactly correct. You may have to make some trade-offs. In doing so, keep in mind that the human eye is not equally sensitive to all colors and all color differences. It is more important to get the red and green primaries right than it is to get the blue one right. Furthermore, it is also more important to get correct hues than it is to get correct saturation.

6.3 Video Processing

The term *video processing* refers to the alteration of the digital video signal. As the digital video travels through your system from the source (disc or file) to the display it is altered by your video equipment. For example, HD video (1920×1080 pixels) is scaled to fill the entire screen of a 4K display (3840×2160 pixels). Or, as I described in the previous section, the gamma and black level of the digital video are changed to create the optimal viewing experience with your particular display in the ambient light conditions of your room. Generally speaking, both your source components and your display apply certain processing steps to the digital video signal. If your video is routed through your A/V controller or A/V receiver, then this component might also add some processing steps. Some video processing techniques are absolutely necessary to show the digital video in the best possible way, while other techniques are fancy gimmicks that intrude on the goal of faithfully reproducing the video. In this section, I will show you which video processing features you should definitely use and which ones you should switch off.

Some video processing techniques can be performed by either the source component or the display. For example, the black and white level may be set on either your Blu-ray player or on the display. Other processing techniques are specifically tailored to one of these devices or may be available on both but be drastically different in their implementation. For example, the conversion of interlaced video to progressive video, called *deinterlacing* can typically be performed by your Blu-ray player as well as your display. However, one of these devices may do a much better job. There are ways to test the quality of certain video processing techniques as I will explain in the upcoming sections. You can use such tests to decide which component in your system is best suited to handle a certain video processing job.

The calibration of your display, as described in section 6.2, involves adjusting some controls, such as gamma, color, and white level. These controls are usually present on both your display and your source component. Thus, the question arises which of these components is best suited to handle the video processing associated with these controls. In other words, are you going to adjust the controls on your Blu-ray player or the controls on your display during the calibration process? For two reasons, I recommend you use the controls on your display, and leave the controls on the source component in their default or neutral position. First, setting the controls on your display ensures that all other components you have connected to the display benefit from the calibration, though you might have to copy the calibrated settings that you have obtained to each different input on the display, as most displays have independent picture control memories for each input. Second, the source component may not be using high enough precision in its video processing circuits. Such a lack of precision may result in irreversible changes to the video signal, such as clipping or compression of several levels (Spears and Munsil, 2013).

Video processing is nothing more than the manipulation of the numbers representing the digital video. The accuracy of such manipulations depends on the number of bits used to perform the necessary calculations. Since the number of bits is always fixed, round-off errors occur during these calculations (Jack, 2007; Pohlmann, 2011). These errors can accumulate from one processing step to the next and degrade the picture quality. Therefore, it is always best to use the largest bit depth that your equipment supports. All digital video equipment supports 24-bit processing (8 bits for each RGB or $Y'C_BC_R$ channel), because this is the minimum standard used for SD and HD video (ITU-R BT.601, 2011; ITU-R BT.709, 2015). Many source components and displays support 30 bits (10 bits per channel) or even 36 bits (12 bits per channel). It is recommended that you use these higher bit modes. Displays automatically switch to the appropriate mode, but on your source components you have to activate it in the setup menu. The setting you are looking for is usually in the HDMI submenu and labeled 'deep color' or simply '36 bits'. Enabling this option will not magically improve the colors of the video, but it ensures that the video processing steps are performed with maximum precision. Using maximum precision all the way up to the display avoids banding and posterization (see section 3.2.1) and preserves smooth color transitions and gradients.

> **Recommendation 6.32** *Configure your source components to process and output digital video with a bit depth of 30 bits (10 bits per channel) or 36 bits (12 bits per channel).*

6.3.1 Scaling

Your display should show the entire video image without cropping or zooming in. This may seem obvious, but in fact it is not. Displays are often in a mode called *overscan* where the image is slightly zoomed in such that on all four sides parts of the image are missing. Overscan was a useful feature during the days of analog television. On an analog set there was some uncertainty as to how the picture would be positioned on the screen. Overscan created a useful margin at the sides of the picture to deal with this uncertainty. On today's digital displays overscan is no longer needed. As a matter of fact it is highly undesirable. You want to be able to see the entire picture so the overscan mode of your display needs to be turned off. Look in your display's setup menus for a setting called 'overscan', 'screen size', or 'display area'. Try the various modes and find out which one shows the entire image (Spears and Munsil, 2013). You can use the test patterns listed in table 6.12 to check this. When viewing these

Table 6.12 Test patterns for checking scaling, overscan, and one-to-one pixel mapping.

AVS HD 709	Basic Settings	Sharpness and Overscan
	Misc. Patterns	Resolution
DVE HD Basics	Basic Patterns	Overscan
		Pixel Phase
	Advanced Patterns	Vertical Multiburst
DVS UHD	Basic Setups	Sharpness and Overscan
	Miscellaneous Setup Patterns	Resolution Patterns
HD Benchmark	Advanced Video—Setup	Framing
	Advanced Video—Evaluation	Image Cropping
		Luma Multiburst
		Scaling

patterns, make sure that your source component and display are configured for the right aspect ratio (16:9), otherwise the image might still be missing some parts (see section 5.1.1).

The goal is to show the video image on the display with a so-called *one-to-one pixel mapping* where every pixel in the digital video image corresponds to exactly one pixel on the display (EBU-Tech 3320, 2017). A one-to-one pixel mapping requires that the native resolution of your display matches the resolution of the video image. This is the case, for example, if you show HD video on a display with 1920 × 1080 pixels. A one-to-one pixel mapping also requires that the display shows the entire image without overscan. You can use the test patterns listed in table 6.12 to check whether your display is showing the video image with a one-to-one pixel mapping. The test patterns contain fine-grained lines and checkerboards in which the pixels alternate between black and white. If these fine-grained areas look evenly gray, your display has a one-to-one pixel mapping. If these areas show interference or otherwise blotchy patches instead, your display is scaling the image in some way.

Recommendation 6.33 *Configure your display such that it shows the entire video image with a one-to-one pixel mapping and without any overscan.*

On some displays it is not possible to achieve a one-to-one pixel mapping without overscan. In this case it is always better to choose the display mode that provides a one-to-one pixel mapping and accept some cropping on the sides of the picture. With a one-to-one pixel mapping the image will be at its sharpest and all its details will be preserved. Therefore, a small amount of cropping on the sides of the image will be less problematic overall than digital scaling of the image. In case you need to accept some cropping on the sides, make sure that the image is perfectly centered both horizontally and vertically. If your display has picture positioning controls, use these to center the image (Spears and Munsil, 2013).

Scaling of digital video is best avoided, but when the native resolution of the display doesn't match the resolution of the digital video source, it is unavoidable. A typical example is displaying HD video (1920 × 1080 pixels) on a 4K display (3840 × 2160 pixels). The display has twice as many lines and columns as the source material, thus every pixel in the source has to be mapped to four pixels on the display. The calculation of the additional pixels can be done in different ways. The mathematical term for it is *interpolation* (Poynton, 2012). Figure 6.13 shows three different interpolation techniques that are often used for upscaling digital video. The simplest technique is *nearest neighbor* in which the adjacent pixels are just replicated. More sophisticated is *linear interpolation* in which the luma (Y') and chroma values (C_B and C_R) of the adjacent pixels are used to predict the luma and chroma for the pixel in between. Figure 6.13 illustrates how this works for the luma value. A line is fitted through the luma values of the two adjacent pixels and that line is used to calculate the optimal luma value for the pixel in between. The interpolation procedure for the chroma values is similar. One step further is to fit a curve through the adjacent pixels instead of a line as is done in *cubic interpolation*. This results in a more accurate prediction of the luma and chroma values for the pixel in between. Since an image has two dimensions, interpolation to compute the additional pixels has to be done along both the horizontal and the vertical dimension. These two dimensions can be combined and that has led to the terms *bilinear* and *bicubic* interpolation (Poynton, 2012).

Image upscaling is best done using either bilinear or bicubic interpolation. The nearest neighbor approach is best avoided, because it creates jagged diagonal lines that look like tiny staircases. Video components differ in the technique they use for image upscaling and hence the resulting image quality may vary. For example, your Blu-ray player may do a better job upscaling the SD video from a DVD to HD video, than your display does. If this happens to be the case, you should configure your Blu-ray player to always output HD video instead of

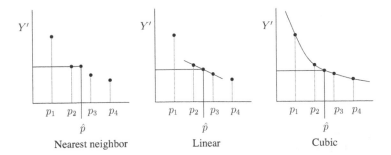

Figure 6.13
Different interpolation techniques for upscaling video images. Interpolation is used to compute the luma value Y' of a new pixel $\hat{p}$ positioned between two existing pixels p_2 and p_3. The nearest neighbor approach just copies the value from the nearest pixel. The linear interpolation technique computes the value for $\hat{p}$ by fitting a line through the two adjacent pixels. The cubic interpolation technique computes the value for $\hat{p}$ by fitting a curve through the nearby pixels.

Table 6.13 Test patterns for setting the sharpness control and for checking the smoothness of diagonal lines.

AVS HD 709	Basic Settings	Sharpness and Overscan
	Misc. Patterns—Various	Star Chart
DVE HD Basics	Basic Patterns	Overscan
	Advanced Patterns	Luminance Zone Plate
DVS UHD	Basic Setups	Sharpness and Overscan
	Miscellaneous Setup Patterns	Resolution Patterns
HD Benchmark	Video Calibration	Sharpness
	Advanced Video—Evaluation	Luma Wedge
		Luma Zone Plate

the native resolution of the source material. Try out different settings for the HDMI outputs of your source components and evaluate the smoothness of diagonal lines in the resulting image. You can use the sharpness test patterns listed in table 6.13 for this purpose.

6.3.2 Sharpness

Displays have a sharpness control that is used to enhance the high-frequency detail in the video image. The sharpness control does not add any detail to the image, it just makes certain edge boundaries appear sharper. This works by increasing the contrast in the immediate vicinity of the edge. Making one side of the edge a bit lighter and the other a bit darker increases the perceived sharpness of the edge. Some amount of edge enhancement is necessary to make the video picture appear pleasantly sharp. We all prefer sharp and clearly defined images over soft and blurry ones. The proper amount of edge enhancement strongly depends on the resolution of the source material and on the viewing distance. For example, on a 4K display, UHD video material inherently has more fine detail than upscaled SD video has. The upscaled video could benefit from some edge enhancement to make it appear sharper. Viewing distance is important too. The further away you sit from the screen, the less sharp the picture appears. Of course, the quality of your eyesight also plays a role, but in general at a larger viewing distance you need more artificial sharpening to make the image pleasantly sharp.

Too much artificial sharpening is bad. It introduces ugly artifacts in the picture: edges start to have halos and ghost lines; diagonal lines are no longer smooth and look like tiny staircases; and smooth areas close to an edge might start to show ripples. On most displays the default setting of the sharpness control is way too high. It should be turned down to avoid introducing distortion and artifacts.

Figure 6.14
Examples of different
sharpness control settings.
The *middle part* shows
a black line on a gray
background at the proper
sharpness setting. In the
***upper part* sharpness is set**
too high resulting in a white
halo around the black line.
In the *lower part* sharpness
is set too low resulting in
the line becoming fuzzy.

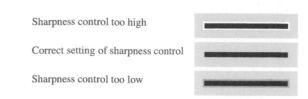

Sharpness control too high

Correct setting of sharpness control

Sharpness control too low

You can use the test patterns listed in table 6.13 to properly set the sharpness control. These patterns consist of multiple black lines on a gray background. The idea is that you set the sharpness control as high as possible without introducing white halos around the lines. As you turn up the sharpness control of the display, you should start to see faint halos around the lines. Turn sharpness up until the screen looks harsh and oversharpened when viewed from your regular seating position, then turn the sharpness control down again until all the lines and curves look smooth again. You have to strike the right balance. Setting sharpness too low will result in edges getting blurred and fuzzy. Setting sharpness too high results in undesirable white halos around edges. The setting of the sharpness control is illustrated in figure 6.14.

Another test pattern to check out is the zone plate. It consists of a series of concentric circles. If the sharpness control is set too high, the circles will show stairstepping and strong interference patterns (moiré) will be present. You should turn the sharpness control up and down and see how the interference patterns change (Spears and Munsil, 2013).

> **Recommendation 6.34** *Use a test pattern that has black lines on a gray background to adjust the sharpness control of your display. Set the sharpness control as high as possible without introducing white halos around dark lines.*

In addition to edge enhancement, displays and also many source components employ other more advanced techniques to increase the apparent sharpness of images in particular areas of fine textured detail such as wood, grass, hair, and skin texture. Enhancing such textures makes an image more lifelike and creates a sense of depth. Manufacturers use different names for these controls; look through the setup menu for items that are labeled 'detail enhancement', 'depth enhancement', 'reality creation', 'super resolution', or 'resolution remaster'. If your video components have such controls it is worthwhile testing out their different settings. Again, do not set these controls too high as that may introduce artifacts and result in harsh and oversharpened images.

6.3.3 Output Color Space

Your display creates an image using different subpixels for the red, green, and blue primary colors. To activate these subpixels your display needs the *RGB* components for each subpixel. Well, actually it uses the $R'G'B'$ gamma corrected ones (see figure 3.7). However, digital video is not stored in the *RGB* format, but as separate luma and chroma channels. As I explained in section 5.1.4, the luma channel Y' represents the luminance information of the image, and the two chroma channels C_B and C_R represent the color information. This $Y'C_BC_R$ video format is used by lossy video codecs to reduce the number of bits needed to represent the image. These codecs exploit the fact that the human eye is more sensitive to changes in brightness than to changes in color. The codecs store the luma channel Y' at its full resolution, but they reduce the resolution of the chroma channels. This data-reduction technique is called *chroma subsampling*. With respect to this subsampling there are three

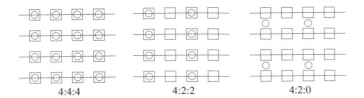

4:4:4   4:2:2   4:2:0

Figure 6.15
Chroma subsampling. A *box* indicates the center of a luma Y' sample. A *circle* indicates the center of a chroma sample pair C_B and C_R. Note that the chroma samples for 4:2:0 are between the scan lines. They are computed as the average of the two surrounding lines.

$Y'C_BC_R$ formats that are widely used and standardized. They are denoted by three digits separated by a colon as follows:

- 4:4:4—No chroma subsampling. For every pixel, there is a Y' value, a C_B value, and a C_R value.
- 4:2:2—Horizontal chroma subsampling. For every four pixels on a line, there are four Y' values, two C_B and two C_R values for the even lines and two C_B and two C_R values for the odd lines. For example, a 1920×1080 HD video image is stored using 1920×1080 Y' values, 960×1080 C_B values, and 960×1080 C_R values.
- 4:2:0—Horizontal and vertical chroma subsampling. For every four pixels on a line, there are four Y' values, two C_B and two C_R values for the even lines, and zero C_B and zero C_R values for the odd lines. For example, a 1920×1080 HD video image is stored using 1920×1080 Y' values, 960×540 C_B values, and 960×540 C_R values.

Figure 6.15 clarifies this somewhat awkward notation. Digital video stored on Blu-ray discs and DVDs is always in the 4:2:0 format. Most digital video files are also in the 4:2:0 format.

In order to display digital video, the full chroma resolution is needed. Therefore, the subsampled 4:2:0 format from the digital video source needs to be upsampled. This upsampling can be handled by your display, or your source component. If you route the video through an A/V receiver or A/V controller, that device may also be used for chroma upsampling. On many source components you can set the HDMI output to either 4:2:0, 4:2:2, or 4:4:4. If you choose to set it to 4:2:0 or 4:2:2, then your display still needs to upsample the chroma to 4:4:4. You would expect that it doesn't matter which device performs the chroma upsampling, but there are large quality differences. After the chroma has been upsampled to the 4:4:4 $Y'C_BC_R$ format, one final step remains: the conversion from $Y'C_BC_R$ to $R'G'B'$. Again this conversion can be handled by your display, your source component, or your A/V receiver or controller. And again, they may all perform differently. Therefore, it is worthwhile to determine which device in your system performs best.

Choosing which device handles the chroma upsampling and/or the conversion to $R'G'B'$ is commonly referred to as choosing a *color space*. Most source components have an item in their setup menu labeled 'HDMI color space', or 'HDMI output'. You can set it to:

- *4:2:0 color space:* The device doesn't perform any upsampling. It outputs the digital video in $Y'C_BC_R$.
- *4:2:2 color space:* The device only upsamples the chroma from 4:2:0 to 4:2:2. It outputs the digital video in $Y'C_BC_R$.
- *4:4:4 color space:* The device upsamples the chroma from 4:2:0 to 4:2:2, and subsequently to 4:4:4. It outputs the digital video in $Y'C_BC_R$.
- *RGB color space:* The device upsamples the chroma from 4:2:0 to 4:2:2, and subsequently to 4:4:4. It also converts the $Y'C_BC_R$ to $R'G'B'$ and outputs the digital video in $R'G'B'$.

The 4:2:0 option is only available if your components support HDMI version 2.0 or higher; lower versions of HDMI require a video signal that is upsampled to at least 4:2:2.

Figure 6.16
Four different output color space configurations when a source component is directly connected to the display. In the *top row* the output of the source component is set to 4:2:0; in the *next row* to 4:4:2; then to 4:4:4, and in the *bottom row* it is set to $R'G'B'$.

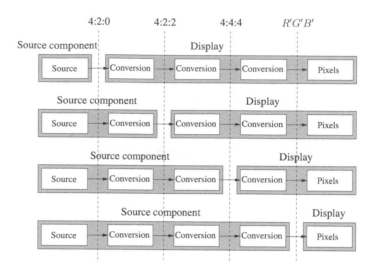

No matter how you set the output color space of your source components, in the end the digital video has to be converted to $R'G'B'$. The choice of the color space of your source component only determines which device handles the upsampling and conversion. Figure 6.16 shows the different possibilities when you connect your source component directly to your display. Which of these possibilities yields the best results depends on the particular implementation used in your player and display. It also depends on the way that the player and display interact. The internal video processor of some devices can only handle one particular format. If you feed it another format, it will first convert it back to the format it supports. For example, some video processing chips only support 4:2:2 as an input. If you feed it 4:4:4 or $R'G'B'$, the chip will first convert the input signal back to 4:2:2, and only then apply its particular processing steps. This conversion introduces an extra processing step that can potentially degrade the image. The only way to find out how to optimally configure the color spaces on your devices is to test all the possibilities listed in figure 6.16 and evaluate the image quality (Spears and Munsil, 2013).

If you route the video from your source component through your A/V controller or receiver, things get a bit more complicated, because now you have to choose a color space for the output of your player and also for the output of your A/V controller or receiver. You need to test and evaluate all the different combinations of color space settings on these two devices as there is no other way to tell what will work best. It is often better to avoid routing your digital video through you A/V controller or receiver, because you eliminate the possibility that the controller or receiver degrades the quality of the digital video. For this reason, many Blu-ray disc players have two HDMI outputs; one HDMI output is used to send the digital video directly to the display, while the other HDMI output is used to send the digital audio to the A/V controller or receiver (see figure 5.16).

You can use the HD Benchmark test disc to determine the optimal color space settings of your video components. The producers of this disc, Spears and Munsil, have devised special test patterns to evaluate certain image-quality aspects associated with chroma upsampling and $Y'C_BC_R$ to $R'G'B'$ conversion. These patterns are listed in table 6.14. The main idea is to systematically test all the different combinations of color space settings (see figure 6.16). For each combination, you look at the test patterns and judge the image quality. The main things to look for are:

1. *Chroma alignment:* The luma and chroma channels should be perfectly aligned. Mistakes or shortcuts in the chroma upsampling algorithm can cause misalignment.
2. *High-frequency detail:* High-frequency chroma bursts (patterns of fine horizontal or vertical lines) should have clear, bright colors that look identical to the colors in the

low-frequency bursts. Problems in the 4:2:2 to 4:4:4 conversion show up as muted vertical lines in the horizontal bursts, while problems in the 4:2:0 to 4:2:2 conversion show up as horizontal muted lines in the vertical bursts.

3. *Upsampling quality:* The chroma channels should be upsampled using a bilinear or bicubic algorithm (see figure 6.13). These algorithms produce smooth and clean chroma transitions. The inferior nearest neighbor upsampling method should be avoided, because it creates jagged color contours.

4. *Smooth diagonals:* Diagonal chroma bursts should be smooth without any jaggedness or stairstepping. The smoothness of the diagonals depends on the upsampling method used. The nearest neighbor method produces jagged diagonals, while the bilinear and bicubic methods yield smooth ones. Another reason for jagged diagonals is the *chroma upsampling error* (CUE) explained in box 6.3.

On their website Spears and Munsil (spearsandmunsil.com) explain the use of the test patterns for color space selection in elaborate detail. You can also download a *color space evaluation form* from their website. This form is a great help in evaluating the different color space settings. In addition to the chroma upsampling quality, Spears and Munsil also recommend that you check for the following possible differences between the color space settings:

1. *Smooth ramps:* Grayscale and color ramps should look smooth and even, with no bands or streaks anywhere along it.

2. *Clipping:* None of the $Y'C_BC_R$ or $R'G'B'$ should be clipped.

3. *Color conversion:* The colors should be decoded according to the Rec. 709 specification for HD video (ITU-R BT.709, 2015), and not according to the Rec. 601 specification for SD video (ITU-R BT.601, 2011).

4. *Chroma range:* The entire chroma range should be used. Some color conversion chips erroneously limit the chroma range which causes subtle color decoding errors.

Recommendation 6.35 *Use the HD Benchmark test disc to determine the output color space of your Blu-ray disc player (4:2:0, 4:2:2, 4:4:4, or R'G'B') that results in the best-quality chroma upsampling.*

It is important to calibrate your display (see section 6.2) before you start judging the test patterns for color space evaluation. You need to perform the display calibration separately for each of the different color spaces: 4:2:0, 4:2:2, 4:4:4, and $R'G'B'$. Once you have set the display controls correctly for each different color space, you can start the evaluation and use the *color space evaluation form* to register if a particular color space setting passes or fails the image-quality tests listed above. The color space setting that passes the largest number of these tests is the one you should use. Some differences are subtle. So it helps to sit close to the screen, much closer than your normal viewing distance, during this evaluation.

Table 6.14 **Test patterns for evaluating color space conversion.**

HD Benchmark	Advanced Video—Evaluation	Color Space Evaluation
		Clipping
		Dynamic Range RGB
		Dynamic Range YCbCr
		Chroma Alignment
		Chroma Multiburst
		Chroma Wedge
		Chroma Zone Plate
		CUE

BOX 6.3 CHROMA UPSAMPLING ERROR (CUE)

The chroma upsampling error (Spears and Munsil, 2003) results in colored streaky horizontal lines on diagonal edges. This error occurs in interlaced video when the chroma upsampling from 4:2:0 to 4:2:2 or 4:4:4 is done incorrectly. Correct chroma upsampling requires the use of two different algorithms: one for interlaced video that originates from progressive video, and another one for video that is interlaced by origin. It is important to distinguish between these two types of video. As I explain in more detail in the upcoming section 6.3.4, when a progressive source (such as film) is converted to interlaced video, the two resulting fields can be used to form one frame. The reason is that the two fields were originally captured at approximately the same time instant. In this case chroma upsampling should be performed using all four chroma samples from both fields. On the other hand, in video that was originally captured in interlaced format, the fields should be treated separately. The reason is that the two fields were captured at two different time instances and therefore may be completely different. In this case chroma upsampling should be performed for each field separately. The difference between the two different ways of upsampling is illustrated in figure 6.17.

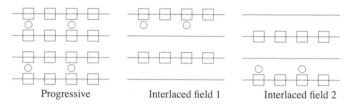

Progressive   Interlaced field 1   Interlaced field 2

Figure 6.17
Chroma subsampling 4:2:0 in interlaced video. A *box* indicates the center of a luma Y' sample. A *circle* indicates the center of a chroma sample pair C_B and C_R. If the interlaced video originates from progressive source material, the two fields can be combined into one progressive frame and all four chroma samples can be used to interpolate. If, on the other hand, the video is interlaced by origin, two fields should be kept separate and only two chroma samples should be used for the interpolation.

6.3.4 Deinterlacing

Modern displays work in progressive scanning mode. Recall from section 5.1.3 that in progressive video the pixel lines are shown one after another, while in interlaced video the even and odd lines are shown in an alternate fashion. Modern displays are able to show interlaced video, but only after it has been converted to a progressive format. The conversion from interlaced to progressive scanning is called *deinterlacing*. Displays and source components contain video processing chips that are able to accomplish this conversion. These chips are called *deinterlacers*.

Interlaced video is produced by an analog or old digital video camera. In the past many TV shows and documentaries were shot in an interlaced video format. Everything that has been shot with a modern digital camera is in a progressive format. The progressive format is preferred over the interlaced one, because it results in a higher vertical resolution. In progressive video all lines in the image are used at the same time to form a frame. By contrast, a frame in interlaced video is made up of two separate fields. The first field only contains the odd lines, while the second field only contains the even lines. These fields are captured and displayed sequentially in time, so there is a slight delay between them. In NTSC video this delay is 1/60 s, resulting in a frame rate of 30 fps. In PAL video this delay is 1/50 s, resulting in a frame rate of 25 fps. As long as the camera doesn't move and everything in the scene is static, there is no difference between a frame of interlaced video and a frame of progressive video. They are exactly the same. However, when the camera moves or when objects in the scene are moving,

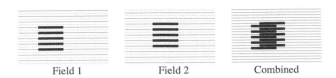

Field 1   Field 2   Combined

Figure 6.18
Field tearing in interlaced video. A black square moves from left to right on the screen. Due to the slight delay between the capture of the two interlaced fields, the square appears in two different positions on the screen. When the two fields are combined to form a complete frame, the displacement of the black square results in jagged edges.

the frames differ. In a progressive frame the even and odd lines are captured at approximately the same time instant. They go together quite well. By contrast, the interlaced frame consists of two separate fields that were captured at two different time instances. As a result, fast moving objects appear at different positions in the two fields. When you combine the two fields the odd and even lines do not always go well together and the edges of moving objects become jagged (Poynton, 2012). This effect is called *field tearing* and is illustrated in figure 6.18.

The simple deinterlacing method of combining two interlaced fields is called *weave* or *field replication* (Poynton, 2012). It only works well if the camera and scene are static. In all other cases motion artifacts occur in the reassembled frame. An alternative deinterlacing method uses interpolation to compute the missing lines in each field. This method is simply called *interpolate*. In this case no motion artifacts occur, but the vertical resolution is cut in half and is only restored by approximating the missing lines using one of the interpolation techniques from figure 6.13. As a result the image appears less sharp and important detail is lost. It would be best if the deinterlacer switches between the weave and interpolate methods depending on the content of the fields. This is in fact what good deinterlacers do, but there are different ways to accomplish this. Let's start with the most simple method and then move on to the more sophisticated techniques.

Interlaced video doesn't need to be inherently interlaced. It can be derived from a progressive source. The most common example is when the original material was shot on film and later converted into interlaced video and put on DVD. Film is inherently progressive: the entire frame is captured at the same time instant. However, DVDs always contain interlaced video, either in the NTSC format (60 fields or 30 frames per second) or in the PAL format (50 fields or 25 frames per second). The DVD standard was developed when analog televisions and displays still reigned supreme. The DVD standard leaves no room for progressive video—only interlaced formats are allowed (Taylor et al., 2006).

A good deinterlacer recognizes when interlaced video originates from a progressive source, and it utilizes this fact to reconstruct the progressive frames with the weave or field replication method. To understand how this works, I first need to explain how film is converted to interlaced video using a process called *telecine*.

Film is typically shot at 24 fps, but the interlaced NTSC format requires 30 fps or equivalently 60 fields per second. The telecine method used for the conversion from 24 fps progressive video to 30 fps interlaced video is called *2:3 pulldown* or *3:2 pulldown*. This method is illustrated in figure 6.19. Each film frame (X) is separated into two fields: one containing the odd lines (X1) and one containing the even lines (X2). The first film frame (A) is shown as two fields (A1 and A2) at a rate of 60 fields per second. This creates a slight time difference, because the film frame would take up 1/24 s, while the two interlaced fields together only take up 1/30 s. The next film frame (B) is also shown as two fields (B1 and B2), but to make up for the time difference the first field (B1) is repeated once. Thus, two film frames are represented by five fields. The five fields at a rate of 60 fields per second, take up 5/60 = 1/12 seconds, which is exactly equal to the time it takes to show two film frames: 2/24 = 1/12 s. The 2:3 pulldown method introduces a time irregularity in the interlaced video: the first film frame takes up 2/30 s while the second film frame takes up 3/30 s. The alternation between these two different timings causes a motion artifact called *judder*. This artifact is especially visible when the camera pans slowly from one side to the other: the image doesn't move smoothly as in the original film material, but it appears to have a certain jerkiness (Poynton, 2012; Taylor et al., 2009).

Figure 6.19
2:3 pulldown. The four
film frames on the *left* at
24 fps are converted into
the interlaced video fields
on the *right* at 30 fps.
The *broken lines* indicate
the time irregularity that
is introduced by this
conversion. The *letters* in
the video fields indicate the
corresponding film frame.
The *numbers 1 and 2* refer
to the first and second field
of interlaced scanning.

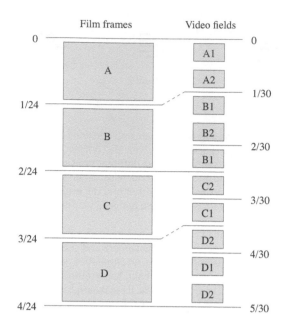

The interlaced PAL format requires 25 fps or equivalently 50 fields per second. The telecine method that converts 24 fps progressive video to 25 fps interlaced video is called *2:2 pulldown*. It is a rather simple method in which each film frame is shown as two fields. To make up for the 4% time difference between 24 fps and 25 fps, the audio track is played at a slightly increased playback speed. This causes a small pitch shift that can be electronically compensated (Poynton, 2012; Taylor et al., 2009). The *2:2 pulldown* method introduces no timing irregularities and thus doesn't result in judder. An alternative pulldown method that avoids the audio pitch shift completely is to repeat one field after every 12 film frames, such that the sequence of fields becomes: 2:2:2:2:2:2:2:2:2:2:2:3. This method is known as *12:1 pulldown* (Jack, 2007). This method introduces a slight timing irregularity, which is less noticeable than the irregularity of the 2:3 pulldown method for interlaced NTSC video.

A deinterlacer can perfectly reconstruct the original progressive frames from the interlaced fields, but it needs to be able to figure out which fields should go together. There are two methods to identify progressive frames in interlaced video: 1) flag reading and 2) cadence reading. Flags are used by a video encoder to indicate that a certain frame in an interlaced video sequence is made up of two fields that originate from the same film frame. When the flag is set the deinterlacer simply combines the two fields to reconstruct the original progressive frame. When the flag is not set, the deinterlacer should probably use an interpolation technique. In practice flag reading is not very reliable, because the flags are not used consistently and on some material they are not used at all. A deinterlacer that solely relies on the correct flags will never be able to reconstruct the progressive frames from such material. In such cases the second method, cadence reading, would be a better choice. A deinterlacer that uses cadence reading constantly compares the data in several successive fields. By identifying repetitive fields the deinterlacer is able to recognize the 2:3 cadence. A good cadence-reading deinterlacer should also be able to recognize a 2:2 cadence and be able to deal with problems and glitches in the cadence. Correctly identifying a 2:2 cadence is much harder and requires more sophisticated methods, because there are no repetitive fields in such a cadence (Munsil and Florian, 2000; Taylor et al., 2006).

The best deinterlacer continuously examines the content of several successive fields in the video stream, and based on that decides how to combine the different fields. When it detects progressive frames, it goes into the so-called *film mode* and merges the fields such that the progressive frames are perfectly reconstructed. When it doesn't detect progressive frames it goes into *video mode*. In this mode it is assumed that each field belongs to a different frame. Since only half of the lines of each frame are known, the deinterlacer has to figure out how to compute the missing lines. Several interpolation techniques can be used for this purpose. The simplest method is to just repeat each line (nearest neighbor interpolation). A bit better is to use bilinear or bicubic interpolation. More sophisticated interpolation methods go beyond such intrafield interpolation and use data from previous and subsequent fields. The best deinterlacers are able to segment each field into areas that are stationary and areas that are changing and treat them differently. In *motion adaptive deinterlacing* the static areas are combined using the weave method, while for the changing areas interpolation is used. In *motion compensated deinterlacing*, the deinterlacer calculates motion vectors for the changing screen areas and uses these vectors to adapt its interpolation direction (Jack, 2007; Poynton, 2012; Taylor et al., 2006). These sophisticated techniques are also referred to as *edge adaptive deinterlacing*.

Modern displays and source components are often equipped with advanced deinterlacing chips. They automatically detect the optimal deinterlacing mode from the video sequence. On some older equipment you may need to manually activate automatic deinterlacing in the setup menu. Unfortunately, the menu labels differ among players and can be quite cryptic. Look for 'auto', 'film mode', or 'cinema conversion'. Be sure to check out the user manual or search the Internet for the correct setting on your player.

> **Recommendation 6.36** *If your video equipment has an item in the setup menu for the deinterlacing mode or progressive mode, set it such that film mode with cadence detection is engaged.*

Deinterlacing can be performed by your source component, your display, or even by your A/V controller or receiver. Good equipment reviews (see table 6.1) usually report on the quality of the deinterlacing. Typically three situations are tested: 1) 2:3 pulldown of film-based material; 2) 2:2 pulldown of film-based material; and 3) interpolation of video material, often called 'jaggies' or 'motion adaptive (MA)' test. If you want to perform your own deinterlacing tests you can use the test patterns listed in table 6.15. Typically, you configure your Blu-ray player to output progressive video, and thus the deinterlacing is done in the player and not in the display. In general, Blu-ray disc players tend to have more advanced video processing capabilities than most displays do. So it is best to let your player perform the deinterlacing. If you want to test the deinterlacing performance of your display you can temporarily change the output setting of your Blu-ray disc player such that the interlaced video reaches the display.

One more thing, the most common format used on Blu-ray discs is progressive video with 24 fps. Modern video equipment should be able to pass this material unaltered. However, some devices may actually convert the 24 fps video to a 60 fps 2:3 pulldown version. You do not want that, because it introduces judder. On some Blu-ray disc players you may need to go into the setup menu and properly configure the output to pass 24 fps video. You can use test patterns from the HD Benchmark disc (Video Processing—24p) to check whether your

Table 6.15 Test patterns for checking deinterlacing.

AVS HD 709	Misc. Patterns	Progressive Motion
		Interlaced Motion
		Deinterlacing
		Numbered Frames
		Numbered Fields
HD Benchmark	Video Processing	Source Adaptive
		Edge Adaptive

equipment is set up properly for 24 fps progressive video. Some Blu-ray disc players have an additional feature that you can activate to convert the 60 fps 2:3 pulldown that results from playing a DVD into progressive video with 24 fps. This feature is called 'DVD 24p conversion'. On DVDs that originate from a progressive source, it restores the original 24 fps film frame rate and eliminates the judder from the 2:3 pulldown process, which of course is a good thing.

6.3.5 Motion Resolution

On most video displays moving objects may appear less sharp than stationary ones. This phenomenon is called *motion blur*. It is particular prevalent on LCD, LCoS, and OLED displays. These types of displays use a *sample-and-hold* method to display the video images: each video frame remains on the screen until it is time for the next one. In the time period between two frames the image on the screen doesn't change; it is stationary. For static scenes this is fine, but with moving objects it causes problems. When an object moves, your eyes are going to track the object across the screen. This is a natural response of the human visual system that allows you to see details in moving objects. Your eyes try to smoothly track the object along its trajectory. However, due to the sample-and-hold nature of the display, the moving objects do not smoothly move in time. They are fixed in one position and when the next video frame appears on the screen, they jump to the next fixed position. Since your eyes track the object faster than the refresh rate of the display, the moving objects appear to be blurred. There is a loss of detail and sharpness compared to the stationary parts of the scene. In other words, the temporal definition of the video image is reduced (see section 3.2.2). The apparently reduced resolution is called the *motion resolution*. It is measured in the number of visible lines. As sensitivity to motion blur differs from person to person, motion resolution is a highly subjective measure.

The key to reducing motion blur is to increase the refresh rate of the display. With a higher refresh rate, the time that each video frame is held stationary on the screen is reduced. There are essentially three methods to increase the refresh rate: 1) *black insertion*, 2) *frame doubling*, and 3) *frame interpolation*. Modern displays use one of these techniques or a combination of them to increase their refresh rates from 60 Hz to 120 Hz, 240 Hz, or higher (Fleischmann, 2017).

Black insertion reduces motion blur by inserting completely black frames between the existing video frames (Hong et al., 2005). The additional black frames simply reduce the time that each image is held stationary on the screen. With these added black frames it is also necessary to increase the refresh rate of the display to at least 120 Hz. Otherwise the blackness between the original frames would induce visible flicker. The main drawback of black insertion is that it reduces the perceived brightness of the display, sometimes significantly. The reason is that the human visual system integrates the luminance output over short periods of time, and due to black insertion the display is completely black between two successive frames.

On LCDs, black insertion can be done by flashing the backlight on and off. This technique is known as *backlight flashing* or *backlight scanning* (Feng, 2006; Fisekovic et al., 2001). The liquid crystals in the display have a finite response time that lies typically between 3 and 16 ms, depending on the specific type of crystals used (VA panels are the fastest, while IPS panels are the slowest). The slower liquid crystal panels cannot be driven at a very high refresh rate: refreshing every 16 ms only yields a rate of 62 Hz. The LEDs in the backlight can be flashed much faster, within a fraction of a millisecond, and thus can reach much higher refresh rates.

Frame doubling reduces motion blur by simply duplicating frames and doubling the frame rate. Since the duplicated frames are exactly the same, this method must be combined with black insertion in order to reduce the time that the image on the screen remains the same. One way to achieve frame doubling is to flash the backlight two times for each original frame. The disadvantage of frame doubling is that it introduces judder (Jack, 2007; Roberts, 2002). As illustrated in figure 6.20, the frame that was duplicated shows a moving object in exactly the

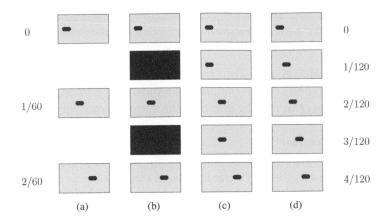

Figure 6.20
Different methods that reduce motion blur. A ball moves with constant speed from the left side of the screen to the right. Each *column* shows a different method to display the moving ball. Each *row* corresponds to a time instant. The time instances are indicated on the far *left* and *right* in seconds. *(a)* Displayed at 60 fps. *(b)* Displayed at 120 fps with black frame insertion. *(c)* Displayed at 120 fps by showing each frame twice. *(d)* Displayed at 120 fps using motion prediction to interpolate frames.

same position as in the original frame. However, you expect to see it in a different position: it should have moved a bit further during the time between the two frames. This disruption of the expected movement results in a certain jerkiness and is the cause of the perceived judder.

Frame interpolation reduces motion blur by inserting new frames that are computed from existing past and future frames. Advanced temporal interpolation techniques are used that separate each frame in static areas and moving objects. The direction and speed of the moving objects are estimated from the existing frames and used to compute the new frames (Choi et al., 2007; Kang et al., 2007). Motion-compensated frame interpolation (MCFI) results in extremely smooth and fluid motion and also significantly reduces motion blur. Moving objects appear much sharper. However, the interpolation is not always able to do a perfect job, and sometimes visual artifacts may occur in the interpolated frames. These small glitches only appear for a fraction of a second and are usually most noticeable with fast camera pans.

Motion-compensated frame interpolation reduces motion blur without introducing judder. In fact, it also removes most of the judder inherent in 24 fps film-based material. The resulting ultrasmooth motion is not to everyone's liking, because it destroys the cinematic look of the movie. In movie theaters every frame of the original 24 fps film is shown twice to avoid flicker. This frame doubling increases the frame rate to 48 fps, but at the same time introduces a small amount of judder. Lots of people associate this judder with a signature cinematic film look. Smoothing the motion and removing the judder with frame interpolation makes the movie appear like it has been shot using a video camera. The popular name for this effect is the *soap opera effect*. This name stems from the fact that most daytime soap operas are shot on video instead of film and therefore don't have the distinctive judder of film. Not everyone is bothered by the soap opera effect. Some people simply don't notice it. Others even like it, because it significantly increases the clarity of the entire image during camera pans and reduces motion blur on fast moving objects.

Display manufacturers use proprietary hybrid techniques to reduce motion blur in which they combine black insertion with motion-compensated frame interpolation. They offer different settings for motion control in the display's setup menu. Which of these modes you prefer depends on your sensitivity to motion blur and your tolerance of the soap opera effect. Your optimal setting is a matter of personal taste. The different options in the setup menu correspond to different trade-offs between these two. Unfortunately, manufacturers all use different names for their settings that control motion blur. Controls with names like 'smooth' or 'judder' are usually biased towards frame interpolation, while controls with names like 'clear', 'impulse', 'blur', or 'cinema' often apply black insertion. The best way to find out what the different motion control modes on your particular display do is to try them all out, one by one, and evaluate their effect on the video image. The motion test patterns of the HD Benchmark test disc (Video Processing—Motion) are great for evaluating the resulting motion resolution. In addition to these test patterns you should also watch some film-based material to judge the extent of the soap opera effect that each mode creates.

Recommendation 6.37 *Try out the different motion processing modes of your display, and choose the one that suits your personal taste with respect to motion blur and the soap opera effect.*

6.3.6 Image Enhancement

Many displays and source components provide additional image enhancement settings in their setup menu, besides the ones already discussed above. The most common enhancements are the ones designed to change contrast, change colors, and reduce image noise. Most of these special enhancements change the original source material such that it deviates from the creator's original intentions. Thus to achieve accurate reproduction they are best turned off. Some enhancements are specifically designed to improve low-quality video material but actually degrade the image quality of high-quality material like Blu-ray discs. While you might want to activate them occasionally to improve your viewing experience of a particular low-quality piece, these enhancements are not suitable for general viewing. So don't forget to turn them off after you have finished watching a low-quality source (Spears and Munsil, 2013).

Many contrast enhancements try to improve shadow detail by changing contrast based on the average brightness of the picture. Their names and implementations differ. Examples are: 'adaptive contrast', 'contrast enhancement', 'shadow detail', 'black correction', and 'auto black level'. Such enhancements completely mess up the display's gamma curve and are thus better left turned off.

Recommendation 6.38 *Turn off all automatic contrast enhancements on your display and source components.*

Dynamic contrast is another contrast enhancement that is found on many displays. It is best turned off. Dynamic contrast automatically controls the brightness of the display and makes dark scenes much darker. On LCD displays this is accomplished by dimming the backlight. On projectors it is done by controlling a mechanical iris or aperture that sits right behind the projector's lens. Varying the size of this iris controls the total light output of the projector (Brennesholtz and Stupp, 2008). Dynamic contrast can significantly increase sequential contrast as it makes dark scenes darker. However, it does nothing to improve simultaneous contrast as the entire screen, including the brighter parts, are made dimmer. On some projectors the iris does not adapt fast enough and lags a bit behind. This creates a visible effect known as *image pumping* or *breathing*.

Recommendation 6.39 *If possible, turn off automatic backlight adjustment and automatic iris adjustment on your display.*

One exception to the general advice to turn off all contrast enhancements is the local dimming feature found on high-end LCD displays (see section 6.1.2). Local dimming should always be turned on, because it significantly improves both sequential and simultaneous contrast. Many displays let you control the amount of local dimming that it applies. You should adjust this amount such that you achieve the darkest blacks without seeing any noticeable halos or blooming around smaller bright parts in a dark scene. The visibility of halos and blooming strongly depends on the amount of ambient light in the room. Therefore, you should adjust the amount of local dimming while watching your display in your typical ambient light conditions. You can use the uniformity test patterns of the HD Benchmark test disc (Advanced Video—Evaluation) to fine-tune the local dimming settings.

> **Recommendation 6.40** *Turn on local dimming on LCD displays. Adjust the amount of local dimming such that you achieve the darkest blacks without seeing any noticeable halos or blooming around smaller bright parts in a dark scene.*

Turn off all automatic color enhancements on your display and source components. I already touched upon this subject in section 6.2.4, but it is worth repeating. Artificial color enhancement messes up the accurate reproduction of colors on your display. Therefore, disable all items in the display's setup menu that have names like 'color enhancement', 'vivid color', 'live color', 'skin tone', and 'color remapping'. Some HDR-capable displays have a mode in which standard dynamic range (SDR) video is automatically converted into HDR video. Such a conversion messes up both the contrast and the colors of the original material, so it is also best left turned off. Example items to look for in the setup menu include 'HDR+', 'HDR tone management', 'extended dynamic range', and 'UHD color'.

> **Recommendation 6.41** *Turn off all automatic color enhancements on your display and source components.*

Noise reduction is a common image enhancement feature found on most displays and source components. It can be used to reduce the amount of noise in low-quality source material. However, it is better not used on high-quality material, as it may cause a loss of details. Only use noise reduction when you need it, and don't leave it on by default. Noise reduction comes in two flavors: analog noise reduction and digital noise reduction. The analog variant is aimed at reducing the noise that has invaded the picture during capture and production. Analog noise reduction reduces the roughness of the picture and eliminates most of the flickering dots that are mostly visible in plain backgrounds. The setup menu item for this type of noise reduction is often simply called 'noise reduction'. The other variant, digital noise reduction, is aimed at reducing compression artifacts in digital video. These artifacts are visible as small rectangular tiles in smooth backgrounds and as small white dots in the neighborhood of sharp edges. The setup menu item for this type of reduction is often labeled 'MPEG noise reduction' or 'Mosquito noise reduction'.

> **Recommendation 6.42** *Only use noise reduction on your display and source components when you need it on low-quality material. Don't leave it on by default.*

7
AUDIO EQUIPMENT

Good sound requires high-quality loudspeakers, amplifiers, and DACs. While some obsessive audiophiles claim that everything matters, these three parts of the audio chain have the most influence on the resulting sound quality. The quality of the loudspeakers is by far the dominant factor, followed by the quality of the amplifier and the way that the amplifier and loudspeakers play together. Sound-quality differences between DACs do exist but are much less pronounced today. Technology has advanced to a point where high-quality DACs are becoming mainstream even in modestly priced equipment.

Not all loudspeakers sound the same. There is a huge difference in sound rendition among the different models. The perfect loudspeaker doesn't exist. Rendering all the nuances of real-life sounds in a believable way is quite a difficult task to perform. Even with today's marvelous technology it is all about compromise. Though loudspeaker design is based on engineering principles, it is still for a large part art. Designers have to make different engineering trade-offs, and their choices result in a huge variety in loudspeaker designs (Newell and Holland, 2007; Winer, 2018).

Different amplifiers also do sound different, especially when combined with certain loudspeakers (Cordell, 2011; Duncan, 1997). It is important to drive your loudspeakers with an amplifier that is a good match. Some loudspeakers sound fabulous when driven by a high-end amplifier, but when you connect them to a less powerful receiver those same loudspeakers

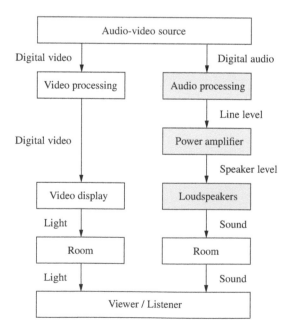

may sound a lot less satisfying. Never buy an amplifier or loudspeaker without listening to it. Most high-end audio stores have a dedicated listening room in which you can audition audio equipment that you consider buying. Always audition the combination of amplifier and loudspeaker and evaluate the whole. Thus, when you are out there listening to new loudspeakers, make sure that you drive your prospective loudspeakers with the same amplifier model that you have at home. Some stores even allow you to take the loudspeakers home and audition them in your own listening room (Harley, 2015).

As explained in section 2.1.2, a basic audio system consists of a receiver, five main loudspeakers, and one or preferably two subwoofers. The receiver contains all the DACs and the power amplifiers that drive the main loudspeakers. Most subwoofers have their own built-in power amplifier. It is important that all components in your audio system are of comparable quality. It doesn't make sense to spend an exorbitant amount of money on top of the line loudspeakers, when you are using a mediocre receiver to drive them. The amount of money that you spend on the different components should be balanced. That being said, money is not always a good indicator of sound quality. Some components are created purely to maximize manufacturer's profit while others are made to sound their best for a given amount of money. Quality differences can be huge, so be on the lookout for some real bargains out there. As a rule of thumb, companies that specialize in audio equipment build higher quality components than companies that manufacture a wide range of electronic devices. It is also important to realize that the look and feel of a component doesn't always reflect its sound quality. Some manufacturers of dedicated high-end audio equipment use beautiful designs and materials to create components that have an elegant appearance and a luxury feel. The gorgeous packaging of these components typically costs a lot of money and rarely contributes to the sound quality of the electronics that are inside (Harley, 2015; McLaughlin, 2005).

7.1 Loudspeakers

A loudspeaker converts the electrical signal from the amplifier into sound. The word 'loudspeaker' can either refer to a loudspeaker driver or to a loudspeaker system. A *loudspeaker driver* is a single transducer that converts electrical energy into acoustical energy. A *loudspeaker system* is the combination of one or more drivers mounted in an enclosure or cabinet (Small, 1972b). In this section I will focus on loudspeaker systems with multiple drivers that are intended to reproduce a large part of the audible frequency spectrum. In other words, I focus on the front and surround loudspeakers. Subwoofers are dealt with in section 7.3.

High-performance loudspeakers consist of multiple drivers, because no single driver can reproduce the entire audible frequency range with sufficient quality. The reproduction of low frequencies requires a driver with a large diaphragm while the accurate reproduction of high frequencies requires a driver with a small diaphragm. At the low-frequency end the driver needs to be large in order to achieve a high enough SPL. The ear is less sensitive at low frequencies compared to mid and high frequencies, and as a consequence more air needs to be moved in order to make the low-frequency sound loud enough (see figure 2.24). At the high-frequency end the driver needs to be small in order to radiate sound over a wide enough angle. A loudspeaker driver becomes less directional and radiates sound over a larger angle when it is small compared to the wavelength of the sound that it reproduces (see figure 2.28). Thus, the size of the driver influences both the attainable output power and the directivity in a certain frequency range (Colloms, 2018; Newell and Holland, 2007).

The majority of small loudspeaker systems uses two separate drivers: one driver for the low and mid frequencies and one driver for the high frequencies. Such a loudspeaker system is called a *two-way system*. The low/mid-frequency driver, also known as the *woofer*, typically has a diameter of 120–180 mm (5–7 inch) and the high-frequency driver, also known as the *tweeter*, often has a diameter of 25 mm (1 inch). An electrical filter, called a *crossover filter*, separates the audible frequency range into two parts such that each driver receives the appropriate frequencies. Typically, the crossover point lies in the range of 2–4 kHz (Colloms,

Figure 7.1
Three different loudspeaker
designs with three drivers.
On the left, a two-way
design in which both
woofers reproduce the
same frequency range. *In
the middle*, a 2 1/2-way
design in which one woofer
reproduces both the mid
and low frequencies,
while the other woofer
only reproduces the low
frequencies. *On the right*,
a three-way design where
each driver reproduces its
own part of the frequency
range.

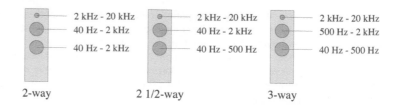

2018). Frequencies below the crossover point go to the woofer, and frequencies above the crossover point to the tweeter. Larger three-way loudspeaker systems are also very common. They have separate drivers for the high-, mid-, and low-frequency ranges where each driver reproduces only part of the entire frequency spectrum. Some four-way systems also exist. Adding more than four drivers is not very common as it adds complexity to the design. Note that you cannot tell if a speaker is two-way or three-way from just counting the number of drivers. Many larger loudspeakers have one high-frequency driver and two woofers that reproduce the same frequency range. While such a loudspeaker has three separate drivers, it is still a two-way design. The designation 'two-way' refers to the number of frequency bands created by the crossover filter and not to the number of drivers. To make matters even more complicated, a loudspeaker that has two woofers can also be designed as a *2 1/2-way system*. In such a system, one woofer reproduces both the mid and low frequencies, while the second woofer only reproduces the low frequencies. Such a design is neither a two-way nor a three-way system, because of the partial overlap between the frequency ranges of both woofers (Colloms, 2018). Figure 7.1 shows an example of three different loudspeaker designs with three drivers.

The three basic parts of any practical loudspeaker system are the drivers, the cabinet, and the crossover filter. These three parts need to be designed such that the overall sound of the loudspeaker is well-balanced without any specific frequencies or frequency ranges standing out. While the individual drivers must be chosen such that their individual sound characteristics can be integrated into a pleasing whole, the design of the crossover filter and the design of the enclosure also play a vital role. It is not only the drivers that matter. The drivers, the crossover filter, and the enclosure must be designed together as a system. Let's now have a more in-depth look at each of these three parts.

7.1.1 Drivers

The most widely used loudspeaker driver is the *moving-coil driver*, also called the *dynamic driver*. Although many other transducer types exist, the moving coil remains the most popular one among loudspeaker manufacturers because of its effectiveness, economy, wide dynamic range, high-power handling, and relatively simple design (Colloms, 2018; Harley, 2015).

The moving-coil loudspeaker produces sound waves by rapidly moving a stiff diaphragm back and forth. This diaphragm is attached to a cylinder onto which a coil of wire is wound. This coil and the attached diaphragm move back and forth when an alternating current from the power amplifier is applied to the coil. This back and forth movement is the result of two magnetic fields interacting. According to Faraday's law of induction, the alternating current applied to the coil generates an alternating magnetic field around the coil. This alternating field interacts with a static magnetic field generated by a permanent magnet and as a result the coil is pulled back and forth.

Figure 7.2 shows a schematic diagram of a moving-coil loudspeaker driver. The coil, also called the *voice coil* is wound onto a lightweight cylinder that is suspended freely in a cylindrical *gap* of the *magnet assembly*. Attached to the magnet assembly is a permanent *magnet* that generates a static magnetic field across the cylindrical gap. The combination of the voice

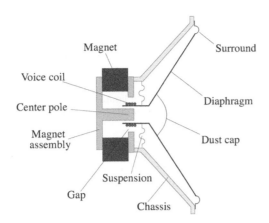

Figure 7.2
Parts of a moving-coil
loudspeaker.

coil and the permanent magnet is called the *motor system*. The motor system is the driving force that moves the diaphragm back and forth. The diaphragm is attached to the cylinder onto which the voice coil is wound and to the *chassis* that keeps the whole assembly together. The *suspension* connects the cylinder to the chassis such that the voice coil remains centered in the gap. The *surround* connects the outer edge of diaphragm to the chassis. A *dust cap* prevents the ingress of dust into the cylindrical gap (Newell and Holland, 2007).

The moving-coil loudspeaker driver comes in many varieties. Loudspeaker designers choose different shapes, sizes, materials, and constructions depending on the application and performance requirements. Each driver design presents its own unique engineering and performance compromise (Colloms, 2018).

For the reproduction of low and mid frequencies, it is customary to use a moving-coil loudspeaker with a cone-shaped diaphragm. The diameter of these cone-shaped loudspeakers varies from 50 mm to 500 mm (2 inch to 20 inch). Typically, the larger the diameter, the louder the loudspeaker can reproduce the lowest frequencies. The cone needs to be lightweight and rigid. A light cone can respond fast, stop quickly, and accurately follow the changes in the electrical signal from the power amplifier. A lightweight cone improves the clarity of the reproduced sound. You can hear the individual notes including their attack and decay (see section 3.4.2). The cone also needs to move back and forth as a perfectly stiff surface without wobbling or flexing, because if it breaks up into a nonuniform movement its frequency response will be less than ideal. Break-up will cause phase cancellations among different parts of the cone's surface, which introduces unwanted colorations such that certain frequencies may stand out more than others and the accurate reproduction of timbre is hampered (section 3.4.1). To avoid cone break-up and maintain cone rigidity, different cone materials and sandwich constructions are being used. Cones can be made of paper, plastics, polymers, metal alloy, Kevlar, and carbon fiber. Many designers avoid the use of larger drivers and favor the use of multiple small drivers, because they feel they can control them better (Colloms, 2018; Newell and Holland, 2007).

For the reproduction of high frequencies, it is customary to use a dome-shaped diaphragm. Dome loudspeaker drivers work on the same principle as their cone-shaped counterparts, but the voice coil has the same diameter as the diaphragm. Domes are the most widespread type of high-frequency drivers. While they can be made in different sizes the majority of them have a diameter of 25 mm (1 inch). Their frequency response may extend far beyond audibility, up to 80 kHz. The dome diaphragm is usually made from a material with a high strength to weight ratio, such as plastic, impregnated cloth, aluminum, titanium, beryllium, or in some rare cases diamond (Colloms, 2018; Newell and Holland, 2007).

7.1.2 Crossovers

The crossover filter splits the entire frequency range into a number of separate parts and feeds them to the appropriate drivers. An example of a three-way crossover filter is shown in figure 7.3. Every high-quality loudspeaker needs a crossover filter, because no single loudspeaker driver can provide a flat frequency response over the entire audible frequency range. Every driver suitable for serious listening has a limited frequency range that it can produce without any distortion or coloration at a sufficiently loud level.

Most loudspeakers include a *passive crossover filter*. Such a filter splits the signal from the power amplifier into multiple frequency ranges and feeds these ranges to the appropriate drivers. A passive crossover filter is part of the design of a loudspeaker system; it is specifically designed for the system and built into the loudspeaker cabinet. An alternative to the passive crossover filter is the *active crossover filter*. Such a filter splits the frequency range before the audio signal is amplified. The use of an active filter requires a separate amplifier for each loudspeaker driver. Figure 7.4 shows the different setups for a passive and an active crossover

Figure 7.3
Example of a three-way crossover filter. Each loudspeaker driver receives the frequency range that it is able to reproduce with sufficient quality.

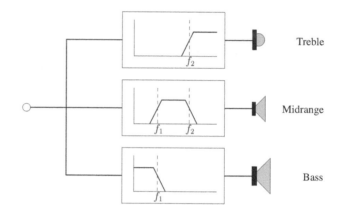

Figure 7.4
Different crossover filter configurations. *At the top*, a single power amplifier with a passive crossover filter. *At the bottom*, an active crossover filter with three power amplifiers, one for each frequency band.

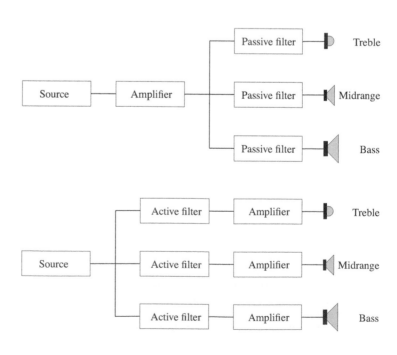

filter. Although an active filter has certain advantages, it is not frequently used due to the added complexity. The majority of loudspeaker systems on the market uses a passive crossover. With one notable exception: subwoofers. A subwoofer has its own built-in amplifier that is fed from the active crossover filter that is integrated into the A/V controller or receiver.

A crossover filter is characterized by its *crossover frequency* and its *filter order*. The crossover frequency is the frequency at which the transition from one frequency range to the next frequency range occurs. The order describes the steepness of the transition. Figure 7.5 illustrates these two concepts. The higher the order, the steeper the slopes in the transition region. A first-order filter has a slope of 6 dB/octave, a second-order filter 12 dB/octave, a third-order one 18 dB/octave, and a fourth-order one 24 dB/octave. The choice of the filter order is an important design decision. The use of a first-order filter results in a large transition region between the different drivers. The considerable overlap in the frequencies that the drivers reproduce can cause problems. The combined sound produced by both drivers can result in destructive interference and an irregular frequency response. The use of higher-order filters considerably reduces the overlap between the drivers and consequently also minimizes the undesirable acoustic interaction between the drivers. However, higher-order filters have a much poorer transient response than first-order filters. First-order filters are much better at retaining clarity and definition in the reproduced sound. As with so many of the choices in loudspeaker design, the choice of the filter order is an engineering trade-off. Second-order filters are a popular choice for small two-way loudspeaker systems. Third-order filters are also extensively used in larger loudspeaker systems (Colloms, 2018; Harley, 2015; Newell and Holland, 2007).

The design of a passive crossover filter is not an easy task. The filter has to perform many duties. Besides the separation of the frequency ranges, it also has to properly balance the relative loudness of the different drivers and ensure that the combined impedance of the

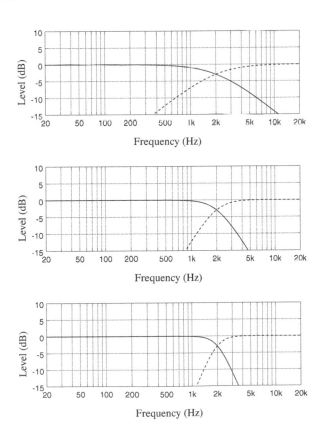

Figure 7.5
The filter order of a crossover filter determines the slope of the frequency roll-off and the size of the transition region in which the frequency ranges overlap. *From top to bottom:* a first-order, second-order, and third-order crossover filter. In all cases the crossover frequency is 2 kHz. As the order increases the slope becomes steeper and the transition region becomes smaller.

drivers results in an easy load for the power amplifier. A skillful crossover design results in a good integration of the sound of the individual drivers; not only right in front of the loud-speaker system, but also within a certain horizontal and vertical angle. The directivity of the loudspeaker system (see figure 2.28) should smoothly change as the frequency of the sound increases without any abrupt changes when one driver takes over from the other. Small movements of the listener's head to the sides or up and down should not result in noticeable changes in the sound of the loudspeaker system (Colloms, 2018; Newell and Holland, 2007).

7.1.3 Enclosures

The enclosure of the loudspeaker system has an enormous influence on its sound quality (Colloms, 2018). There are three ways that the enclosure modifies the overall sound of the loudspeaker system. First, it determines the lowest frequency that the loudspeaker system can reproduce. Recall from section 2.3.3 that low frequencies are radiated in all directions. An enclosure around the low-frequency driver is needed to prevent the sound radiated from the back of the driver canceling the sound radiated from the front. The type of enclosure and its size will determine the lowest frequency that the loudspeaker is able to reproduce with suf-ficient loudness. Second, the enclosure vibrates in response to the movement of the drivers. The panels of the enclosure will dissipate some of the vibrational energy from the drivers. The damped vibration of the panels will result in some unwanted acoustic output. In many loudspeakers much of the coloration in the reproduced sound is due to the quality of the enclosure and not due to the quality of the drivers or the crossover filter. Third, the shape of the enclosure can cause irregularities in the mid- and high-frequency response due to diffrac-tion of sound waves (see section 2.3.3). Diffraction occurs around the edges of the cabinet and around any protruding parts on the front panel in which the drivers are mounted. For this reason most loudspeaker systems have a smooth front panel and rounded cabinet edges.

The most commonly used enclosure types are the *closed-box* and the *bass-reflex*. The closed-box loudspeaker is sometimes also referred to as *sealed-box* or *infinite baffle.* The bass-reflex loudspeaker is also known as *vented-box*. Two other types that you might encounter are the *band-pass* and *transmission-line* enclosures. Figure 7.6 shows these four types of enclosures.

The closed-box seals the loudspeaker driver in an airtight cavity (Small, 1972a, 1973a). The sound energy radiated from the back of the driver is trapped inside the cabinet and can't interfere with the sound radiated from the front of the driver. The air trapped in the enclo-sure acts as a spring. The resonance frequency of this spring determines the lowest frequency that the closed-box loudspeaker system can reproduce. The smaller the box the higher the resonance frequency. For this reason large loudspeaker enclosures are needed for good low-frequency reproduction. Below the lowest frequency the response of the closed-box system falls off at 12 dB per octave.

The bass-reflex enclosure has an opening through which the sound radiated from the back of the driver is channeled out of the enclosure (Small, 1973b, 1973c, 1973d, 1973e; Thiele, 1971a, 1971b). The opening is usually fitted with a port tube. This port delays the sound from the rear of the driver such that this sound is shifted in phase when it leaves the enclosure and doesn't cancel the sound radiated from the front of the driver. The air trapped in the port acts as a

Figure 7.6
Different types of
loudspeaker enclosures: *(a)*
closed-box; *(b)* **bass-reflex;**
(c) **band-pass; and (d)**
transmission line.

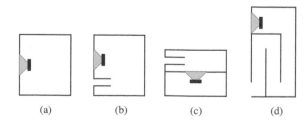

(a)　　　　(b)　　　　(c)　　　　(d)

mass that resonates with the spring created by the air inside the cabinet. The port resonance contributes to the overall sound production of the loudspeaker in two important ways. First, above the resonance frequency of the spring, it greatly reduces the diaphragm excursion needed from the driver and increases the acoustic efficiency. In other words, when driven by the same amplifier power, a bass-reflex system plays louder than a closed-box system. Second, below the resonance frequency of the spring, the port still radiates sound and extends the frequency response to a lower frequency. As a result, at a fixed enclosure size the bass-reflex system can reproduce lower frequencies than a closed-box system. However, below the lowest frequency the response of the bass-reflex system falls off at 24 dB per octave—twice as fast as the closed-box system does.

Some bass-reflex systems use a *passive radiator* instead of a port (Small, 1974a, 1974b). A passive radiator, also known as an *auxiliary bass radiator* (ABR), is a driver without a magnet and voice coil. It is just a diaphragm suspended in a chassis. It usually has the same diameter as the driver. On the outside of the loudspeaker cabinet it can look exactly the same as the main driver. The passive radiator performs the same function as the port in a normal bass-reflex enclosure. Its main advantage is that there is no air flow. The air flow in a bass-reflex port can sometimes generate unwanted sounds, such as blowing, port resonances, and leakage of high-frequency sounds from the cabinet. These unwanted sounds can have a significant influence on the overall sound quality (Colloms, 2018).

A band-pass enclosure consists of two compartments with the driver mounted on the division between them (Geddes, 1989). The rear of the driver is enclosed in a sealed compartment, while the front of the driver faces a compartment with a port. This port provides the only radiation of sound to the outside. The name 'band-pass' refers to the fact that the frequency response of this type of loudspeaker is limited at both the high and the low end of the frequency spectrum. The roll-off at both ends tends to be 12 dB per octave. Band-pass enclosures are mostly used for low-quality subwoofers. The main reason is that for a given output capability the band-pass enclosure can be made physically much smaller than a closed-box or bass-reflex system. However, the low-frequency sound produced by band-pass enclosures often sounds less satisfying as it tends to be thick and boomy (Colloms, 2018).

A transmission-line enclosure consists of a very long duct filled with sound-absorbing material. This duct functions as a transmission line for the sound from the rear of the loudspeaker driver. If the line is infinitely long, all rear sound from the driver is absorbed and the transmission-line system can play very low frequencies. In practice the line needs to be finite and as a result the system partially functions as a bass-reflex system. The line needs to be as long as the lowest wavelength of the sound that needs to be absorbed. This results in very long lines that are often folded into a labyrinth to keep the cabinet size manageable. Nevertheless, transmission-line systems tend to be very large and that is the main reason why they are almost never used (Colloms, 2018; Newell and Holland, 2007).

All enclosure types suffer from panel vibrations and enclosure resonances (Iverson, 1973; Tappan, 1962). When the driver moves back and forth it will excite the panels of the enclosure into vibration. The vibrating panels radiate sound that is concentrated in a narrow frequency band around the frequency at which the panel resonates. This unwanted sound leads to coloration of the overall sound output of the loudspeaker. It changes the delicate timbres of musical instruments and voices (Harley, 2015). In some loudspeakers, the panels continue to vibrate after the loudspeaker driver has ceased moving. This so-called *hangover* imparts an unpleasant ringing to the reproduced sound and reduces its clarity (Tappan, 1962).

Loudspeaker designers use different techniques to reduce the enclosure resonances to inaudible levels. They make the enclosure of a material with a high stiffness-to-weight ratio: thick MDF (medium density fireboard) panels or high-quality hardboard are most often used. More exotic materials such as molded plastic, steel, marble, and concrete have also been used, and some manufacturers even use sandwich wall constructions with damping materials such as lead sheets and fibrous bitumen. In addition, loudspeaker enclosures are often internally braced, such that the panels are subdivided into smaller unequal areas. This reduces the panel vibrations and also staggers the resonance frequencies of the different areas. Some manufacturers use curved panels that are much stronger than flat panels and therefore less prone to resonate (Colloms, 2018; Newell and Holland, 2007). You can easily evaluate the construction quality of a loudspeaker cabinet with a simple test. Gently knock on the enclosure walls and listen to the resulting sound. A well-made resonance-free enclosure will produce a dead thump while a poorly damped enclosure will ring and sound hollow (Harley, 2015; Tappan, 1962). Another indication of a well-damped enclosure is its weight. The heavier the better.

The shape of the enclosure can also affect the sound quality of the loudspeaker. At the edges of the cabinet the radiated sound encounters a discontinuity that leads to diffraction of the sound (see section 2.3.3). The use of rounded corners can mitigate the effect to some extent. The frequencies at which diffraction occurs depend on the relative size of the cabinet with respect to the wavelength of the sound. For small stand-mounted loudspeakers diffraction may occur from 1 kHz to 2 kHz, whereas for larger floor-standing loudspeakers it may occur from 200 Hz to 800 Hz. The accurate reproduction of high frequencies may also be influenced by diffraction around surface irregularities such as the driver's mounting bolts and the grille frame. Diffraction can result in a significant loss of stereo image stability (Colloms, 2018).

Most loudspeakers have a fabric or metallic grille that protects the drivers and screens them from view. The grille can have a considerable effect on sound quality. The best grille is no grille at all. Grilles are never completely acoustically transparent and will absorb and reflect some of the sound energy. The frame of the grille can become a diffraction source producing further coloration of the sound (Colloms, 2018; Harley, 2015; Newell and Holland, 2007). Many commercial loudspeakers sound quite superior with their grilles removed. They have a better frequency response in the 2–20 kHz range and provide better imaging (Colloms, 2018).

Recommendation 7.1 *Remove the grilles from your loudspeakers for an improved reproduction of high frequencies.*

7.1.4 Performance

High-quality loudspeakers should provide neutral sound for any type of reproduction, be it music, speech, or a movie soundtrack. This is easier said than done. Loudspeakers are complex devices. They are sufficiently complex that accurate and reliable prediction of their sound radiation is difficult, even with advanced computer and measurement technology (Newell and Holland, 2007). High-quality loudspeakers should of course meet certain technical requirements, but the final arbiter of their sound quality is a listening test. It is widely agreed that it is absolutely necessary to listen to a loudspeaker to assess its sound quality, looking purely at technical measurements never suffices (Colloms, 2018; EBU-Tech 3276, 1998; ITU-R BS.1116, 2015; Toole, 1985, 2018). Therefore, you should never select and buy loudspeakers without performing an extensive listening test to evaluate their sound quality.

Listening tests involve a certain degree of subjectivity. Therefore, it is important that you audition prospective loudspeakers yourself before buying them. Only you can decide whether you like what you hear. However, keep in mind that you want to select a loudspeaker that sounds as neutral as possible. Loudspeakers that overlay the reproduced sound with a certain sonic quality of their own will be boring to listen to in the long run. And, more importantly, they don't reproduce the sound in a way that the artist and producer intended. So, always strive for accuracy when selecting loudspeakers.

Recommendation 7.2 *Before you buy loudspeakers, always listen to them to evaluate their sound quality. Select loudspeakers that sound as neutral as possible.*

Listening tests are also extensively used by reviewers in audio magazines and websites. Their reviews can be a good starting point for narrowing down your search for new loudspeakers, but you should also treat these reviews with caution. The majority of reviewers evaluate the loudspeakers using only a rudimentary listening test without the proper controls to make such a test both reliable and repeatable. For example, it is no secret that knowledge of the loudspeaker under evaluation influences the perceived sound quality. It has been shown that it is nearly impossible to objectively evaluate a product if you know its price, brand, or reputation (Toole, 2018). To eliminate such psychological bias the listening test has to be performed blind without any a priori knowledge of the loudspeaker under evaluation. Other variables that influence the outcome of a listening test and that should be controlled include the choice of listening room, listeners, music, sound levels, positions of loudspeakers and listeners, and the design of the listening procedure (Gabrielsson and Lindström, 1985). Listening tests conducted by the audio press almost never have the appropriate controls and as a result the reliability of these tests can be questioned. It is not uncommon that uncontrolled listening tests by audio reviewers result in a large variation of opinion on a particular loudspeaker (Toole, 1985, 2018). So, be careful when interpreting the results of such tests. It is always better to trust your own ears when you are selecting loudspeakers that you are going to buy.

To perform your own listening tests on loudspeakers, I recommend that you follow the general guidelines that I laid out in section 3.5. In addition, you should pay attention to the guidelines that follow. Since the sound of the loudspeaker is greatly influenced by their positions in the room, it is important to place all the loudspeakers that you are comparing in the exact same position in the room, one after another. It is equally important that during the comparisons you sit at the exact same spot in the room (Toole, 2018). Of course, it would be best to listen to the loudspeakers in your own listening room, but this is not always feasible. When listening to loudspeakers at the dealer, make sure that you sit at the same distance from the loudspeakers as you sit from them at home. Also pay attention to the height of the chair that you sit on. This height should also be the same as the height of your chair at home, because loudspeakers have different tonal balances at different listening heights, especially in the high-frequency region (Harley, 2015). It is important to drive the loudspeakers that you are comparing with the same power amplifier or receiver that you have at home. The interaction between the amplifier and the loudspeaker influences the reproduced sound. Some loudspeakers are a better match to a certain amplifier than others (Harley, 2015). Finally, make sure that the amplifier is connected directly to the loudspeakers and not through a switch box. Some stores use a switch box to be able to easily switch between different loudspeakers. Avoid the use of such a box as it can change the sound (McLaughlin, 2005).

Recommendation 7.3 *When evaluating and comparing loudspeakers at the dealer, make sure that during the listening tests:*

- *The loudspeakers under evaluation are placed at the same positions in the room.*
- *You sit at the same position in the room for each evaluation.*
- *You sit at the same distance from the loudspeakers as you do at home.*
- *You sit in a chair that has the same height as the one you use at home.*
- *The loudspeakers are driven by the same amplifier that you have at home.*

As a general guideline, good loudspeakers sound neutral and tend to disappear into the soundstage. The less you are aware of the fact that the music is reproduced by the loudspeakers, the better. The loudspeakers should not attract any attention to themselves and should not stand in the way of enjoying the music (Harley, 2015). I recommend that during your listening tests you critically listen and evaluate all the sound-quality aspects listed in section 3.4 and also use the associated sonic descriptions to consciously characterize the reproduced sound. For easy comparison of the different loudspeakers that you audition, you could even consider assigning a score to each of the quality aspects using a number between 1 and 5.

For the critical evaluation of loudspeakers it is best to use recordings that have a broad and dense frequency content. Such recordings are more likely to uncover audible flaws that are due to unwanted resonances of the drivers or cabinet. Simple vocal and instrumental recordings are not the most revealing as they may fail to excite the troublesome resonances (Toole, 2018). However, certain vocal and instrumental recordings can still be useful to uncover other flaws in the reproduced sound. Male speaking voice is a good test for colorations in the upper bass. The voice should sound natural and not heavy or chesty. Another good test for upper-bass coloration is a descending or ascending line of lower notes played on the piano. You should be able to distinguish every note and none of the notes should jump out; they should all sound even in tone and volume (Harley, 2015). Male and female vocals are a good test for midrange coloration. Many loudspeakers suffer from such annoying colorations. We humans are very sensitive to subtle differences in the sound of the human voice (Lipshitz and Vanderkooy, 1981). Vocals should not sound nasal nor should they sound as if they are produced through cupped hands. They should sound pure and open (Harley, 2015). It is often useful to play loudspeakers at quite low sound levels to uncover the common flaw of an emphasized upper midrange. At such a low level speech and vocals should still sound natural and lifelike and not unnaturally thin (Colloms, 2018). Excessive vocal sibilants (s, sh, and ch sounds) indicate poor high-frequency reproduction. Although at first it might give you the false impression of increased clarity, the overly emphasized high frequencies will quickly lead to listening fatigue. Another indicator of poor high-frequency reproduction is a grainy or dirty sound to violins and cymbals. The high frequencies should never attract too much attention; they should be integrated with the rest of the music and never sound as a separate component (Harley, 2015).

Recommendation 7.4 *Use recordings with a broad and dense frequency content when performing listening tests to evaluate the sound quality of loudspeakers. In addition, use speech and vocals, solo piano, and violin. Listen for upper-bass, midrange, and high-frequency colorations, and other reproduction flaws.*

Frequency Response

One of the most important technical measurements of a loudspeaker system is its amplitude frequency response. When measured in the right way, the frequency response is a good indicator of the loudspeaker's sound quality. Ideally, a loudspeaker should convert the electrical signal from the power amplifier into sound with an equal sensitivity at all frequencies, but practical loudspeakers have a limited bandwidth. They are both limited in the lowest and the highest frequency that they can produce. Furthermore, the directions in which the loudspeaker radiates sound changes as the frequency increases. At low frequencies the loudspeaker radiates sound in all directions, but as the frequency increases the sound is only radiated in a forward projected beam. This beam gets narrower as the frequency increases (see figure 2.28). Thus, the frequency response of a loudspeaker very much depends on the position from which it is measured. The response contains more high frequencies when measured directly in front of the loudspeaker than when it is measured more towards the side.

A loudspeaker generates a three-dimensional sound field that interacts with the listening room before it arrives at the two ears of the listener. To be able to describe such a three-dimensional field, multiple frequency-response measurements are needed. At frequencies below about 100 Hz the loudspeaker is an omnidirectional radiator and one frequency-response curve is sufficient to describe its sound field. At higher frequencies the sounds are only radiated in certain directions and multiple frequency-response measurements around the loudspeaker are needed to describe the radiated sound field. No single frequency-response

Figure 7.7
Positions of the
measurement microphone
for on-axis and off-axis
frequency response
measurements in both the
vertical (*left*) and horizontal
(*right*) dimension.

curve can sufficiently describe this three-dimensional behavior (Toole, 2018). It is customary to measure the frequency response of a loudspeaker on its main axis (called on-axis response) and off-axis in both the vertical and horizontal dimension. Figure 7.7 illustrates the position of the measurement microphone with respect to the loudspeaker. The on-axis response is usually measured from the center of the high-frequency driver. The off-axis response is often measured at regular intervals of 10° or 15° left to right and top to bottom.

When a loudspeaker is measured in a normal room its frequency response, both on-axis and off-axis, is altered by the reflections, the adjacent-boundary effect, and the room resonances (see section 2.3.4). The frequency response of the loudspeaker will be different in different rooms and it is not easy to separate the contribution of the room from that of the loudspeaker. To measure only the response of the loudspeaker it is customary to measure the frequency response in an *anechoic room*. In such a room all surfaces are covered with a thick layer of sound-absorbing material. This material absorbs almost all sound that strikes it and as a result the room no longer alters the frequency response of the loudspeaker (except at very low frequencies that are difficult to absorb completely).

Frequency-response measurements made in an anechoic room provide a useful insight in the sound quality that a particular loudspeaker is capable off. The use of an anechoic room results in frequency-response curves that are independent of the room and are a good description of the capabilities of the loudspeaker itself. The anechoic response curves can be used to predict the sound quality of the loudspeaker when placed in any room. Recall from section 2.3.4 that at mid and high frequencies the sound of a loudspeaker in a room is largely determined by both the direct sound and the early reflections. The anechoic on-axis response of a loudspeaker is a good representation of the direct sound, and the multiple off-axis responses provide information on the frequency content of the early reflections (Toole, 2018).

Frequency-response measurements can be used to recognize true excellence and true inferiority among different loudspeakers. Both the on-axis and multiple off-axis responses are needed for such a judgment. Furthermore, the measurements need to have sufficient frequency resolution (Toole, 1986a, 1986b). High-resolution measurements tend to show lots of small peaks and dips that are mostly due to acoustic interference between the loudspeaker and the measurement microphone. These small fluctuations in the response differ for each position of the microphone and are not really of interest. To get rid of them, it is customary to spatially average several curves. The direct sound of a loudspeaker is best represented by the spatial average of the on-axis response and all the off-axis responses within 15°. This spatial averaged curve is called the *listening window*. It represents the sound that arrives at the listener's ears directly from the loudspeakers. The average of the off-axis responses between 30° and 45° is often used to characterize the sound that arrives at the listener's ears from the early reflections, and the average of the responses between 60° and 75° to characterize the late reflections (Toole, 2018).

The ideal shape of the spatially averaged frequency-response curves is shown in figure 7.8. These curves are based on the average of the frequency-response curves of 11 loudspeakers that had the highest overall fidelity rating in extensive listening tests by Toole (1986b). The ideal listening window curve is a flat and smooth line that extends from the lowest frequency that the loudspeaker is able to reproduce, all the way up to 20 kHz. The off-axis responses at

Figure 7.8
Frequency-response curves
for a high-performance
loudspeaker. The *top* curve
represents the listening
window (on-axis and up
to 15° off-axis), the *middle*
curve the off-axis response
in the range of 30°–45°, and
the *bottom* curve the off-
axis response in the range
of 60°–75°.

Based on Toole (1986b), figure 7d.

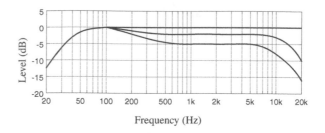

30°–45° and 60°–75° are also smooth. However, they are far from flat and deviate in shape from the on-axis response. The reason for this deviation is the directivity of the loudspeaker (see figure 2.28). At low frequencies the loudspeaker radiates equally in all directions. As a result, the on-axis and off-axis frequency responses are all equal at low frequencies. That is why below 100 Hz the three curves in figure 7.8 lay on top of each other. As the frequency increases beyond 100 Hz, the two off-axis response curves start to differ from the on-axis response. The loudspeaker becomes more directional and radiates less sound towards the sides. As a result the off-axis frequency responses drop in level. The sound level decreases as the off-axis angle increases. Thus, the 60°–75° response has a lower level (–5 dB) than the 30°–45° response (–2 dB) for most of the midrange frequencies. At the highest frequencies, beyond 10 kHz, the loudspeaker becomes even more directional and both off-axis responses rapidly decrease even further. It is important that the 30°–45° off-axis high-frequency response doesn't differ too much from the on-axis response. In a truly excellent loudspeaker it should not deviate more than 4 dB below 10 kHz (Colloms, 2018; EBU-Tech 3276, 1998; ITU-R BS.1116, 2015).

While off-axis high-frequency roll-off is typical, the on-axis response should extend all the way to 20 kHz. Some loudspeakers are able to reproduce even higher frequencies up to about 45 kHz. This might seem beneficial for the reproduction of high-resolution audio (HRA) that can contain frequencies above 20 kHz (see section 5.2.1). However, at such very high frequencies the loudspeaker will be very directional and only radiate these frequencies in a tiny beam around its main axis. Loudspeakers that can only reproduce frequencies up to 20 kHz are still excellent loudspeakers, because it is not very likely that you will be able to hear any difference in the frequency range above 20 kHz. It is much more important that the high frequencies below 20 kHz are reproduced accurately. In fact, a truly natural timbre for the human voice and instruments such as the violin and acoustic guitar can only be achieved when below 20 kHz the on-axis high-frequency response doesn't deviate more than 0.5 dB from the ideal target in figure 7.8. A larger positive deviation makes the loudspeaker sound too bright and harsh. While a larger negative deviation makes the loudspeaker sound dull and muffled (Colloms, 2018). An example of such a negative deviation that results in rolled-off high frequencies is shown in figure 7.9. If you compare this figure with figure 7.8, you will see that such a deviation in the on-axis response also reduces the amount of high frequencies in the off-axis responses.

All loudspeakers are limited in the lowest frequency that they can reproduce. It is not easy to reproduce low frequencies that extend all the way to 20 Hz. Fortunately, if you use subwoofers your main loudspeakers don't need to go that low. Typically the subwoofers take over at a crossover frequency between 80 Hz and 120 Hz. To ensure a smooth transition between the subwoofers and the main loudspeakers, the frequency response of your main loudspeakers should be flat to about one octave below the crossover frequency. Thus, with a typical subwoofer crossover frequency of 80 Hz your main loudspeakers do not need to go any lower than 40 Hz (AES, 2001; EBU-Tech 3276, 1998; McLaughlin, 2005). It is important that the low-frequency response is flat as in figure 7.8. Some small loudspeakers exist that are purposely designed to accentuate the upper bass in an attempt to compensate for their limited ability

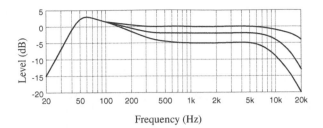

Figure 7.9
Frequency-response
curves for a loudspeaker
with a bump in the bass
response and rolled-off
high frequencies. Compare
these curves with the ones
in figure 7.8.

to play low. They have a distinct bump in their low-frequency response and sound thick and heavy. An example of such a low-frequency response is shown in figure 7.9.

Extensive listening tests have shown that listeners like loudspeakers that have a flat and smooth on-axis frequency response and that also have a smooth off-axis frequency response. The off-axis response is never flat, but the closer it follows the on-axis response, the higher the fidelity rating of the loudspeaker. In other words, listeners like loudspeakers that have a uniform wide dispersion (Toole, 1986b, 2018; Toole and Olive, 1988).

Recommendation 7.5 *Your main loudspeakers should have a flat and smooth on-axis frequency response between 40 Hz and 20 kHz. They should also have smooth off-axis frequency responses, with the 30°–45° response closely following the on-axis response and not deviating more than 4 dB below 10 kHz.*

Unwanted coloration of the reproduced sound remains a major fault with loudspeakers. No loudspeaker has a truly flat frequency response. Smooth deviations are generally acceptable, but abrupt changes are not (Colloms, 2018; Newell and Holland, 2007; Toole, 2018). Upward thrusting peaks in the frequency response are a sign of unwanted coloration. The audibility of peaks depends both on their bandwidth and their amplitude (see figure 3.15). Wide low-amplitude bumps can be as annoying as high narrow peaks. Resonances in the loudspeaker driver and its cabinet are often the cause of audible colorations.

Although high-resolution frequency-response measurements of a loudspeaker are good indicators of its sound quality, they are not very useful in practice. Unfortunately, most manufacturers and reviewers do not provide sufficiently accurate frequency-response measurements. Often they only show the frequency response in numerical form without any curves (Toole, 2018). Sadly, many published specs are incomplete and sometimes even misleading (Winer, 2018). One notable exception is the SoundStage! Network that cooperates with Canada's National Research Council (NRC). They publish detailed and reliable loudspeaker measurements along with their reviews (www.soundstage.com).

Phase Response

Accurate reproduction of an audio waveform requires that both its amplitude spectrum and its phase spectrum are reproduced unaltered. In other words, the relative amplitudes and relative phases of each of the waveform's frequency components need to be preserved. Recall from section 2.3.1 that many different audio waveforms have exactly the same amplitude spectrum; you need both the amplitude and the phase spectrum to be able to uniquely describe the waveform. The frequency response of a loudspeaker, such as the one in figure 7.8, only describes how a loudspeaker changes the amplitude spectrum of the audio signal that it reproduces. For a complete description, the *phase response* of the loudspeaker is also needed. The phase response describes how the loudspeaker changes the phase spectrum of the signal that it reproduces.

Extensive listening tests show that good-sounding loudspeakers need to have a smooth phase response. Surprisingly, the particular shape of the phase response appears to be irrelevant. This means that exact reproduction of the audio waveform is not a requirement. Shifts in phase appear to be inaudible as long as there are no abrupt changes (Newell and Holland, 2007; Toole, 2018). The effect of phase shift is clearly much less relevant than peaks and dips in the amplitude frequency response. Only large and abrupt phase shifts are audible. In general, listeners have great difficulty hearing changes in phase when listening to music reproduced through loudspeaker in a normally reflective room. Although under certain circumstances a phase shift can be audible, its effect is quite subtle. Therefore, accurate phase reproduction is not of any practical relevance (Lipshitz and Vanderkooy, 1981; Toole, 1986b, 1988).

Nonlinear Distortion

A loudspeaker converts an electrical input signal into acoustical output. As long as the loudspeaker behaves like a *linear system,* the transformation from the input to the output can be completely described by its frequency and phase response (see box 7.1). However, real-life loudspeakers don't behave like a linear system. How much the behavior deviates from a linear system is called *nonlinear distortion*. Good loudspeakers have low levels of nonlinear distortion and operate very much like a linear system.

A loudspeaker is usually quite linear for small excursions of its diaphragm. Nonlinear distortion starts to occur as the excursion starts to increase. Reproducing low frequencies and putting out high sound levels requires large excursions. Therefore, nonlinear distortion is worse at the lowest frequencies and at the highest sound levels. Large excursions create nonlinear distortion, because the voice coil has to travel a larger distance through the magnetic gap and moves close to the boundaries of that gap (see figure 7.2). At these boundaries the magnetic field flares out and differs from the homogeneous field further inside the gap. This difference in geometry of the magnetic field causes nonlinear distortion (Gander, 1981).

An important property of a linear system is that its output doesn't contain frequency components that were not present in its input signal. When the input is a sine wave, the output of a

---

BOX 7.1 LINEAR SYSTEM

By definition, a system is a linear system if it satisfies two properties:

1. *Homogeneity:* The amplitude of the output is proportional to the amplitude of the input. Thus, if the input is amplified by a factor k, the output is also amplified by a factor k.
2. *Superposition:* The output to a sum of input signals is the sum of the outputs to each individual input. Thus, if the input $x_1(t)$ yields an output $y_1(t)$ and the input $x_2(t)$ yields an output $y_2(t)$, then the output to the input signal $x_1(t) + x_2(t)$ equals $y_1(t) + y_2(t)$.

The combination of these two properties shows that an input signal $k_1 x_1(t) + k_2 x_2(t)$ produces an output signal $k_1 y_1(t) + k_2 y_2(t)$. These properties of a linear system together with the theory of Fourier analysis (see section 2.3.1) enable us to easily compute the output of a linear system that results from any arbitrary complex input signal. Recall that Fourier analysis allows us to decompose any input signal into a sum of sine waves each having a particular amplitude and phase. The output of the linear system is simply the sum of the outputs that result from each of these sine waves. When the system is nonlinear, this method does not work and the output contains additional frequency components that are more difficult to compute.

linear system is a sine wave that has exactly the same frequency, but possibly with a different amplitude and phase. The changes in amplitude and phase can be completely described by the (amplitude) frequency response and the phase response of the system. A nonlinear system does behave differently: it introduces additional frequency components in the output signal. When the input is a sine wave with a frequency f, the output of a nonlinear system contains frequency components that are multiples of this frequency, that is $2f$, $3f$, $4f$, etc. These components are harmonics of the input frequency, and they are not present if the system behaves linearly. They are jointly referred to as *harmonic distortion*. Furthermore, when the input to a nonlinear system consists of two sine waves with frequencies f_1 and f_2, the output contains frequency components corresponding to the sum and difference of these two frequencies and their harmonics, that is $f_1 - f_2$, $f_1 + f_2$, $2f_1 - f_2$, $2f_2 - f_1$, etc. These additional frequency components are jointly referred to as *intermodulation distortion*.

Nonlinear distortion, both harmonic and intermodulation distortion, is usually specified as a level in dB or as a percentage. The level specification is a negative number that relates the distortion components to the level of the pure sound at 0 dB. For example, a specification of −40 dB means that the distortion components have a level that is 40 dB lower than the undistorted sound. The distortion level can also be expressed as a percentage as indicated in table 7.1. The relationship is simple: each factor of 10 changes the level by 20 dB. Harmonic distortion is usually specified as the *total harmonic distortion* (THD), which is the total amplitude of all the harmonic distortion components relative to the amplitude of the undistorted sound. Intermodulation distortion is more difficult to specify, because it is very dependent on the circumstances. The number of intermodulation components very much depends on the number of frequency components in the input signal, as every component interacts with all the others. A loudspeaker may sound great on relatively simple musical signals, such as a single instrument or voice, while it may sound rather unpleasant with music that contains a dense frequency spectrum, such as orchestral arrangements. Due to this complexity no truly meaningful figure of merit for intermodulation distortion has been established (Newell and Holland, 2007).

Loudspeakers, even high-quality ones, have nonlinear distortion levels that are much higher than those of other components such as power amplifiers. Distortion levels of a few percent are quite common (Moore, 2013; Winer, 2018). It appears that these higher distortion levels are not that important, because they do not correlate well with the subjective sound quality of a loudspeaker (Colloms, 2018; Toole, 2018). The reason is that the audibility of nonlinear distortion components depends on masking. The human ear is not able to detect distortion components if they are being masked by other frequency components (see also

Table 7.1 Relative distortion levels in percentages and decibels (Newell and Holland, 2007).

Decibel	Percentage
0 dB	100%
−10 dB	30%
−20 dB	10%
−30 dB	3%
−40 dB	1%
−50 dB	0.3%
−60 dB	0.1%
−70 dB	0.03%
−80 dB	0.01%
−90 dB	0.003%
−100 dB	0.001%

section 5.2.5). Thus, nonlinear distortion is often partially masked by the music itself. Certain nonlinear distortion components are more easily masked than others. Consequently certain nonlinear distortions are subjectively more objectionable than others. Harmonic distortion is often less objectionable than intermodulation distortion. One reason is that we are used to hearing multiple higher harmonics at once, because musical instruments naturally generate plenty of harmonics that are higher in frequency than the fundamental. Furthermore, the lower harmonics (second and third order) will be masked to some extent by the fundamental frequency. The higher harmonics will be too far from the fundamental and may be more easily audible. Intermodulation distortion is different because it not only generates distortion at frequencies above the fundamental, but also below it. The frequencies below the fundamental are not easily masked, because masking is much more effective in the upward-frequency direction (Moore, 2013). Intermodulation distortion is more offensive to the ear than harmonic distortion, because it adds a dissonant quality to the sound (Newell and Holland, 2007; Winer, 2018).

With modern high-quality loudspeakers it is very rare that harmonic and intermodulation distortion levels are so high that they influence the subjective sound quality (Toole, 2018). Most loudspeakers have nonlinear distortion levels that are below the maximum values given in table 7.2. Some smaller loudspeakers may struggle to keep up at frequencies below 100 Hz, but this is not really an issue if you use subwoofers to reproduce the lowest frequencies.

Recommendation 7.6 *Your main loudspeakers should have low nonlinear distortion levels that do not exceed the values given in table 7.2.*

Another nonlinear distortion effect that can occur in loudspeakers is *frequency modulation* (FM) distortion, also called *Doppler distortion*. When a loudspeaker diaphragm reproduces low frequencies and high frequencies at the same time, the high frequencies may be shifted upward and downward as the diaphragm moves alternately toward and away from the listener. In practice there is not much to worry about. Two-way and three-way loudspeaker systems split the entire frequency range over multiple drivers and as a consequence don't suffer from audible FM distortion (Allison and Villchur, 1982; Colloms, 2018).

Maximum Sound Level

Your loudspeakers should be able to generate high sound levels of at least 96 dB at the listening position. An SPL of 96 dB corresponds to the required peak level for playing music and movies at the recommended average reference level for home listening (see box 3.2). If you want to achieve an SPL that corresponds to cinema reference level, then your loudspeakers need to be able to generate a peak SPL of 105 dB, but the majority of people find that way too loud for home listening.

Due to the inverse-square law (see section 2.3.3) the SPL decreases with increasing distance. Therefore, if you want to achieve an SPL of 96 dB at the listening position, your loudspeakers should be able to generate more than 96 dB depending on the distance between the

Table 7.2 Nonlinear distortion specifications for loudspeakers. At 90 dB SPL the distortion levels should not exceed these values (Colloms, 2018; EBU-Tech 3276, 1998; ITU-R BS.1116, 2015; Newell and Holland, 2007).

	Excellent	Good
Below 100 Hz	–30 dB (3%)	–26 dB (5%)
100 Hz–250 Hz	–40 dB (1%)	–30 dB (3%)
250 Hz–20 kHz	–50 dB (0.3%)	–40 dB (1%)

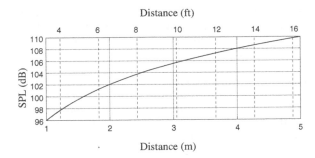

Figure 7.10
Maximum loudspeaker
SPL needed to achieve
96 dB SPL at the listening
position as a function of the
distance (in meters or feet)
between the loudspeaker
and the listener.

BOX 7.2 SPL AT THE LISTENING POSITION

Due to the inverse-square law, the SPL will decrease 6 dB every time the distance between the listener and the loudspeaker is doubled:

$$SPL_D = SPL_0 - 20\log_{10}(D)$$

where D is the distance (in meters) between the loudspeaker and the listener. Indoors, the inverse-square law is a conservative estimate: the SPL at a certain distance will be a bit higher than this law predicts, because the presence of early reflections slightly increases the perceived sound level.

---

loudspeakers and your prime listening position. Figure 7.10 shows the exact amount. This figure is based on a simple calculation presented in box 7.2. For example, at a typical listening distance of 3 m (10 ft) your loudspeakers should be able to produce approximately 106 dB of peak SPL.

> **Recommendation 7.7** *Your main loudspeakers should each be able to produce a peak SPL of at least 96 dB at the listening position.*

Loudspeaker manufacturers specify the maximum SPL that their loudspeakers can handle without getting damaged. Good loudspeakers are able to handle an SPL of 110 dB or more. Producing a high SPL requires a high electrical power consumption. Since most loudspeakers are only a few percent efficient, a large amount of the input power is dissipated as heat in the voice coil. Prolonged and excessive heating of the voice coil can damage the loudspeaker (Colloms, 2018). Heating of the voice coil also results in unwanted *dynamic compression* or *power compression* where the high temperature of the voice coil leads to a reduced SPL (Gander, 1986). The increased temperature of the coil lowers its resistance, which in turn reduces the efficiency of the loudspeaker. The multiple drivers in a loudspeaker will have different levels of dynamic compression and as a result, different portions of the frequency spectrum are affected by different amounts, and the frequency balance of the loudspeaker changes. In addition, dynamic compression creates the impression of strain (Harley, 2015; Toole, 2018). Loudspeaker manufacturers try to combat dynamic compression by using drivers with oversized voice coils that dissipate the heat more easily.

7.1.5 Front Loudspeakers

The three front loudspeakers (left, center, and right) create the frontal soundstage. They do not need to reproduce the lowest frequencies, because the subwoofers take care of them. Therefore, the front loudspeakers do not need large diameter drivers, which also means that their cabinet size can be kept relatively small. As a matter of fact, the majority of front loudspeakers are modestly sized bass-reflex designs. The greatest advantage of the bass-reflex design over the closed-box is that it can achieve the same SPL with a significant reduction of diaphragm excursion. This reduction helps to maintain low values of nonlinear distortion (Small, 1973c; Thiele, 1971b).

Some people believe that the larger the loudspeaker and the more drivers it has the better it is. This is certainly not the case. A small two-way loudspeaker is likely to sound much better than a similarly priced four-way floor-standing one. First, at the same price point, the two drivers employed in the smaller loudspeaker can be made of a much higher quality than the four drivers required for the larger loudspeaker (Harley, 2015). Second, the larger the cabinet, the more difficult and expensive it is to control the panel vibrations of the cabinet. The unwanted sound output that results from such vibrations is a major discriminating factor in loudspeaker sound quality. Taller floor-standing cabinets may suffer from height-related panel resonances and internal standing waves. Small cabinets have fewer problems. In most commercial designs cabinet coloration is usually worse with larger cabinets unless they are very carefully designed (Colloms, 2018; Harley, 2015).

Recommendation 7.8 *Prefer a small front loudspeaker over a larger one if they both have the same retail price. The smaller one will likely have a better sound quality.*

The mid/bass drivers in your front loudspeakers shouldn't be too small. The front loudspeakers have to reproduce all the frequencies above the subwoofer crossover frequency (above 80 Hz). Small drivers will have trouble reproducing sufficient output power in the crossover range creating a significant dip in the overall frequency response (McLaughlin, 2005). Furthermore, a small driver will require a larger diaphragm excursion, which will increase nonlinear distortion.

Recommendation 7.9 *Choose front loudspeakers having a low/mid-frequency driver that has a diameter of at least 10 cm (4 inch).*

In some loudspeaker designs the high-frequency driver is mounted in a recessed cone in the front panel. This recessed cone acts as a *waveguide* that restricts the dispersion of the high-frequency driver such that it better matches the directivity of the midrange driver at the crossover frequency. The diameter of the midrange driver is large compared to the wavelength at the crossover frequency and therefore it has a narrow dispersion. By contrast, the high-frequency driver is small compared to this wavelength, and hence it has a wide dispersion. The use of a waveguide helps to avoid a discontinuity in directivity in the frequency region around the crossover frequency; it helps to achieve a smooth off-axis frequency response (Harley, 2015).

Most front loudspeaker designs consist of two or three drivers that reproduce different parts of the frequency spectrum (see section 7.1). The particular arrangement of the drivers influences the frequency response and the directivity of the loudspeaker. As stated in section 7.1.4 good-sounding loudspeakers have a smooth off-axis frequency response. The sound radiated off-axis determines to a large extent the sound quality of the early reflections. It has a major influence on the overall sound quality of the loudspeaker (Toole, 2018; Queen, 1979). The smoothness and uniformity of the off-axis frequency response depends on the physical arrangement of the drivers. The most common design is to mount the drivers in a vertical in-line formation. This particular arrangement results in symmetrical directivity in the horizontal plane such that the off-axis sound radiated towards the left and the right has the same frequency content. Such a symmetrical horizontal directivity is important for stereo image stability (Colloms, 2018). Unfortunately, the vertical in-line formation results in an asymmetrical directivity in the vertical plane. Since the drivers are separated vertically, the frequency content of the sound radiated towards the ceiling differs from the sound radiated towards the floor. Furthermore, the vertical separation of the drivers also results in a nonuniform dispersion pattern that changes with frequency. Figure 7.11 shows an example of such a pattern. At many off-axis positions the distance to the drivers differs. At such an off-axis position the sound from both drivers arrives with a different time delay. This time delay causes a phase shift at those frequencies that have a wavelength that is of the same order of magnitude as the vertical separation between the drivers. The amount of phase shift determines how the sound from the drivers combines. At the crossover point the sound from both drivers overlaps in frequency and due to the phase shift the combined sound can be reinforced or weakened. When the path-length difference results in a 180° phase shift the sound at that particular frequency is canceled. The frequency-dependent reinforcement and weakening results in an irregular and nonsmooth off-axis frequency response.

Some loudspeaker designs employ a *concentric driver* in which the high-frequency driver is mounted inside a larger midrange driver (see figure 7.12). Examples are the Tannoy Dual Concentric driver and the KEF Uni-Q driver. The big advantage of these designs is that the acoustic centers of the drivers coincide and as a result the directivity is smooth and uniform in both the horizontal and vertical dimension. The asymmetrical and irregular directivity that is a common drawback of traditional vertical in-line arrangements can be completely avoided. In addition, the midrange cone acts as a waveguide to the high-frequency driver such that the directivity of both drivers matches at the crossover frequency. Concentric drivers are able to produce a wide symmetrical dispersion and a smooth on- and off-axis frequency response.

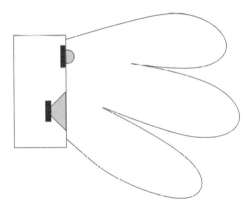

Figure 7.11
The commonly employed vertical in-line arrangement of loudspeaker drivers results in an irregular and nonsmooth off-axis frequency response in the vertical dimension around the crossover frequency. At many off-axis positions the distance to the drivers differs, because they are vertically separated. This difference in path length results in a frequency-dependent phase shift. Several lobs occur, because at certain off-axis positions the sound from both drivers is in phase while at other positions it is out of phase partly canceling the total radiated sound.

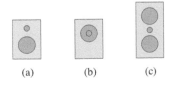

Figure 7.12
Different driver arrangements: *(a)* common two-way loudspeaker; *(b)* concentric loudspeaker; and *(c)* D'Appolito loudspeaker.

Their good directivity improves the sound quality of the early reflections and also the image definition in the frontal soundstage (Colloms, 2018; Kessler and Watson, 2011).

Another design that avoids asymmetrical vertical directivity is the *D'Appolito* driver configuration, also referred to as *M-T-M* (see figure 7.12). In this design the high-frequency unit (T for tweeter) is vertically centered between two midrange drivers (M). The two midrange drivers produce the same frequency range to achieve symmetry in the vertical dimension. Although the dispersion is symmetrical it is still irregular, because path differences between the drivers still give rise to phase shifts at certain points in space. As with the traditional vertical in-line arrangement, the D'Appolito arrangement still produces irregular lobbing in the vertical dimension as in figure 7.11. The important difference is that the lobs are symmetrical.

Since for most loudspeaker designs (including the D'Appolito driver configuration) the vertical directivity differs from the horizontal directivity it is never a good idea to place loudspeakers horizontally. When placed horizontally, the irregular lobbing will cause the sound of the loudspeaker to change with sideways movements. Multiple listeners seated in a row will experience a slightly different sound, and the sound might even change when the listeners move their heads sideways (Newell and Holland, 2007).

> **Recommendation 7.10** *Do not place loudspeakers horizontally, unless they are properly designed for it.*

Despite the fact that a horizontal arrangement of drivers creates an irregular dispersion pattern, many dedicated center loudspeakers employ such an arrangement. Since the center loudspeaker is placed below the video screen, aesthetic considerations dictate a low-profile horizontal loudspeaker cabinet. A horizontal D'Appolito M-T-M driver configuration (see figure 7.13a) is most commonly used, because it has symmetrical directivity in the horizontal dimension. However, it suffers from lobbing that leads to a dip in the frequency response often in the 1 kHz region (Colloms, 2018). A better center loudspeaker design employs an additional midrange driver that is placed below the high-frequency driver, as shown in figure 7.13b. Such a three-way design has no directivity problems in the crossover region between the high-frequency and midrange drivers, because these drivers are stacked vertically (Toole, 2018). It also doesn't suffer from directivity problems in the crossover region between the midrange and bass drivers. The reason is that the crossover frequency lies at a much lower frequency that has a corresponding wavelength that is much larger than the horizontal separation between the different drivers. At these low frequencies the slight separation between the drivers can be neglected, because it doesn't cause any significant phase shift.

> **Recommendation 7.11** *Choose a center loudspeaker that has a driver arrangement that is horizontally symmetrical. Vertical driver arrangements (shown in figure 7.12) provide the best results. Alternatively, a horizontal design with a vertically stacked high-frequency and midrange driver (shown in figure 7.13b) can be used.*

For the best results the center loudspeaker should be the exact same model as your other two front loudspeakers (Harley, 2015). Using three identical front loudspeakers is the ideal

(a)   (b)

Figure 7.13
Different driver arrangements for center loudspeakers: *(a)* common center loudspeaker and *(b)* alternative center loudspeaker with improved horizontal dispersion.

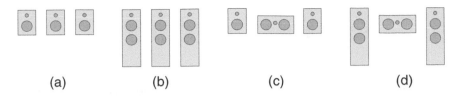

Figure 7.14
Different front loudspeaker arrangements: *(a)* three identical stand-mounted loudspeakers; *(b)* three identical floor-standing loudspeakers; *(c)* two stand-mounted loudspeakers with a center loudspeaker; and *(d)* two floor-standing loudspeakers with a center loudspeaker.

situation (see figure 7.14). It ensures a smooth and continuous sound field across the front where the sound does not change abruptly if it moves from one side to the other. If the center loudspeaker is a different model than the left and right loudspeakers then it probably sounds a bit different and this sonic difference may introduce a discontinuity in the frontal soundstage. However, it is not always practical to use three identical loudspeakers. The placement of the center loudspeaker underneath the video display may dictate another cabinet shape and size than that used for the left and right front loudspeakers. If you need your center loudspeaker to be different, make sure that it is timbre-matched with the other loudspeakers; the differences in frequency response should preferably not exceed the value of 1 dB in the frequency range of at least 250 Hz to 2 kHz (ITU-R BS.1116, 2015). Many loudspeaker manufacturers provide specific loudspeaker lines with different models (small, large, dedicated center) that work well together. The loudspeakers in such a model line all use the same type of drivers; they only differ in the number of drivers used, diameters of the midrange and bass drivers, cabinet size, and cabinet shape. Try to avoid mixing models from different model lines or from different manufacturers (Briere and Hurley, 2009; Harley, 2015; McLaughlin, 2005).

> **Recommendation 7.12** *Prefer a center loudspeaker that is identical to the other two front loudspeakers. Alternatively, make sure that the center loudspeaker is a timbre-matched model from the same manufacturer and model line.*

Even with sonically matched or identical front loudspeakers there might still be some difference in the sound emerging from the center loudspeaker. One reason is that the center loudspeaker is positioned in the middle of the room at a different distance from the sidewalls than the left and right loudspeakers. The match of the left and right loudspeakers is usually better than the match of the center and the left or right loudspeaker (Mäkivirta and Anet, 2001). A better match can be achieved by controlling the sidewall reflections (see section 8.2.1), but this can have the undesirable effect of making the reproduced sound dull and less spacious.

7.1.6 Surround Loudspeakers

Surround loudspeakers come in two different types: *direct-radiating* and *multidirectional* (Harley, 2015; Holman, 2008). As the name implies, direct-radiating surround loudspeakers radiate their sound directly towards the listener. Their design equals that of the front loudspeakers; they consist of multiple drivers mounted on the front side of the loudspeaker cabinet. Multidirectional surround loudspeakers radiate their sound into multiple directions. The most common design is the *bidirectional* or *dipole* loudspeaker. This loudspeaker consists of two sets of drivers mounted at opposite sides of the enclosure. These loudspeakers are meant to be placed at the sides of the room such that they fire sound towards the front and the back of the room (see figure 7.15).

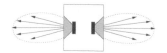

Figure 7.15
Two different types of surround loudspeakers. *On the left*, dipole surround loudspeaker that radiates sound towards the front and the back of the room. *On the right*, direct-radiating surround loudspeaker that radiates sound directly towards the listener.

Dipole surround loudspeakers were introduced at a time when surround sound consisted of a single signal that was fed to multiple surround loudspeakers. At that time surround sound mainly consisted of ambient sound that should envelop the listener and not be localizable. The main idea was to add complexity to the sound field by bouncing the surround sound off the front and back walls such that the sound arrives at the listener from multiple angles. The indirect sound also made it difficult for the listener to localize the surround loudspeakers as the source of the ambient sound (Holman, 1991, 2008).

Nowadays multidirectional surround loudspeakers such as the dipole loudspeaker should no longer be used. They have become obsolete as surround sound technology has improved (Toole, 2018). Modern surround sound consists of multiple individual surround channels that are used to create directional sound effects as well as ambient sound that envelops the listeners. Dipole surround loudspeakers are not good at delivering directional sound effects, because they do not radiate direct sound towards the listener. Maybe more surprisingly, they are also not good at creating envelopment. Envelopment is created by sounds in the frequency range of 100–1000 Hz that arrive from the sides 80 ms or more after a similar sound has arrived from the front (see section 3.4.5). Dipole loudspeakers generate sounds that arrive from the front and the back, not from the sides. Sounds arriving from the front and the back are much less effective in creating envelopment than those arriving from the sides (Hiyama et al., 2002; Toole, 2018). Furthermore, the directivity of a dipole loudspeaker is very inconsistent with frequency and thus not amenable to delivering high sound quality. In conclusion, a conventional direct-radiating loudspeaker with a smooth off-axis frequency response and a wide dispersion is a much better choice for the surround loudspeakers (AES, 2001; Zacharov, 1998).

Recommendation 7.13 *Use conventional direct-radiating loudspeakers as your surround loudspeakers. Do not use multidirectional loudspeakers such as dipoles.*

The surround loudspeakers can be separated into two categories, the ones installed at ear-level and the overhead loudspeakers used for Dolby Atmos and DTS:X. To avoid any discontinuities in the sound field, all ear-level surround sound loudspeakers should be identical. Similarly, all overhead surround loudspeakers should be identical. Mixing different loudspeaker models is not recommended. In the ideal situation all the loudspeakers in a surround system are identical, including the front, ear-level surrounds, and overhead surrounds. However, often the overhead loudspeakers are architectural loudspeakers without a cabinet meant to be built into the ceiling, while the other loudspeakers are free-standing enclosures. In that case, make sure that the overhead loudspeakers are timbre-matched to the other surround loudspeakers and to the front loudspeakers. Manufacturers often provide different loudspeaker lines that consist of several models that work well together. If you stick to the models in one of these lines, your loudspeakers should all sound sufficiently similar.

Recommendation 7.14 *In the ideal situation all your surround loudspeakers are identical and they are also identical to the front loudspeakers. Since this requirement is often too stringent, at least aim for the following:*

- *All ear-level surround loudspeakers should be identical.*
- *All overhead surround loudspeakers should be identical.*
- *Ear-level and overhead surround loudspeakers should be timbre-matched.*
- *Surround loudspeakers and the front loudspeakers should be timbre-matched.*

All the surround loudspeakers should have wide dispersion such that they cover the entire audience (see figure 4.25). The ear-level surround loudspeakers should have a uniform horizontal directivity of at least 30° (15° in either direction). The overhead surround loudspeakers should preferably have an even wider dispersion, 45° or more over the frequency range from 100 Hz to 10 kHz (www.dolby.com).

In some smaller rooms it is more convenient to use in-wall loudspeakers as ear-level surround loudspeakers, because they take up less space in the room. In-wall speakers can also be chosen for aesthetic reasons; they can be installed inconspicuously and draw less attention to themselves. However, they do have some drawbacks. First, in-wall loudspeakers do not have an enclosure; they use the inside of the wall as an enclosure, which makes their sound somewhat unpredictable. Second, they prevent you from experimenting and tuning the sound by moving them around; once installed they remain in a fixed position. And finally, they can cause vibrations in the wall that result in unwanted sound (CinemaSource, 2002; Rushing, 2004).

7.1.7 Loudspeaker Stands

Small front and surround loudspeakers should be elevated off the floor and installed at ear level (see section 4.2.3 and 4.2.4). They should not be placed on pieces of furniture, because the vibrations of the loudspeaker will couple to the furniture and make it radiate unwanted sound. Loudspeakers should only be mounted on dedicated loudspeaker stands. The construction of the stand can affect the reproduced sound.

Loudspeaker stands should be as nonresonant as possible and mechanically stable and rigid. You should avoid flimsy, wobbly, lightweight stands. The most popular stands consist of three to four hollow-section welded steel columns connected to a large base area to improve stability. Broad columns are to be avoided in favor of narrow ones, because the stand should be as acoustically transparent as possible and not introduce unwanted reflections or diffraction. The columns of some stands can be filled with dry sand, lead shot, or steel shot. Such a filling increases the mass of the stand and makes it less prone to vibration. For increased vibration reduction, you can also mass load the bottom panel of the stand by attaching a damping layer of bitumen or lead sheet to its underside (Colloms, 2018; Harley, 2015; Newell and Holland, 2007).

Recommendation 7.15 *Mount small loudspeakers on a rigid, nonresonant, and mechanically stable loudspeaker stand. The stand can be mass loaded to make it less prone to vibrations.*

Loudspeaker stands should be rigidly coupled to the floor to prevent them from rocking back and forth. Under heavy transients cabinet vibrations can make the stand bounce on its support and degrade the quality of the reproduced sound. Rigid coupling avoids rocking and considerably improves clarity, definition, and imaging. The best way to couple the stand to the floor is to use three or four hardened steel spikes with narrow points. These spikes are meant to penetrate carpet and connect the stand to the underlying floor boards to create a vibration-free coupling. On hard-surfaced floors, such as polished wood, the spikes connect directly to the floor. To avoid damaging the floor's surface, spikes are often supplied with small metal discs that can be placed underneath the spikes. Adjust the spikes such that the stand doesn't rock and make sure that the weight is carried by all three or four spikes. Use a bubble level to check if all the spikes are installed at the same height (Colloms, 2018; Harley, 2015).

Recommendation 7.16 *Use hardened steel spikes with narrow points to rigidly couple the loudspeaker stand to the floor.*

The coupling between the loudspeaker and the stand is also important. Here you want to suppress the transmission of the resonances of the loudspeaker cabinet. The most effective suppression is achieved by using small lumps of compressed mastic, Blu-Tack, or a similar sticky gumlike material. One thin pad in each corner of the top plate of the stand provides a good coupling and a measure of adhesion that locks the stand and loudspeaker together (Atkinson, 1992; Colloms, 2018; Harley, 2015).

Recommendation 7.17 *Use four small lumps of compressed mastic, Blu-Tack, or a similar sticky gumlike material to couple the loudspeaker to the stand.*

7.2 Power Amplifiers

The power amplifier turns a line-level analog audio signal of about 1 or 2 volts into a larger signal of several tens of volts to drive the loudspeakers. The two basic building blocks of a power amplifier are its *output stage* and its *power supply*. The output stage takes the small line-level input signal and uses the energy from the power supply to amplify the input signal and turn it into a larger output signal. In other words, the line-level input signal is used to modulate the energy from the power supply. Therefore, the quality of the output signal not only depends on the quality of the electronics that make up the output stage, but also on the quality of the amplifier's power supply circuits.

As the name suggests, a power amplifier delivers electrical power to the loudspeakers. Electrical power P is defined as the voltage V multiplied by the amount of electrical current I, that is, $P = V \cdot I$. Power is measured in watts (W). Power amplifiers that are intended to be used at home usually have power ratings from 20 W to 150 W, with some high-end models putting out 250 W or more. The more powerful an amplifier is, the higher its output voltage and the higher the amount of current that it can deliver. Consequently, powerful amplifiers need to have a mighty power supply. Two important parts of the power supply are the *transformer* and the *reservoir capacitors*. The transformer changes the electrical supply voltage (120 V or 230 V) in a lower voltage at which the amplifier operates. The transformer is a large metal block or cylinder inside the chassis. The larger it is, the more current it can deliver. In high-performance amplifiers the transformer is usually the heaviest part. The reservoir capacitors store electrical energy. They release their energy to smooth out the sinusoidal variations in the supply voltage and to briefly deliver high peak currents during large transients in the audio signal. Again, the larger the capacitors, the more current they can deliver (Cordell, 2011; Duncan, 1997; Harley, 2015).

High-quality amplifiers are usually large and heavy (Harley, 2015). Not only do they have a large and heavy power supply, they also have a large heat sink to get rid of the excess heat that builds up in the output stage. Most amplifiers are not very efficient, which means that a relatively large amount of their electrical energy is not converted to output power, but dissipated as heat. Getting rid of this heat is an important design consideration; it requires large heat sinks that increase the cost and size of the amplifier (Cordell, 2011). The price of high-quality amplifiers remains high, partly because amplifiers do not follow the trend of miniaturization and mass-production techniques that are widely used to produce all kinds of other electronic devices (Newell and Holland, 2007).

7.2.1 Power Requirements

How much amplifier power do you need? It is important to be able to answer this question, because the cost of an amplifier is proportional to its output power. An oversized amplifier can be a waste of money. While on the other hand, an underpowered amplifier will prevent your loudspeakers from sounding their best: the reproduced sound will be constricted and uninvolving with an overall sense of strain during climaxes (Harley, 2015).

Obviously, the louder you want to play your music or movies, the more amplifier power you need. A peak SPL of 96 dB is a good target to aim for. It is the peak SPL that corresponds to the recommended reference SPL for home listening (EBU-Tech 3276, 1998; ITU-R BS.1116, 2015) as explained in box 3.2. Note that you need to achieve this peak SPL at the listening position, and must therefore take into account the distance from the loudspeakers. Figure 7.10 shows the SPL that your loudspeakers need to produce in order to achieve 96 dB at a certain distance. To summarize, the farther away you sit from the loudspeakers, the more SPL your loudspeakers need to produce and hence the more amplifier power you need.

The most important factor that determines how much amplifier power you need is the *sensitivity* of your loudspeakers. The sensitivity specification determines how efficient your loudspeakers are in converting the electrical power supplied by the amplifier into acoustical output. The standard way to specify sensitivity is to measure the SPL in dB that the loudspeaker produces at a distance of 1 m when driven by an amplifier signal of 2.83 V (Colloms, 2018). A moderately efficient loudspeaker has a sensitivity of about 88 dB. Thus, such a loudspeaker produces 88 dB SPL at 1 m when the output of the amplifier is 2.83 V. To calculate the corresponding amplifier power, we need to know the *impedance* of the loudspeaker. The impedance describes the load that the loudspeaker produces for the amplifier; it describes how voltage and current are related and is measured in ohms (denoted as Ω). The impedance of a loudspeaker varies with frequency, but for power calculations it is customary to use the *nominal impedance,* which is a kind of average impedance. Most commercial loudspeakers have a nominal impedance of 8 Ω. With a simple calculation that is shown in box 7.3, it follows that a voltage of 2.83 V applied to a loudspeaker with a nominal impedance of 8 Ω results in exactly 1 W of power. To summarize, when you apply 1 W of amplifier power to a loudspeaker with a sensitivity of 88 dB SPL and a nominal impedance of 8 Ω, it produces an SPL of 88 dB at a distance of 1 m.

You can determine the amount of amplifier power that you need from the sensitivity of your loudspeakers and the listening distance. You can simply read off the required power in figure 7.16, or alternatively use the equation in box 7.4. For example, if you sit 3 m (10 ft) from your loudspeakers and they have a sensitivity of 88 dB SPL (at 1 m/2.83 V) you will

---

BOX 7.3 AMPLIFIER POWER

Electrical power P is defined as $P = V \cdot I$. With Ohm's law, $V = I \cdot R$, the power can also be expressed as

$$P = \frac{V^2}{R}$$

where R is the nominal impedance of the loudspeaker.

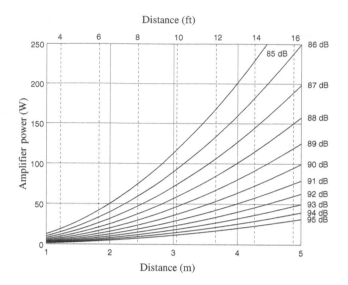

BOX 7.4 AMPLIFIER POWER REQUIREMENTS

The sound pressure level produced by a loudspeaker with a nominal impedance of 8 Ω is given by:

$$SPL = S - 20\log_{10}(D) + 10\log_{10}(P)$$

where S is the sensitivity of the loudspeaker in dB (at 1 m/2.83 V), D is the distance in meters between the loudspeaker and the listener, and P is the electrical power rating of the amplifier in watt.

need approximately 60 W of amplifier power to be able to produce the desired peak SPL of 96 dB at your listening position. Note that figure 7.16 is only valid for loudspeakers with a nominal impedance of 8 Ω. Some loudspeakers exist that have a nominal impedance of only 4 Ω. They draw twice as much current from the amplifier when the amplifier signal is 2.83 V (because $I = V/R$). Consequently, they require twice the amplifier power. Thus, if your loudspeakers have a nominal impedance of 4 Ω, you need to double the power obtained from figure 7.16.

Recommendation 7.18 *The amount of amplifier power that you need depends on the sensitivity of your loudspeakers and the listening distance. Use figure 7.16 to determine the amount of power that you need to be able to produce a peak SPL of 96 dB at the listening position.*

From figure 7.16 it follows that a drop of 3 dB in loudspeaker sensitivity requires a doubling of amplifier power in order to still be able to generate the same SPL at the listening position. For example, at a listening distance of 3 m (10 ft), we need 60 W of amplifier power for a loudspeaker with a sensitivity of 88 dB SPL (at 1 m/2.83 V), but already 120 W for a loudspeaker with a sensitivity of 85 dB SPL (at 1 m/2.83 V). Or to put it differently, if you want your loud-speakers to play louder, every increase of 3 dB in SPL requires a doubling of amplifier power. This is an important point as it puts the amplifier power figures into perspective. The SPL difference between a 60 W amplifier and a 75 W amplifier is not going to be all that great. A noticeable increase of 3 dB already requires 120 W, and the next increase of 3 dB requires 240 W. Recall from section 2.3.2 that in the mid- and high-frequency region where the ear is most sensitive, an SPL increase of 10 dB gives you the impression that the sound is twice as loud. Such an increase of 10 dB would mean that the amplifier power needs to be increased tenfold (see table 2.5). Thus, to be able to play your music or movie twice as loud you would need to replace your 60 W amplifier by an amplifier that puts out a whopping 600 W.

The maximum power output of an amplifier is limited by the largest voltage it can produce and the maximum amount of current it can provide. The ideal power amplifier is a *voltage source*, which means that it produces an output voltage that is proportional to its input volt-age and that it can provide an unlimited amount of current. Ideally, the output voltage is inde-pendent of the loudspeaker's impedance. This would imply that the maximum power that the amplifier delivers to a 4 Ω loudspeaker is twice that of an 8 Ω loudspeaker (see also box 7.3). In practice, amplifiers cannot do this due to their limited current capabilities; power into 4 Ω is typically between 1.95 and 1.65 times the power into 8 Ω (Duncan, 1997; Self, 2009). The amount of current that an amplifier can deliver is important, because loudspeaker impedance varies with frequency. Figure 7.17 shows a typical example of loudspeaker impedance as a function of frequency for a two-way loudspeaker system. The two peaks correspond to the resonance frequencies of the two drivers. In some loudspeakers the impedance can drop way below the nominal impedance of 8 Ω at certain frequencies. The lower the impedance the more current the amplifier needs to provide. Peak currents for different loudspeakers can be between 3.8 and 6.6 times larger than that drawn by an 8 Ω load (Otala and Huttunen, 1987). Many receivers and stand-alone power amplifiers are not able to deliver that much current. The ability to deliver large amounts of current is a major differentiating factor among power amplifiers. The design of the power supply will have a big influence on amplifier performance. Inadequate current capabilities and a power supply voltage that sags under heavy loads can impair sound quality (Cordell, 2011). As I have said before, capable amplifiers are often large and heavy; they have big power transformers and reservoir capacitors (Harley, 2015).

When a power amplifier reaches its maximum output it starts *clipping*: it can no longer produce the peaks of the audio signal. Figure 7.18 shows what happens if a sine wave gets clipped. When the amplifier reaches its maximum output voltage the peaks of the sine wave get stuck at this maximum voltage. This is called *hard clipping* or *peak limiting*. Often before this happens the amplifier will go into *soft clipping* or *saturation* where the tops of the sine wave get distorted. Both types of clipping lead to nonlinear distortion. As with loudspeakers

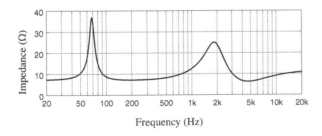

Figure 7.17
Typical example of loudspeaker impedance for a two-way system.

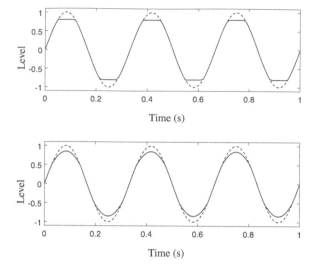

Figure 7.18
Examples of a clipped sine wave. The *broken line* is the original sine wave while the *solid line* is the clipped signal. *At the top,* hard clipping where the peaks of the sine wave are cut off. *At the bottom,* soft clipping where the peaks of the sine wave are distorted in shape.

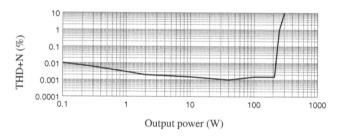

Figure 7.19
Example of nonlinear distortion in an amplifier as a function of output power when the input is a 1 kHz sine wave. The vertical axis represents the total harmonic distortion (THD) plus noise (THD+N). At low power levels the THD is very low and stays close to the noise floor. At a certain power level the distortion starts to rise rapidly. The maximum power that the amplifier can deliver is often defined as the power at which the distortion reaches 0.1%.

(see section 7.1.4) amplifiers behave like a linear system for small values of their input signal. As the input gets larger and the amplifier starts to run out of power at its output, the amplifier becomes nonlinear. The maximum power that an amplifier can deliver is defined as the power for which its nonlinear distortion is less than a certain amount (Self, 2009). One way to determine the maximum power is to use a sine wave of 1 kHz as an input signal and measure the total amount of harmonic distortion (THD) at the output as the power of the sine wave is increased. An example of such a power sweep is given in figure 7.19. At low power levels the THD is very low and stays close to the noise floor. Like all analog audio components, thermal noise in the electrical circuits will prevent the amplifier from being completely silent when no input signal is present. At a certain power level the THD starts to rise rapidly. The maximum power that the amplifier can deliver is often defined as the power at which the distortion reaches 0.1%. Some amplifier manufacturers are known to artificially boost their power ratings by stating the maximum power at increased distortion levels of 1% or even 10%. Power ratings with a distortion of 0.1% or less provide a more honest statement.

There are many different ways to specify the maximum power rating of an amplifier. When comparing different amplifiers it is important to know what to look for. Some manufacturers are known to be creative with their power ratings to make them more impressive while other manufacturers provide a more conservative and honest rating (Harley, 2015; McLaughlin, 2005). First, the power should be specified per channel and not as a total sum. Second, the nonlinear distortion at the maximum power level should not exceed 0.1%. Third, the power bandwidth should be 20 Hz to 20 kHz. The reproduction of low frequencies requires more power than the reproduction of mid and high frequencies. Therefore, a power specification for a single sine of 1 kHz will always be higher than for a more realistic audio signal that covers the entire 20 Hz to 20 kHz frequency range. Fourth, the power rating should be for continuous power, sometimes incorrectly called *rms* (root mean square) power. Amplifiers are able to generate more power for short time bursts than for sustained periods of time. The peak-power rating of an amplifier is always larger than its continuous power rating. The difference mainly depends on the amount of energy that can be stored in the reservoir capacitors of the amplifier's power supply. Fifth, the power rating should be specified for the situation in which all channels of the amplifier are driven. This is especially important in multichannel amplifiers with five or seven channels. These amplifiers can often put out more power when only one or two channels are driven, because in that case the power supply has to work less hard. Finally, the power specification should be for a nominal load of 8 Ω. Some manufacturers specify their power ratings only at 6 Ω or 4 Ω. Such specifications are always more impressive as an amplifier will deliver more power at these lower impedances. However, it becomes impossible to compare the power rating with other competitive models specified at the recommended nominal load of 8 Ω.

> **Recommendation 7.19** *When comparing power specifications of multichannel amplifiers make sure that the power is listed as follows:*
>
> - *watts per channel (not the total sum),*
> - *with no more than 0.1% of THD (not 1% or 10%),*
> - *over a bandwidth of 20 Hz to 20 kHz (not at 1 kHz, not from 40 Hz to 20 kHz),*
> - *continuous power (not peak power),*
> - *with all channels driven (not just one or two channels), and*
> - *into a nominal load of 8 Ω (not 6 Ω or 4 Ω).*

Some people wrongly assume that the maximum-power rating of an amplifier should always be less than the power rating of the loudspeakers. They worry that a too powerful amplifier will easily blow up the loudspeakers. In fact, it is the other way around: a less powerful amplifier is more likely to destroy your loudspeakers. When an amplifier runs out of power it will distort the output signal by clipping its peaks (as shown in figure 7.18). If clipping occurs only briefly, you will not notice it. However, if clipping is frequent or sustained for a longer period of time, your loudspeaker will quickly be destroyed. Loudspeakers are able to handle surprising amounts of power provided that the amplifier is not clipping. Therefore, it is better to use an amplifier that is rated for more power than the loudspeakers can handle (Harley, 2015; Winer, 2018). Most loudspeakers are rated at 100 W or more. That being said, determining the maximum power rating of a loudspeaker is not an exact science. The maximum rating is meant to describe how much power the loudspeaker can take before it is damaged by overheating or by an excessively large diaphragm excursion. It is quite complicated to come up with a reasonable estimate due to the unpredictable transient nature of music and sound in movies (Colloms, 2018).

7.2.2 Amplifier Types

Amplifiers come in different types depending on the implementation of their output stages. The design of the output stage has a large influence on the performance and cost of the amplifier. The output stage consists of several output devices that operate at high power levels and often generate lots of heat. Three types of output devices that are commonly used are: 1) transistors, 2) IC (integrated circuit) amps, and 3) vacuum tubes. Most amplifiers use transistors. Some lower-end amplifiers use IC amps that consist of several transistors integrated on a small chip. Although the quality of IC amps has improved over the years, low distortion and high power still requires output stages that are built around discrete transistors (Self, 2009). Some audiophile amplifiers use vacuum tubes. These amplifiers have a completely different sound than the ones built from transistors. The tubes generate a fair amount of harmonic distortion that gives tube amplifiers their characteristic sound. Some audiophiles prefer this distorted sound; it sounds more pleasing to them. However, pleasing as it might seem to some, the fact remains that tubes add their own sonic signature that impairs the accuracy of the reproduction (Self, 2009; Winer, 2018).

The output stage can be designed in different ways. A system of amplifier class designations describes the different circuit topologies. Not all amplifier classes are suitable for high-performance sound reproduction. The classes that are most often used are A, AB, B, D, G, and H. The higher classes provide an improvement in efficiency where the amplifier wastes less power and produces less heat. In some designs ultimate sound quality is traded for a higher power efficiency to reduce the cost of the amplifier and to make better use of energy resources.

Class A

In a class A amplifier all transistors in the output stage are continuously delivering current. This is in contrast with a class B amplifier (to be explained next) where one set of transistors operates on the positive parts of the audio signals and another set on the negative parts of the signal. In a class A amplifier there is no switching between different transistors as in class B. The lack of switching is the biggest advantage of a class A amplifier; it avoids the introduction of a type of nonlinear distortion called *crossover distortion* where a small discontinuity occurs in the output signal as one set of transistors takes over from the other (Cordell, 2011; Duncan, 1997; Self, 2009).

Class A amplifiers are very inefficient. This is their biggest disadvantage. At its best only 50% of the energy it consumes is used for amplification, the other 50% is dissipated as heat. In practice, the efficiency tends to be even lower: 35% or less. Class A amplifiers are impractical to build; they have high hardware and energy costs, heavy weight, and generate lots of heat. Therefore, their power rating is practically limited to about 100 W (Duncan, 1997).

Class B and AB

In a class B amplifier the transistors in the output stage remain on for only half of the signal cycle; one set of transistors amplifies the positive parts of the audio signals while another set amplifies the negative parts. The hand-off from one set to the other is abrupt and can lead to crossover distortion (Cordell, 2011). Crossover distortion introduces a discontinuity in the amplified signal that can be audibly unpleasant (Newell and Holland, 2007).

Class B amplifiers are more efficient than class A amplifiers. The efficiency of a class B amplifier can approach 78.5%, which is a lot more than the maximum efficiency of 50% for class A devices (Duncan, 1997).

Pure class B amplifiers are not much used in high-quality sound reproduction. The most commonly used amplifier is a class AB amplifier. Class AB is not a class in itself but a combination of class A operation and class B operation. These amplifiers behave largely like a class B amplifier, but some overlap between the operation of the different sets of transistors is allowed, such that the transition from one set to the other is a bit more smooth. For small output signals the amplifier effectively operates in class A; for larger outputs it operates in class B. In class AB operation crossover distortion can be kept under control, but temperature variations in the output transistors can affect sound quality. Temperature variations should be minimized, because they influence the transition from class A to class B operation (Cordell,

2011). For this reason some people let the amplifier warm up for several minutes, before they start listening to music.

Class D

The output stage of a class D amplifier consists of transistors that are alternately switched fully on or off. These amplifiers work on an entire different principle than class A, B, and AB amplifiers. While in these other classes the transistors supply current that is proportional to the input signals, the transistors in a class D amplifier are either on or off; they act as switches that connect the output of the amplifier directly to the positive and negative power supply rails. The audio information is contained in the duration of the switching intervals. The input signal is converted into a series of pulses by a process called *pulse-width modulation* (PWM). The train of pulses determines the switching intervals of the output transistors. The variation in the switching intervals results in variations of the average output voltage such that this average voltage is proportional to the input signal. The switching occurs at a very high frequency, often in the range of 50–1000 kHz. A low-pass filter at the output of the amplifier averages the high-frequency pulses and extracts the average voltage to drive the loudspeakers (Cordell, 2011; Self, 2009).

Because of their specific way of operating class D amplifiers are often called *switching amplifiers* or *PWM amplifiers* (Harley, 2015). Sometimes they are also called *digital amplifiers*, but this is not a correct term. While some class D amplifiers may be driven directly by a digital signal, the output stage of a class D amplifier doesn't work with sampled time instances. The output voltages and currents exist at any time instance. Consequently, any disturbance of the input signal or power supply voltage will directly distort the output of the amplifier. In a completely digital system small disturbances will only affect the output if they become larger than the quantization interval. Class D amplifiers do not exhibit such robustness.

Class D amplifiers are very efficient, often between 80% and 95% (Cordell, 2011; Self, 2009). They dissipate little power, because the output transistors are either fully on or off. Almost all the power from the amplifier's power supply is transferred to the loudspeakers. Since Class D amplifiers generate little heat, they do not need massive heat sinks nor massive power supplies. Therefore, they are relatively small and lightweight (Duncan, 1997). This makes them a popular choice for powered subwoofers and multichannel home theater amplifiers.

Despite these advantages of class D amplifiers, they have long suffered from poor sound quality. Imperfections in the high-frequency switching process can generate lots of nonlinear distortion in class D amplifiers (Cordell, 2011). Over the years matters have improved dramatically, but many audiophiles still prefer the sound of the more traditional class AB amplifiers. Class D amplifiers can have a less than optimal sonic performance when they reproduce frequencies above 5 kHz (Duncan, 1997; Newell and Holland, 2007). At low frequencies the performance is up to par with the other amplifier classes and therefore class D amplifiers can without any concern be used in powered subwoofers.

Class G and H

The development of class G and class H amplifiers is another attempt at improving the efficiency of the class AB designs. These amplifiers utilize different power supply voltages to increase efficiency. They operate similarly to class AB amplifiers, but under low power conditions they utilize a lower supply voltage to significantly reduce the power wasted by the output transistors. When higher power conditions arise, they increase the power supply voltage to generate more power. Class G amplifiers switch between different supply voltages, while class H amplifiers use a variable power supply voltage. These amplifiers are more efficient than the traditional AB designs, but still less efficient than the class D designs. Like the class D amplifiers, they are a popular choice for powered subwoofers (Duncan, 1997; Self, 2009).

7.2.3 Performance

Amplifiers differ in their sound quality. As mentioned before, it is important to match the amplifier to your loudspeakers. Not all amplifiers and loudspeakers optimally work together. For starters, the power output and the sensitivity of the loudspeakers should be matched (see figure 7.16), otherwise the amplifier may sound strained and harsh as it goes into clipping and produces too much distortion during the louder parts of the program material. Some people claim that all well-engineered amplifiers sound the same as long as they provide adequate output power with low distortion. However, amplifiers do also sound different due to differences in performance when driving complex loudspeaker loads (Newell and Holland, 2007). Measurements of the electrical performance of an amplifier do not tell the whole story. While measurements are extremely useful to confirm a correct design, they cannot predict all the nuances of the perceived sound quality (Cordell, 2011; Duncan, 1997). Therefore, it is important to audition an amplifier with your own loudspeakers before you buy one. Some dealers allow you to take the amplifier home for auditioning the combination. Alternatively, you could bring your loudspeakers to the store, or if the dealer sells the same loudspeakers use those (Harley, 2015).

Listen to an amplifier at high and low volumes and compare its sound. If the bass is weak at high levels, it is a sure sign that the amplifier is underpowered. Other signs of an underpowered amplifier are a congested sound, sense of strain, and hardening of timbre. If you feel a sense of relief when the volume is turned down then the high frequencies may be too bright and hard. Many amplifiers make the high frequencies forward and unpleasant. Listen to brass instruments, violin, piano, and flute. A good amplifier sounds neutral without any sense of strain or harshness (Harley, 2015).

Recommendation 7.20 *Before you buy an amplifier, always listen to how well this amplifier drives your loudspeakers (in the store or at home). It should sound neutral without any sense of strain or harshness, even at high volumes.*

Besides a thorough listening test, it is also a good idea to verify that the amplifier complies with a number of standard technical performance criteria. The most important one, the power output, has already been described in section 7.2.1. The other technical specifications are listed in table 7.3. You can obtain these specifications from the manufacturer or from an independent technical review (online or in an A/V magazine). Let's look at these specifications in more detail.

Recommendation 7.21 *When you buy an amplifier, make sure it complies with the technical specifications listed in table 7.3.*

Table 7.3 Recommended technical specifications for power amplifiers.

	Excellent	Good
Frequency response 20 Hz–20 kHz	±0.5 dB	±1 dB
Nonlinear distortion	0.01%	0.1%
Signal-to-noise ratio (unweighted)	90 dB	80 dB
Signal-to-noise ratio (A-weighted)	100 dB	90 dB
Channel separation	60 dB	50 dB
Damping factor	50	20
Slew rate	20 V/μs	10 V/μs

Frequency Response

The amplifier should reproduce all frequencies in the audible range with an equal amount of amplification. The standard way to measure this aspect is to determine the frequency response between the amplifier's inputs and its outputs. The ideal frequency response is a ruler flat line between 20 Hz and 20 kHz. Some amplifiers may struggle to provide adequate amplification at the frequency extremes of this range. The frequency response may be down a few dB at 20 Hz and at 20 kHz. Most modern amplifiers have no trouble whatsoever and are only down by 0.5 dB on either side (Self, 2009). Sometimes the frequency response is specified at the standard −3 dB points; for example as 10 Hz to 80 kHz. In this case the specified frequency range should be large enough such that the −3 dB points are far enough from the 20 Hz and 20 kHz frequencies and that the frequency response at 20 Hz and 20 kHz deviates less than 1 dB (Cordell, 2011; Winer, 2018). Some amplifiers have a positive deviation at the high-frequency end of the spectrum; for example +1 dB at 20 kHz. Such a positive deviation should be looked at with extreme suspicion, as it can mean that the amplifier has stability problems at the high-frequency end of the spectrum (Self, 2009). Well-designed amplifiers have approximately the same frequency response under different loads: they don't lose bandwidth when the load changes from 8 Ω to 4 Ω. In some technical reviews the changes in frequency response with different loads are measured and compared (see for examples the amplifier reviews at audioholics.com and soundstage.com).

Nonlinear Distortion

The frequency response of an amplifier doesn't tell the whole story. While an ideal amplifier behaves like a linear system, real-life amplifiers are only approximately linear; they suffer from nonlinear distortion. Recall from figure 7.19 that nonlinear distortion in an amplifier depends on the power output. At low powers the distortion is low and at a certain power level it raises dramatically. During normal operation, the amplifier functions at power levels way below the point where the distortion rises. We are interested in the amount of distortion produced in this region of the power curve. Compared to loudspeakers, the nonlinear distortion that amplifiers produce is way lower. While it is fairly typical for loudspeakers to have nonlinear distortion levels of a few percent (see section 7.1.4), a moderately good amplifier will have distortion levels of 0.1% (−60 dB) and often even below 0.01% (−80 dB) at normal operating powers. This raises the question whether nonlinear distortion in amplifiers is insignificant compared to the much higher levels of distortion that the loudspeakers produce. The answer is that you cannot simply compare the figures. The amount to which nonlinear distortion is subjectively disagreeable very much depends on the nature of the nonlinear distortion. The specific frequency components that make up the distortion are important. For example, second or third order harmonic distortion is far less objectionable than higher-order harmonic distortion. It turns out that nonlinear distortion produced by loudspeakers tends to be benign compared to the nonlinear distortion that an amplifier can produce. Thus, simply comparing distortion levels is misleading (Duncan, 1997; Newell and Holland, 2007).

Nonlinear distortion is difficult to characterize completely. While harmonic distortion can be adequately described by specifying the distortion levels of all the harmonic frequency components that it generates, the levels and frequencies of these components depend on the frequency of the test signal. Most often the total harmonic distortion (THD) is measured using a test signal with a frequency of 1 kHz. However, THD measured with a test signal of a different frequency might yield a different result. It is not difficult to achieve low THD figures at 1 kHz, but achieving low distortion at higher frequencies is a completely different story. A good amplifier might have a THD of 0.005% at 1 kHz, but at the same time have a THD of 0.02% at 20 kHz (Cordell, 2011). To complicate matters further intermodulation distortion is even more difficult to characterize as it involves not only one but two different test frequencies. Two commonly employed tests are the CCIF and the SMPTE intermodulation distortion tests

(Cordell, 2011). The CCIF test, also known as the IMD (ITU-R) test, involves two sines, one at 19 kHz and one at 20 kHz, that have the same amplitudes (ITU-R SM.1446, 2000). The SMPTE test involves two sines, one at 60 Hz and one at 7000 Hz mixed in a ratio of 4:1 (SMPTE RP 120, 2005).

It is nearly impossible to predict the sound of an amplifier from its distortion measurements. There are too many different distortion measurements and it remains unclear how they relate to the subjective sound quality of an amplifier (Duncan, 1997; Newell and Holland, 2007). However, while the measurements are all different, it is important to realize that they are just different ways to characterize the underlying mechanism that is responsible for the nonlinear distortion. In fact, it is virtually impossible to have one type of measured distortion without having another type of measured distortion. Therefore, a very low THD figure for all test frequencies leaves little room for most other distortions to be present. If THD is well under 0.01% (−80 dB) at all test frequencies, the other distortion measurements like CCIF and SMPTE will also be very small (Cordell, 2011). So, don't dismiss the THD measurement all together. A graph showing THD as a function of test frequency is much more revealing of amplifier performance than a single THD number at the typical test frequency of 1 kHz. While low distortion is a good thing to have, it remains unclear how it exactly relates to the perceived sound quality of an amplifier. There is no evidence that an amplifier producing 0.001% THD sounds any cleaner than one producing 0.005% (Self, 2009). In general, there is no point in getting too excited about distortion figures below 0.01%.

Signal-to-Noise Ratio

The signal-to-noise ratio (SNR) describes how quiet an amplifier is. When there is no input signal present the amplifier should not produce any sound. In practice, every amplifier produces a very low hiss that results from electronic noise that all electronic circuits have some amount of. Ideally, the noise in the amplifier should be less than the quantization error in the audio signal from your digital source. This requires an SNR of at least 98 dB for 16 bits and a whopping 146 dB for 24 bits (see table 5.7). Recall from section 5.2.2 that the accuracy of a DAC is also limited by electronic noise. The best DACs can achieve an accuracy of at most 21 bits, which corresponds roughly to an SNR of 128 dB. Currently no power amplifier can match this. It is nearly impossible to find a power amplifier with an SNR higher than 104 dB, which corresponds to about 17 bits in the digital world. Very good amplifiers have an SNR that reaches 90 dB (about 15 bits) when their output voltage equals 2.82 V into 8 Ω (Cordell, 2011). Some manufacturers specify the SNR in dBA; they apply frequency weighting using the A-weighting curve. This weighting curve mimics the sensitivity of the ear at moderate sound levels and emphasizes the contribution of the mid and high frequencies. Compared to the unweighted SNR specification the A-weighted specification will be about 10 dB higher with the same amount of noise. So take care when comparing SNR figures.

Channel Separation

In a multichannel amplifier the signal from one audio channel may leak into another channel. A good amplifier has a *channel separation* of 60 dB or more, which means that the signal from the leaked channel is 60 dB or more below the original signal. The amount of leakage is called *cross talk*; thus another way to say the same thing is to state that the amplifier has a cross talk of −60 dB or less. Cross talk in amplifiers occurs because the high-level amplified signals can easily leak into the adjacent electronic circuits (Duncan, 1997). Proper internal wiring and circuit board layout may help, as well as separate power supplies for each channel. The ultimate channel separation is achieved when one mono amplifier with its own chassis and power supply is used for each of the different channels. However, this is a very expensive option (Harley, 2015). There is no need to go that far, because a carefully designed multichannel amplifier can have a channel separation of 90 dB (Self, 2009). Furthermore, in practice, a separation of 50 dB or 60 dB is more than sufficient and does not yield any audible imperfections.

Damping Factor

An ideal amplifier acts as a voltage source, which means that it supplies its voltage without being influenced by the load that the loudspeaker provides. Practical amplifiers have a certain output impedance that prevents them from acting as a perfect voltage source. The lower the output impedance of the amplifier the closer it comes to achieving its goal of being a perfect voltage source (Greiner, 1980). The effect of a finite amplifier output impedance is explained in more detail in box 7.5.

The output impedance of an amplifier is often indirectly specified as the *damping factor*. This factor is calculated simply as the number 8 divided by the output impedance. The number 8 is used as it is the nominal loudspeaker impedance that is commonly used to specify the amplifier's power rating. Almost all modern power amplifiers have an output impedance of 0.05 Ω or less, which corresponds to a damping factor of 160 or more (Self, 2009).

In general the higher the damping factor, the less influence the loudspeaker load has on the amplifier. This is important as the impedance of the loudspeaker varies with frequency (see figure 7.17). The interaction between the amplifier and the loudspeaker can cause peaks and dips in the frequency response. The interaction can also influence the transient response, especially at low frequencies. The higher the damping factor the better the amplifier is able to control the motion of the low-frequency driver in the loudspeaker. Amplifiers with a high damping factor provide a tighter-sounding bass (Harley, 2015; Newell and Holland, 2007).

Box 7.5 AMPLIFIER OUTPUT IMPEDANCE

Figure 7.20 shows a simplified electrical diagram of an amplifier driving a loudspeaker. The amplifier is represented by an ideal voltage source V_A with a finite output impedance R_A. The amplifier's output voltage V_O is less than the ideal voltage V_A because of the output impedance R_A. The circuit in figure 7.20 acts as a voltage divider. The equation for the voltage V_O can be derived using some straightforward electrical circuit analysis:

$$V_O = \frac{R_C + R_L}{R_A + R_C + R_L} \cdot V_A$$

where R_C is the impedance of the loudspeaker cable, and R_L is the impedance of the loudspeaker. As long as R_A is much smaller than $R_C + R_L$ the amplifier acts as if it is a perfect voltage source and V_O will be approximately equal to V_A. In practice, typical values are $R_C = 0.02\ \Omega$ and $R_L = 6\ \Omega$ (Toole, 2018). Almost all modern power amplifiers have an output impedance of 0.05 Ω or less, in which case $V_O = 0.99 V_A$.

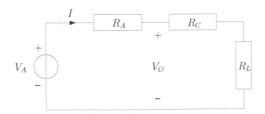

Figure 7.20
Simplified electrical diagram of an amplifier driving a loudspeaker. R_A is the output impedance of the amplifier, R_C is the impedance of the loudspeaker cable, and R_L is the impedance of the loudspeaker.

While a high damping factor is desirable, there is a point of diminishing returns. In fact this point is already reached with a damping factor of about 20 (output impedance less than 0.4 Ω). With a damping factor of 20 and a nominal loudspeaker impedance of at least 6 Ω, no changes are visible in the transient response, and the changes in the frequency response are much less than 1 dB, and then only over a narrow frequency range (Toole, 1975, 2018). Contrary to what some believe, a very high damping factor does nothing to improve the sound. Keeping a safety margin, it is fair to say that a damping factor of about 50 already suffices (Winer, 2018). Clearly, the difference in sound between an amplifier with a damping factor of 100 and another with a damping factor of 2000 is undetectable by human perception (Self, 2009). So don't worry about your amplifier's damping factor too much.

Slew Rate

The *slew rate* of an amplifier is a measure of how fast the output of the amplifier can change. It is an indicator of how well an amplifier can respond to high-level transients. The higher the rated power output of the amplifier, the larger its output voltage is and hence the larger its slew rate must be to keep up with fast-changing audio signals. A minimum slew rate of 5 V/μs is needed for a 100 W amplifier to cleanly reproduce a sine wave with a frequency of 20 kHz. To produce full power at 50 kHz this 100 W amplifier needs a minimum slew rate of 12.5 V/μs (Cordell, 2011; Self, 2009). Adding some safety margin, a slew rate of 20 V/μs should be sufficient for any domestic sound system. High-performance amplifiers can easily achieve slew rates of 50 V/μs or more, providing an even larger safety margin. In practice, any modern amplifier has a sufficiently large slew rate, so you don't need to worry about this particular amplifier specification.

7.3 Subwoofers

A subwoofer is a loudspeaker that is specifically designed to reproduce low-frequency sound. Accurate reproduction of low frequencies at a high enough sound level requires large loudspeaker drivers, large and heavy loudspeaker cabinets, and lots of amplifier power. These are all good reasons for using a dedicated loudspeaker for the reproduction of the lowest frequencies. Though, an even more important reason is that the optimal reproduction of low frequencies requires the subwoofers to be placed in positions in the room that are completely different from the positions of the main loudspeakers, as I explained in section 4.1.

Most subwoofers on the market are *active subwoofers*; they have an integrated power amplifier. These subwoofers are driven with the unamplified line-level signal from the subwoofer output of your A/V receiver or controller. An active subwoofer makes your system *bi-amped:* the amplifier built into the subwoofer takes care of the low bass (below 80–120 Hz) while the rest of the frequency spectrum is amplified using separate power amplifiers or the power amplifiers built into your A/V receiver. The A/V receiver or controller acts as an active crossover filter splitting the entire frequency range into two parts: one for the subwoofers and one for the main loudspeakers.

Bi-amping has the advantage that it reduces the intermodulation distortion in the amplifiers, because they operate over a narrower frequency bandwidth. Bi-amping also increases the ability of the system to play loud without any strain. The reason is that most music is bass heavy and the high-energy bass frequencies will dominate the output of the power amplifier, leaving little to no power for the high frequencies. When the amplifier reaches its maximum output on the bass frequencies, the other frequencies will also get distorted (see also the example in figure 4.2). Using two separate amplifiers avoids such clipping of the higher frequencies and results in an effective increase in headroom (Colloms, 2018).

Amplifiers for subwoofers need to deliver plenty of power. Amplifiers with more than 300 W are not uncommon. At these powers heat dissipation can become an issue with the traditional class A/B amplifier designs. Class D amplifiers with their increased efficiency generate

Figure 7.21
A subwoofer design with two identical bass drivers mounted rigidly back to back to balance their reaction forces. This design can significantly reduce the rocking motion of the enclosure.

significantly less heat and are therefore a popular choice. The sound quality of class D amplifiers at high frequencies is not always that good, but at low frequencies their sound quality is up to par with that of class A/B amplifiers. The general consensus is that sound quality of subwoofer amplifiers is less critical than that of other amplifiers (Self, 2009).

The large powerful drivers that are typical for subwoofers move a lot of air with their large excursions. They exert quite some force on the enclosure, which can result in unwanted enclosure resonances. Such resonances cause the panels of the enclosure to radiate spurious sounds degrading the overall sound quality. Therefore, the subwoofer's enclosure should be solid, rigid, and vibration-free. Some manufacturers reduce the driver-induced enclosure vibrations by using two identical bass drivers mounted rigidly back to back to balance their reaction forces. Such a configuration is shown in figure 7.21. The two drivers are driven such that their frame vibration is in opposition, which greatly reduces the rocking motion of the enclosure (Colloms, 2018). Some manufacturers mount a single driver in the bottom of the cabinet with the driver facing the floor. Such an arrangement is not recommended as it may result in a drumming sound on the floor due to the reaction force of the driver. In a single driver design it is best to mount the driver in one of the side faces of the enclosure (Colloms, 2018).

The size of the subwoofer drivers and the power of the built-in amplifier determine how loud the subwoofer can play and how low it can go. Besides these two criteria it is also important that the subwoofer has a smooth and flat frequency response when placed in the room, that it sounds tight and clean, and doesn't produce any audible nonlinear distortion. The frequency response of the subwoofer is largely determined by its interaction with the room. Performing a listening test at the dealer is therefore only of limited use. It is not a good prediction of how the subwoofer will sound in your own room. It is best to think of a subwoofer as a source of low-frequency energy whose main purpose is to excite the room. The sound that you hear is for a large part determined by the room itself, not by the subwoofer. Looking at the frequency response of the subwoofer is therefore of limited value. Most subwoofers have some electronic filters and equalization built in that allow you to adjust the frequency response of the subwoofer to the response of your room. The best match between the subwoofer and the room involves finding the optimal position of the subwoofer in the room (see section 4.2.6) in combination with some sort of electronic equalization (see section 8.1.3).

7.3.1 Lowest Frequency

How low should your subwoofers go? The lowest audible frequency is 20 Hz, below that sound can only be felt. Thus, the frequency response of a subwoofer placed in your listening room should extend to 20 Hz (meaning that its response is 3 dB down at 20 Hz). However, this doesn't mean that the measured anechoic response of the subwoofer should extend all the way to 20 Hz, because the room will reinforce the lowest frequencies due to the adjacent-boundary effect (see section 2.3.4 and 4.2.6). An example of how the adjacent-boundary effect extends the bass response down to lower frequencies was given in figure 4.40. Due to this in-room response extension a subwoofer that has an anechoic frequency response that extends to about 30 Hz can often still produce frequencies down to 20 Hz when placed in a typical room.

Sufficient low-frequency extension is a very important factor in overall sound quality. The lower your system can go the better it will sound. Bass, on its own, accounts for 25–30% of the overall sound quality rating (Olive, 2004a, 2004b). Extensive listening tests have shown that listeners like low bass, not more bass. In other words, reproducing lower frequencies is more important than boosting the bass (Toole, 2018).

> **Recommendation 7.22** *Your subwoofers should be able to reproduce low frequencies down to 20 Hz when placed in your listening room. This usually means that the anechoic frequency response should extend to at least 30 Hz.*

Bass sound quality is not only influenced by the lowest frequency that the subwoofer can reproduce, but also by the steepness of the roll-off below this lowest frequency. A low-frequency response with a relatively high −3 dB cutoff frequency, for example 60 Hz, and a slow roll-off is preferable to a response that extends to 45 Hz, but which rolls off quickly. An overall exten-sion with a slight roll-off to 35 Hz appears to be more important (Colloms, 2018). The point at which the response of the subwoofer is down to −10 dB seems to be a more reliable indicator of the subjective fullness and extension of the bass (Harley, 2015).

The steepness of the low-frequency roll-off depends on the design of the subwoofer enclo-sure. There is a big difference between the closed-box and the bass-reflex enclosure (see also section 7.1.3). The closed-box design rolls off at 12 dB per octave while the bass-reflex design rolls off twice as fast at 24 dB per octave. Figures 7.22 and 7.23 show the frequency responses of a number of differently tuned closed-box and bass-reflex designs. These different designs correspond to different choices of the enclosure size. The smaller the enclosure, the higher the −3 dB cutoff frequency. If you compare the closed-box design with $Q = 0.7$ and the B4 bass-reflex design, you will see that they have the same −3 dB cutoff frequency, but a very different −10 dB cutoff frequency due to their difference in roll-off steepness.

Subwoofers with a closed-box enclosure are preferred over bass-reflex designs, because by comparison the closed-box produces more bass below its −3 dB cutoff frequency. The major drawback of the closed-box design is that it is much less efficient than a comparable bass-reflex design. When driven by an equal amount of amplifier power, the bass-reflex design puts out 3 dB more sound output (Small, 1973c). Therefore, bass-reflex designs tend to be preferred for large venues like cinemas and concert halls. For your home theater or listening room you are usually better off choosing a closed-box design that has more low bass.

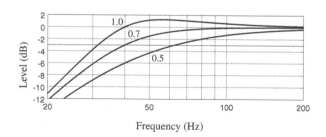

Figure 7.22
Frequency response of a closed-box loudspeaker system for three different values of Q. The closed-box designs are characterized by the Q factor or 'quality factor'. A higher Q corresponds to a larger enclosure with a lower cutoff frequency (Small, 1972a, 1973a).

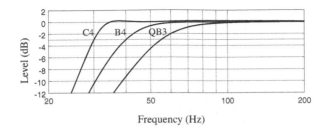

Figure 7.23
Frequency response of a bass-reflex loudspeaker system for three different alignments: QB3, B4, and C4. There are many possibilities, but these three designs are more or less standardized: C4 is the fourth-order Chebyshev equal-ripple alignment, B4 is the fourth-order Butterworth maximally flat alignment, and QB3 is the quasi-third-order Butterworth alignment (Small, 1973b, 1973c; Thiele, 1971a, 1971b).

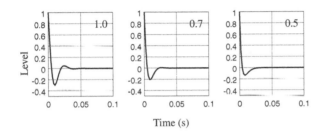

Figure 7.24
Transient response of a closed-box loudspeaker system for three different values of Q.

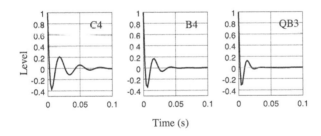

Figure 7.25
Transient response of a bass-reflex loudspeaker system for three different alignments: QB3, B4, and C4.

Recommendation 7.23 *Prefer subwoofers that have a closed-box enclosure over those with a bass-reflex enclosure.*

Another advantage of the closed-box design is that it has a better transient response. A loudspeaker is a resonant system that vibrates in response to an electrical signal. In terms of sound quality it is desirable that after a sudden transient the movement of the loudspeaker diaphragm stops as soon as possible. In practice, it cannot stop instantly and its response to a transient will show some decreasing residual resonance. Figures 7.24 and 7.25 show the transient responses of some different closed-box and bass-reflex designs. Clearly, the closed-box designs have a transient response that decays faster, which means that they have better bass definition and sound tighter and more well-controlled than their bass-reflex counterparts. A poor, that is longer, transient response can result in a booming sound that lacks detail (Harley, 2015). Due to their superior transient response, closed-box subwoofers have a strong following. However, some say that the fast decay times of closed-box subwoofers are irrelevant, as the decay times of room resonances are much longer and therefore dominate the low-frequency sound. Nevertheless, in rooms in which the resonances are well-controlled by sufficient low-frequency absorption and electronic equalization (see section 8.1) the fast decay times of a closed-box subwoofer are still beneficial (Newell and Holland, 2007).

7.3.2 Maximum Sound Level

How loud should your subwoofers be able to play? I explained in section 7.1.4 that your main loudspeakers should be able to generate an SPL of at least 96 dB at the listening position. The desired SPL of 96 dB is the recommended peak level for home listening (see box 3.2). With bass management, your main loudspeakers do not reproduce bass; your subwoofers reproduce the low-frequency content from all the channels: fronts, center, and surrounds. In addition, the subwoofer also has to reproduce the LFE channel (see section 2.1). The LFE channel is mainly used for high-impact low-frequency sound effects. It contains the sounds of explosions, earthquakes, and the like. For this reason surround sound standards prescribe that this channel is reproduced with a gain of 10 dB compared to the other channels (Holman, 2008; ITU-R BS.775, 2012). Thus, the subwoofer must be able to reproduce the LFE channel at an SPL of $96 + 10 = 106$ dB. Some math is required to calculate the SPL that results when the subwoofers are reproducing the LFE channel (at 106 dB) and all the low-frequency content from the other channels (each at 96 dB). The calculations in box 7.6 show that for a 5.1-channel sound system, your subwoofers must be able to generate an SPL of at least 108 dB at the listening position. Unlike the case with the main loudspeakers (see figure 7.10), the subwoofer SPL level doesn't need to be corrected for listening distance. In a typical home theater or listening room, the dimensions of the room are of the same order of magnitude as the wavelengths of the low-frequency sound. In such a case, the low-frequency SPL is approximately the same throughout the whole room and doesn't fall off with listening distance (Fielder and Benjamin, 1988).

Recommendation 7.24 *Your subwoofers should together be able to produce a peak SPL of at least 108 dB.*

BOX 7.6 MAXIMUM SUBWOOFER SPL

The subwoofer has to handle the bass from all the channels, including the LFE channel. The maximum level of each of the main channels is 96 dB and the maximum level of the LFE channel is $96+10 = 106$ dB. The total SPL with bass at maximum level in all channels is given by:

$$SPL_{max} = 10\log_{10}\left(\frac{nI_c + I_{LFE}}{I_o}\right) = 10\log_{10}(n+10) + 96$$

where n is the number of main channels, I_o is the reference intensity, I_c the intensity of one of the main channels, and I_{LFE} the intensity of the LFE channel. Note that the equation follows from the fact that the intensities can be expressed as:

$$I_c = I_o \cdot 10^{96/10}$$
$$I_{LFE} = I_o \cdot 10^{(96+10)/10} = I_o \cdot 10^{96/10} \cdot 10 = I_c \cdot 10$$

Hence, for a 5.1-channel sound system ($n=5$), the maximum subwoofer SPL will be 108 dB.

Producing 108 dB SPL at low frequencies is difficult. It requires multiple large low-frequency drivers with large excursion capabilities (Fielder and Benjamin, 1988). You must move large volumes of air to achieve a high SPL at low frequencies. At the lowest bass frequencies, near the −3 dB cutoff frequency, the SPL of the subwoofer is limited by the maximum volume of air that the driver can displace. Due to natural resonance of the driver in the enclosure, the subwoofer uses relatively little amplifier power at these lowest frequencies. Amplifier power can be a limiting factor in achieving the required SPL at the highest bass frequencies, but not at the lowest frequencies.

The lower the frequency, the larger the volume of air that the subwoofer has to displace in order to generate a high enough sound level. The displacement volume of the driver depends on its diameter and its maximum excursion. Obviously, larger diameter drivers displace more air at a fixed excursion than smaller ones. Large diameter drivers generally also have a larger maximum excursion, but few drivers have a maximum excursion that exceeds 10 mm. Excessive low-frequency peak power can cause the driver to exceed its maximum excursion resulting in mechanical failure. Figure 7.26 shows the excursion needed to generate an SPL of 108 dB with drivers of different diameters. Box 7.7 contains the mathematical equations used to generate this figure. There are four lines in the figure, each one corresponding to a different frequency. The figure clearly shows that it is physically easier to obtain high sound output levels with larger drivers. For example, the figure shows that an 8-inch driver needs a maximum excursion that is about three times larger as a 15-inch driver. Taking 10 mm as a reasonable estimate of the maximum excursion for one driver, you can use figure 7.26 to determine the number of drivers needed to produce 108 dB SPL at different low frequencies. For example, to reach 108 dB SPL at 20 Hz you need at least four 9-inch drivers or two 13-inch drivers. If you increase the lower cutoff of the subwoofer to 30 Hz, the excursion requirements are reduced quite a lot. In this case you only need two 9-inch drivers or just one 12-inch driver.

Recommendation 7.25 *Choose subwoofers with large drivers.*
In total you need at least four 9-inch drivers or two 13-inch drivers.

Producing 108 dB SPL also requires sufficient amplifier power. As I explained in section 7.2.1, the amount of amplifier power depends on the sensitivity of the loudspeaker. However, manufacturers rarely specify the sensitivity of their subwoofers. Since the power amplifier is integrated into the subwoofer, they only specify the amplifier power and sometimes the

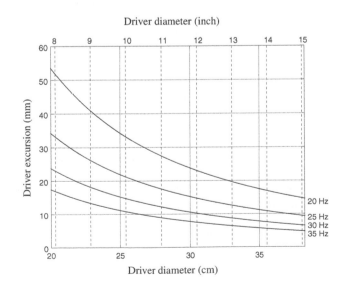

Figure 7.26
Subwoofer driver excursion (in mm) needed to achieve the required maximum SPL of 108 dB in the listening room as a function of the driver diameter (in cm or inch) and the −3 dB cutoff frequency of the subwoofer. See box 7.7 for the calculations and underlying assumptions.

BOX 7.7 SUBWOOFER EXCURSION AND SPL

The acoustic power P_A needed to achieve a specific SPL follows from (Colloms, 2018; Fielder and Benjamin, 1988):

$$SPL = 112 + 10 \log_{10} P_A$$

This equation assumes that the loudspeaker is close to one boundary (the floor) and hence radiates in half space (2π sterradians). Subwoofers are typically placed not only near the floor, but also close to one of the walls. This placement increases the output of the subwoofer by 6 dB, due to the adjacent-boundary effect (see for example figure 4.40). This 6 dB room gain reduces the acoustic power that the subwoofer needs to generate. Thus, in the above equation the SPL can be set to $108 - 6 = 102$ dB. The acoustic power P_A for a closed-box loudspeaker with a typical $Q_{TC} = 1/\sqrt{2} \approx 0.707$ is given by (Small, 1972a, 1973a):

$$P_A = 0.424 \cdot f_3^4 V_D^2$$

where f_3 is the –3 dB cutoff frequency of the loudspeaker and V_D is the peak displacement volume of the driver, which equals

$$V_D = \pi r_D^2 x_{max}$$

with r_D the effective diameter and x_{max} the peak linear excursion of the driver. In the calculations it is assumed that $r_D = 0.85 \cdot r$, where r is the driver diameter specified by the manufacturer. Combining the different equations yields:

$$SPL = 112 + 10 \log_{10} (0.424 \cdot \pi^2 f_3^4 r_D^4 x_{max}^2)$$

---

maximum SPL that the subwoofer can produce. You basically have to trust them on using sufficient amplifier power. You can get a rough idea of the amount of power that you need by looking at the enclosure size. The smaller the enclosure, the more amplifier power you need. Smaller enclosures make the subwoofer less efficient. The underlying principle is called *efficiency-bandwidth-volume exchange* (Small, 1972a). You can pick any two and the third will be dictated; for example a small enclosure that can play very low frequencies (large bandwidth) will need lots of amplifier power. Accepting a higher cutoff frequency (reduced bandwidth) will significantly reduce the required amplifier power. Figure 7.27 shows how subwoofer amplifier power requirements change with cutoff frequency and enclosure size. You can use this figure to estimate the amount of amplifier power that your subwoofer needs to achieve the target SPL of 108 dB. If you have multiple identical subwoofers, you can still use figure 7.27 to estimate the required amplifier power as follows: 1) Use the internal volume of only one of the subwoofers to determine the corresponding amplifier power in figure 7.27. The value you obtain is the total amplifier power for driving all of the subwoofers together. 2) Divide this value by the number of subwoofers, and you get the amplifier power for each of the separate subwoofers.

Recommendation 7.26 *Make sure that the amplifiers in your subwoofers are at least as powerful as indicated in figure 7.27.*

In recent years there has been a tendency to reduce the enclosure size of subwoofers considerably (Colloms, 2018). Understandably, a smaller enclosure is much easier to integrate in the room. The downside is that subwoofers in small enclosures require several hundreds of watts of amplifier power. Consequently, the low-frequency drivers used in these small enclosures

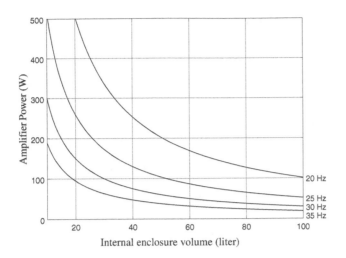

Figure 7.27
Approximation of the
subwoofer amplifier power
needed to achieve the
required maximum SPL
of 108 dB in the listening
room as a function of
the internal enclosure
volume and the −3 dB
cutoff frequency of the
subwoofer. See box 7.8
for the calculations and
underlying assumptions.

BOX 7.8 SUBWOOFER AMPLIFIER POWER

The subwoofer amplifier power P_E required to achieve the target SPL of 108 dB can be calculated from the equation in box 7.4 with $D = 1$ and the sensitivity

$$S = 112 + 10\log_{10}\eta$$

where η is the efficiency of the subwoofer. The efficiency η of a closed-box subwoofer at the highest bass frequencies with a typical $Q_{TC} = 1/\sqrt{2} \approx 0.707$ and a mechanical loss factor of 0.9 is given by (Small, 1972a, 1973a):

$$\eta = 1.234 \times 10^{-6} f_3^3 V_B$$

where f_3 is the −3 dB cutoff frequency of the subwoofer, and V_B is the internal volume of the subwoofer enclosure. Combining the equations yields:

$$SPL = 112 + 10\log_{10}(1.234 \times 10^{-6} f_3^3 V_B) + 10\log_{10}(P_E)$$

Room gain helps to reduce the power requirements a bit. Similar to the calculation in box 7.7 the target SPL is reduced to $108 - 6 = 102$ dB when the subwoofer is placed close to both the floor and one of the walls.

---

must be able to handle large amounts of electrical power (Small, 1973a). Often robust drivers with large oversized magnets and large voice coils are used in such high-powered designs.

Subwoofer enclosure size can be further reduced by using electronic equalization to compensate for the roll-off of the frequency response below the −3 dB cutoff frequency (Greiner and Schoessow, 1983; Small, 1972a; Ståhl, 1981). Such equalization requires even more amplifier power and puts enormous stress on the low-frequency drivers. Thus, this approach requires specially designed robust drivers. Another technique to generate lots of bass with small enclosures involves the use of motional feedback also known as a *servo-driven subwoofer*. In such a subwoofer the driver is coupled to a sensor that measures the excursion of the diaphragm and the electrical signal from this sensor is fed back to the power amplifier to correct the response of the loudspeaker. Again there is no free lunch and such a system also requires generous amounts of amplifier power (Colloms, 2018).

7.4 Cables for Analog Audio Signals

Two types of analog audio cables are being used in your system: loudspeaker cables and analog interconnects. Loudspeaker cables are used to connect the loudspeakers to the power
amplifier or A/V receiver. Interconnect cables are used to connect each output channel of an
A/V controller to the corresponding input channel of a power amplifier, and they are used to
connect the subwoofer output of an A/V receiver or controller to an active subwoofer.

There exists a wide range of specialty analog audio cables ranging from cheap to very expensive ones. The world seems to be divided into those who believe that esoteric audio cables
can make a huge sonic difference and those who think that it is all a lot of nonsense. Regardless of which camp you belong to, it is important to realize that the cables are not the most
important piece of equipment quality-wise. Sound quality is first and foremost influenced by
the quality of your loudspeakers and their interaction with the room. The second important
factor is the quality of your electronics, especially your amplifier and your DACs. Only the
last tiny bit of performance of the entire audio chain is influenced by cables. Their impact
is very small compared to the other factors. A significant improvement in sound quality is
much more likely to come from upgrading your loudspeakers or changing the acoustics of
your listening room.

You should use decent quality cables, but there is no good reason to spend a lot of money
on expensive specialty audio cables with esoteric designs or materials. There is no scientifi-
cally sound evidence to support many of the claims that changing a cable would magically
change the sound of your system (Russell, 2012; Self, 2009; Winer, 2018). Some examples of
far-fetched claims about audio cables are that the cable introduces nonlinear distortion, that
the purity of the copper in the wire affects sound purity, that the cable works best when wired
in one direction, that the cable needs to be broken in for several days, and even that keeping
the cable away from room surfaces will improve the sound (Villchur, 1994).

It is a good idea to label your cables to avoid making connection mistakes. A label on each
end of each cable saves you from the hassle of following cables to the other end to see
which component is connected to it. It is of critical importance that your loudspeaker
cables connect each loudspeaker to the appropriate audio channel on the back of the
power amplifier or receiver. A simple connection mistake completely destroys the imaging and surround capabilities of your system. Similarly, the analog interconnects between
the A/V controller and power amplifier need to be plugged in such that the audio channels
on both components match. Use a simple listening test to check that every loudspeaker
in your system is reproducing the correct audio channel. For this test, you can use the test
tone mode of your A/V controller or A/V receiver. This mode, which is designed to set the
channel balance, will send a noisy signal to each individual loudspeaker, one at a time. It
typically starts with the left front loudspeaker, then moves to the center, the right front,
and on to the surrounds. If anything is out of order you probably have made a mistake connecting your cables. Instead of the test tone mode, you can also use the noisy test signals
on a test disc (see appendix A).

Another useful check involves the polarity of the loudspeaker connections. The terminals on the back of your loudspeakers and amplifier have a red (labeled plus) and a black
(labeled minus) terminal for each loudspeaker. Make sure you connect the red terminal on
the loudspeaker to the red terminal on the amplifier, and similarly for the black ones. Most
loudspeaker cables include a marker on one of the two wires to help you make the correct
connections on both sides. If these connections are accidentally reversed for one of the
loudspeakers, this particular loudspeaker is said to be out of phase. This means that its cone
moves in the opposite direction compared to the cones in the other loudspeakers. Such a
phase shift of 180° reduces the overall bass output of your system and destroys its imaging
capabilities (Harley, 2015; Winer, 2018). Again I recommend a simple listening test to check
that every loudspeaker in your system is connected with the right polarity. For this test you
need noisy test signals that appear to come from the midpoint between two loudspeakers.

Such test signals are included on several audio test discs (see appendix A). If the loudspeakers are connected with the correct polarity, that is if they are in phase, the noise should be a distinct spotlike sound between the two loudspeakers. If, on the contrary, the sound seems hard to pinpoint or seems to come from a space far away, the loudspeakers are out of phase and you need to check the polarity at both ends of the loudspeaker cables. When using identical loudspeakers or loudspeakers from the same brand, they will all be in phase if you connect the red terminals to each other, and the black terminals to each other. But, sometimes if you mix different models from different brands the polarity of the different loudspeakers doesn't match. In this case, you need to reverse the polarity of the offending loudspeaker by switching the red and black connections only at the loudspeaker terminals. Listen to the test signals again and if the imaging of the sound improves, keep the connections this way. Never mind that the red and black terminals no longer line up; always trust your ears.

Recommendation 7.27 *Listen to audio test signals to verify that each loudspeaker is reproducing the appropriate audio channel and that each loudspeaker is connected with the correct polarity.*

7.4.1 Loudspeaker Cables

The ideal loudspeaker cable transmits the electrical signal from the power amplifier to the loudspeaker without altering it. The ideal cable doesn't exist: there will always be some interaction between the amplifier, cable, and loudspeaker. The electrical characteristics of a loudspeaker cable can be completely described by only three parameters: its resistance, capacitance, and inductance (Davis, 1991; Greiner, 1980). Together they form the cable's impedance, which can be thought of as a frequency-dependent resistance. These three parameters depend directly on the geometry of the cable and the materials used for the conductors and insulation (Davis, 1991; Greiner, 1980).

A high-quality loudspeaker cable has a low resistance and a low inductance. The electrical performance of the cable at low frequencies is dominated by its resistance, while its performance at high frequencies is dominated by its inductance. The amount of capacitance is comparatively insignificant (Davis, 1991; Newell and Holland, 2007). Keeping the cable as short as possible ensures both low resistance and low inductance. It is important to keep the two wires in the cable together as close as possible, since the close spacing reduces the inductance. Thus, it is a bad idea to run two single wires each with its own route to the loudspeaker. Increasing the diameter of the wires has the advantage of reducing the resistance of the cable, but it has the opposite effect on the inductance. Increasing the diameter increases the inductance, because of the increased spacing between the centers of the two wires. For each cable, there is an optimum where increased inductance is balanced against reduced resistance (Duncan, 1997; Greiner, 1980; Newell and Holland, 2007).

Recommendation 7.28 *Choose loudspeaker cables that have closely spaced conductors.*

The most important factor for choosing loudspeaker cables is that they should have low resistance. The resistance of a loudspeaker cable causes a voltage loss between the amplifier and the loudspeaker. This loss can vary with frequency, because the impedance of the loudspeaker varies with frequency (see figure 7.17). Frequency variations due to loudspeaker-cable-amplifier interactions are inaudible if they are less than 0.5 dB, which is essentially always the case if the resistance of the loudspeaker cable (for the two conductors together) does not exceed 0.1 Ω (Toole, 2018). This is true for all combinations of modern amplifiers and loudspeakers, as explained in more detail in box 7.9. The resistance of a loudspeaker cable depends on three parameters: the conductivity of the material used, the length of the cable, and its diameter (see box 7.10). Of these parameters the length and diameter are the most important ones. Thus, minimizing loudspeaker cable resistance is easy: use short cables with a large diameter.

BOX 7.9 LOUDSPEAKER-CABLE-AMPLIFIER INTERACTION

From figure 7.20 it follows that the voltage at the loudspeaker terminals equals:

$$V_L = \frac{R_L}{R_A + R_C + R_L} \cdot V_A$$

where V_A is the voltage provided by the amplifier, R_A is the output impedance of the amplifier, R_C is the resistance of the loudspeaker cable, and R_L is the impedance of the loudspeaker. If $R_A = 0$ and $R_C = 0$, the voltage applied to the loudspeaker equals the voltage provided by the amplifier $V_L = V_A$. Therefore, the goal is to make both R_A and R_C small. The question is how small? Since the loudspeaker impedance R_L varies with frequency (see for example figure 7.17). The voltage V_L will also vary with frequency. The maximum voltage $V_{L,max}$ occurs when the loudspeaker impedance reaches its highest value $R_{L,max}$. The minimum voltage $V_{L,min}$ occurs when the loudspeaker impedance reaches its lowest value $R_{L,min}$. This variation in loudspeaker impedance causes an undesirable variation in frequency response. The magnitude of this variation in dB is given by:

$$20\log_{10}\left(\frac{V_{L,max}}{V_{L,min}}\right) = 20\log_{10}\left(\frac{R_{L,max}(R_A + R_C + R_{L,min})}{R_{L,min}(R_A + R_C + R_{L,max})}\right)$$

To keep this variation between the thresholds of audibility, it should be less than 0.5 dB. For a typical consumer loudspeaker the impedance as a function of frequency can vary between 3 Ω and 40 Ω. Using these values $R_{L,min}$ and $R_{L,max}$, the previous equation can be used to compute the maximum value of $R_A + R_C$ that keeps the frequency response variations below 0.5 dB. Doing the math shows that $R_A + R_C$ should be less than 0.19 Ω. Any modern solid-state amplifier has a damping factor much higher than 100 (see section 7.2.3). Assuming a damping factor of 100, the amplifier's output impedance becomes $R_A = 8/100 = 0.08\Omega$. Therefore, no audible frequency variations occur as long as the resistance of the loudspeaker cable R_C does not exceed 0.1 Ω. This number applies to all combinations of a modern solid-state amplifier with a typical high-end loudspeaker.

BOX 7.10 CABLE RESISTANCE

The electrical resistance of a wire equals:

$$R = \frac{\rho\ell}{A} \tag{7.1}$$

where ρ is the volume resistivity (Ωm), ℓ is the length (m), and A is the cross-sectional area (m²). For copper wires, $\rho = 1.72 \cdot 10^{-8}$ Ωm. Note that a loudspeaker cable has two wires, hence to compute the total resistance of a cable the result from equation (7.1) must be multiplied by two. The cross-sectional area A can be computed from the diameter of the copper conductor:

$$A = \frac{\pi d^2}{4}$$

Note that the resistance of stranded wire should be computed from the diameter of an equivalent solid conductor. Due to the interstices between the multiple conductors, the physical diameter of stranded wire is approximately 1.15 times larger than the diameter of solid wire (Barnes, 2004). The diameter (m) of a solid conductor specified according to the AWG system equals:

$$d = 127 \cdot 10^{-6} \cdot 92^{(36-G)/39}$$

where G is the wire size in AWG.

Table 7.4 The maximum recommended length for stranded loudspeaker cables in standard IEC wire sizes (IEC 60228, 2004). The resistance per meter is given for two conductors. The maximum length keeps the resistance of the loudspeaker cable below 0.1 Ω. The diameter in this table equals the physical diameter for stranded wire.

IEC Wire Size (mm²)	Diameter (mm)	Resistance (Ω/m)	Max. Length (m)
0.50	0.9	0.0688	1.5
0.75	1.1	0.0459	2.2
1.00	1.3	0.0344	2.9
1.50	1.6	0.0229	4.4
2.50	2.1	0.0138	7.2
4.00	2.6	0.0086	11.6
6.00	3.2	0.0057	17.5
10.00	4.1	0.0034	29.4

Table 7.5 The maximum recommended length for stranded loudspeaker cables in American Wire Gauge (ASTM, 2008). The resistance per meter is given for two conductors. The maximum length keeps the resistance of the loudspeaker cable below 0.1 Ω. The diameter in this table equals the physical diameter for stranded wire.

Wire Size AWG	Area (mm²)	Diameter (mm)	Resistance (Ω/m)	Max. Length (m)
20	0.52	0.9	0.0665	1.5
18	0.82	1.2	0.0418	2.4
16	1.31	1.5	0.0263	3.8
14	2.08	1.9	0.0165	6.1
12	3.31	2.4	0.0104	9.6
10	5.26	3.0	0.0065	15.4
8	8.37	3.8	0.0041	24.4

The longer your loudspeaker cables, the larger their diameter must be to keep their resistance below 0.1 Ω. Tables 7.4 and 7.5 list the maximum recommended cable lengths for stranded wire of different diameters (see box 7.10 for the calculations). In the United States the wire size is specified in AWG (American Wire Gauge). The lower the AWG number, the larger the diameter. You can of course use larger diameter cables than the ones listed in tables 7.4 and 7.5; it will further reduce the resistance of the cable. However, do not use cables of too large a diameter, as the increased spacing between the two conductors will result in an increased cable inductance and that may cause a loss of high frequencies. The cable acts as a low-pass filter attenuating the higher frequencies in the audio signal. This attenuation is likely to be less than 0.5 dB at 10 kHz and less than 1 dB at 20 kHz for most ordinary loudspeaker cables that do not go beyond 4 mm² (or lower than 12 AWG). Psychoacoustic data on just noticeable differences in frequency responses predicts that the high-frequency attenuation due to the inductance of the loudspeaker cables are almost always below the threshold of audibility (Davis, 1991; Villchur, 1994).

Some people claim that in large diameter cables the so-called *skin depth* becomes an issue. Skin depth refers to the fact that as the frequency of an electrical signal increases, the signal travels closer to the surface of the conductor. It only penetrates to a certain depth and does not utilize the whole cross-sectional area. Skin depth only has a minor effect at audio frequencies (Greiner, 1980). Again, if you do not go beyond 4 mm² (or lower than 12 AWG) everything should be fine.

Recommendation 7.29 *Keep the electrical resistance of your loudspeaker cables below 0.1 Ω by choosing thicker cables for longer lengths such that you do not exceed the maximum recommended lengths listed in tables 7.4 and 7.5. Keep the inductance of the cables low by avoiding cables that have a cross-sectional area that exceeds 4 mm² or that have an AWG number lower than 12.*

While loudspeaker cables should be as short as possible, it is important to keep the lengths (and diameters) of the cables for the left, center, and right front loudspeakers the same. Any difference in cable length or diameter may result in a slight difference in frequency response. Such a difference is usually subtle, but you should try to minimize any differences between the frontal loudspeakers to keep the frontal soundstage as consistent as possible. There is no need to use the same length of cables between the front and the surround loudspeakers. In fact, it is better not to. The cables to the surround loudspeakers are usually required to be much longer, and this would mean that the length of the cables for the fronts would become undesirably long. Again, it is good practice to keep the cables for each pair of left and right surround loudspeakers the same length (Colloms, 2018; Harley, 2015).

Recommendation 7.30 *Keep your loudspeaker cables as short as possible, but also keep the cables for the left, center, and right front loudspeakers the same length and diameter. Similarly, keep the cables for each pair of left and right surround loudspeakers the same length and diameter.*

Loudspeaker wires are usually made of copper or silver, and both materials are more or less equally suitable. Silver has a slightly better conductivity than copper, and thus results in a somewhat lower resistance. However, this difference in resistance is not really significant. The real advantage of silver is that silver oxide has the same resistance as silver. Over time both silver and copper will oxidize when in prolonged contact with air. While silver wires maintain their resistance, copper wires will suffer from an increased resistance. Nevertheless, most loudspeaker wires are made of copper, because silver is more expensive than copper and hard to work with (Lampen and Ballou, 2015). Copper wires are usually made from *oxygen-free copper* (OFC), which only has 50 ppm (parts per million) of oxygen compared to 250 ppm of oxygen for normal copper. This reduction of oxygen content slows down the formation of copper oxides within the wires (Harley, 2015).

Since bare wire will oxidize and degrade the connection it is best to use air-sealed connectors on both ends of your loudspeaker cables. The best-quality connectors are gold-plated ones. The thin plating of gold will prevent the connectors from oxidizing over time. Of course, the gold plating should not only be used on the cable's plugs, but also on the terminals of the amplifier and the loudspeakers (Harley, 2015; Whitlock, 2015; Winer, 2018). Two commonly used types of loudspeaker connectors are the *spade lug* and the *banana plug*. Both types provide a high-quality connection with a sufficiently large contact area and high contact pressure.

Recommendation 7.31 *Use loudspeaker cables that are terminated with air-sealed connectors that are gold-plated.*

Many high-end loudspeakers have two sets of binding posts that can be used for *bi-wiring* where two sets of loudspeaker cables are used to connect the loudspeaker to the amplifier. Figure 7.28 shows the difference between an ordinary connection and bi-wiring. One cable is connected to the part of the crossover filter that feeds the high-frequency driver and this cable only carries the high frequencies. The other cable is connected to the low-frequency part of the crossover filter and consequently carries only the lower frequencies. It is claimed that bi-wiring yields a small but worthwhile improvement in sound quality (Colloms, 2018;

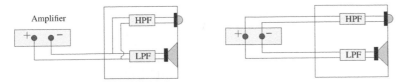

Figure 7.28
Two different ways to connect a loudspeaker to the power amplifier. *On the left,* an ordinary connection with one loudspeaker cable (that consists of two wires) running between the loudspeaker and the amplifier. *On the right,* a bi-wired connection using two loudspeaker cables: one running from the loudspeaker's high-pass filter (HPF) to the power amplifier and another one running from its low-pass filter (LPF) to the amplifier.

Harley, 2015; Newell and Holland, 2007). However, a scientific explanation of why this is the case is currently lacking. So, not everyone is convinced of its merits.

7.4.2 *Interconnects*

Interconnect cables carry line-level signals with a relatively small voltage of approximately 1 volt. This delicate audio signal needs to be properly protected against electromagnetic interference (EMI), also known as radio frequency interference (RFI). All electrical devices that generate changing electrical currents and voltages also generate EMI. This includes your audio and video equipment, but also cell phones, Wi-Fi devices, computers, microwave ovens, light dimmers, drills, ventilators, vacuum cleaners, etc. Typical problems stemming from EMI are increased noise levels, subtle changes in the clarity of sound, or an audible hum. There are two ways that EMI can enter your system through the conductors in your cables (Barnes, 2004; Fause, 1995; Whitlock, 2015): 1) *Electric-field coupling:* Any conductor that has an alternating voltage applied to it creates an electrical field that radiates out. The higher the voltage, the stronger the field. 2) *Magnetic-field coupling:* Any conductor that carries an alternating current creates a magnetic field that radiates out at right angles (90°) to the current flow. When a second nearby conductor cuts through the magnetic field lines, a voltage is induced in this conductor. The higher the current in the first conductor, the stronger the induced voltage in the second one.

To protect against unwanted EMI analog interconnects are constructed as *coaxial cables* in which one or more conductors are centered inside a hollow tubelike conductor that carries the return current (see figure 7.29). The hollow conductor wrapped around the other conductors works as a shield against electric field coupling. It protects the analog signal from outside electrical interference and at the same time also keeps the signal from leaking out and interfering with other nearby cables (Lampen and Ballou, 2015). Coaxial cable is not usually used for loudspeaker cables, because it increases the capacitance of the cable (Winer, 2018) and that increase can have an unwanted effect on the interaction between the loudspeaker and the power amplifier. Furthermore, the relatively high voltages and currents in the loudspeaker cables make electric field shielding unnecessary. Loudspeaker cables carry audio signals that range up to about 60 volts. A small disturbance of about 0.01 volt has almost no perceptible impact on a loudspeaker cable (less than 0.02%), but it makes a much bigger difference on an interconnect cable carrying a line-level signal of only 1 volt (1%).

Interconnect cables come in two varieties: *unbalanced* and *balanced* (Whitlock, 2015). Unbalanced interconnects are more common, as most consumer equipment is fitted with unbalanced connections. The unbalanced connection consists of one coaxial cable for each audio channel. The coaxial cables are terminated at each end with an *RCA plug* that consists of a single pin surrounded by a metal ring. The tubelike conductor of the coaxial cable serves two purposes: 1) it provides the 0 V reference point for the audio signal, and 2) it is connected to ground to shield the center conductor from nearby disturbing electric fields. The shield

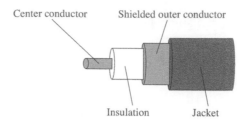

Center conductor   Shielded outer conductor

Insulation   Jacket

Figure 7.29
Coaxial cable.

suppresses noise from electric fields, but it doesn't prevent noise coupling from nearby magnetic fields. Unbalanced connections are effective for short cables that carry line-level audio
signals. They are less suited for long cables and cables that carry low-voltage signals from
microphones (Fause, 1995).

Balanced interconnects are used in professional audio equipment and in a few high-end consumer products. The balanced connection uses coaxial cables with three conductors: two
center conductors and a shield surrounding these two conductors. The cables are terminated
at each end with an *XLR plug* that consists of three pins or sockets surrounded by a metal
ring. The two center conductors carry the audio signal, and the shield is connected to ground
to protect the two center conductors from nearby disturbing electric fields. The two center
conductors carry two identical signals, but with opposite polarity. These signals are fed into
a differential amplifier that only passes the difference between the two signals. This clever
trick makes the balanced connection immune to noise interference from both nearby electric
and magnetic fields. Since the noise pick-up is the same in both conductors, the differential
amplifier rejects the noise. This noise rejection capability is called *common-mode rejection*
(Whitlock, 2015).

To minimize EMI it is good practice to organize and separate your cabling according to their
type: loudspeaker cables, analog interconnects, digital interconnects, and power cables
(Barnes, 2004; Fause, 1995; Harley, 2015; Whitlock, 2015). The different types of cables carry different types of signals that might interfere electromagnetically. Digital interconnects are much
less susceptible to electromagnetic noise than analog interconnects and loudspeaker cables.
Due to the robust nature of digital signals, a small disturbance typically has no effect at all on
the digital data that is being transmitted. However, digital cables are our worst high-frequency
noise sources. The rapid switching between ones and zeros creates a lot of high-frequency electromagnetic radiation that can couple into nearby analog cables. Power cables also generate
significant electromagnetic noise, because they operate at higher voltages and currents. Power
cables of digital equipment can generate a lot of high-frequency noise (up to about 30 MHz),
because such equipment often employs a switching power supply (Barnes, 2004).

Keeping different types of cables separate reduces their mutual electromagnetic coupling.
The main strategy is to increase the distance between different types of cables. The magnetic field rapidly decreases with distance as it follows an inverse-square law: doubling the
distance results in four times less magnetic field strength. Another way to reduce magnetic
field coupling is to let different types of cables meet each other at right angles. At right angles
there will be no magnetic coupling because the cables don't intersect each other's magnetic
field lines. Yet another way to minimize magnetic field coupling is to bundle similar cables,
the reason being that if two identical conductors are exposed to identical magnetic fields,
these conductors will have identical voltages induced which cancel each other. Bundling similar cables places their conductors at approximately the same distance from the source of
the magnetic field, thus minimizing its effect. It is therefore a good idea to bundle all analog
interconnects between any two components. Likewise, the power cables of all components

should be bundled, because bundling will tend to average and cancel the radiated magnetic fields (Whitlock, 2015).

> **Recommendation 7.32** *Organize your cables in the following way to keep potential electromagnetic interference to a minimum:*

- *Separate and bundle your cabling according to their type: loud-speaker cables, analog interconnects, digital interconnects, and power cables. Cable bundles should only contain cables of a single type.*
- *Keep analog interconnects away from digital interconnects and power cables. Separate them as much as possible, but at least 25% of their parallel distance.*
- *Keep loudspeaker cables away from digital interconnects and power cables. Separate them as much as possible, but at least 2.5% of their parallel distance.*
- *Always keep a minimum distance of at least 2.5 cm (1 inch) between the different bundles.*
- *If possible, cross cables from different bundles only at right angles.*
- *Keep all cables as short as possible. Wind any excess cables into a figure-8 pattern.*

Both balanced and unbalanced connections are also susceptible to *common-impedance coupling*. Such coupling occurs when the current from a noise source and the audio signal share a return conductor. A common scenario is that two grounded devices, for example your A/V receiver and subwoofer, have slightly different ground voltages. This difference causes a current flow between the two devices. This current flows in a so-called *ground loop* that consists of two parts: 1) the safety-ground connection between the two devices and 2) the shield of the audio interconnection cables. In extreme cases the noise current in such a ground loop can reach 100 mA and cause a clearly audible hum or buzz. Another kind of ground loop can occur between two ungrounded devices. Capacitive coupling between the power supply and the chassis may cause a small voltage difference between the chassis of the two devices. This voltage difference causes a noise current to flow from one device to the other through the shield of the interconnection cable. Such a current can reach 0.75 mA. Common-impedance coupling can be reduced in two ways: 1) minimize the interchassis currents and 2) reduce the resistance of the shields of the interconnect cables (Whitlock, 1996). The interchassis current depends for a large part on properly designed equipment (Whitlock, 2015).

High-quality interconnect cables have heavy gauge shields with a low shield resistance. Heavy-braided copper shields are preferred over foil and thin wires. A heavy-braided shield has a lower resistance and thus minimizes common-impedance coupling (Whitlock, 2015). The shield should also cover the cable for at least 95% to achieve adequate shielding against radio frequency interference from nearby noise sources. A single braid can cover at most 95%, while a dual braid shield can cover as much as 98%. One hundred percent coverage is not physically possible with braided shields, only with foil sheets. The combination of a thin foil sheet with braided copper combines the best of both worlds (Lampen and Ballou, 2015).

> **Recommendation 7.33** *Use interconnect cables with heavy-braided copper shields, and keep them as short as possible.*

Coaxial interconnect cables can have a very high capacitance if they are long. This capacitance will interact with the output impedance of the signal source and act as a low-pass filter that attenuates high frequencies in the audio signal. A typical balanced or unbalanced interconnect will have a capacitance of about 160 pF/m (50 pF/ft). Combined with a typical output impedance of 1 kΩ, an interconnect cable with a length of 15 m (50 ft) would attenuate the response at 20 kHz only 0.5 dB (Whitlock, 2015). Interconnects used between power amplifiers and A/V controllers are typically much shorter, and their high-frequency attenuation will be insignificant. High-frequency attenuation is also not a problem in interconnects used for subwoofers. While these cables can be quit long, they only carry a low-frequency audio signal (below 120 Hz).

As with loudspeaker cables, it is important that the RCA and XLR connectors used have a low contact impedance and do not oxidize over time. It is therefore recommended using air-sealed connectors that are gold-plated.

Recommendation 7.34 *Use interconnect cables with air-sealed RCA or XLR connectors that are gold-plated.*

7.5 Audio Processing

Your A/V controller (or the controller built into your receiver) forms the heart of your system. You use it to set the master volume and to switch between different content sources. But, that is not all. It also performs some essential audio processing tasks, notably: surround decoding, bass management, level and delay compensation for the different loudspeakers, and optionally dynamic range control and equalization. You will need to configure the audio processing in your controller to suit your particular room and loudspeaker setup. This is usually done through a series of graphical menus that the controller can display on your video screen. Some controllers assist you in setting up your system and have an automatic calibration mode. It works as follows: you position a measurement microphone that came with the device at your prime listening position. The controller emits a series of audible test signals. The microphone picks up these signals and sends them to the controller where they are automatically analyzed to provide the right menu settings for bass management and for level and delay compensation. If you use the automatic calibration, it is important to manually check the settings that were chosen. They should correspond to the settings for bass management, delay, and level that are recommended in this section. Manually correct the settings if necessary.

Many controllers take calibration one step further and use automatic equalization to achieve a flat frequency response in your room. The measurement microphone is used at multiple positions in the room to measure the combined frequency response of the room and your loudspeakers. Based on these measurements equalization is used to correct certain room acoustical problems, such as the adjacent-boundary effect and room resonances (see section 2.3.4), and also to correct frequency-response anomalies in your loudspeakers.

The audio processing in your controller is completely performed in the digital domain. *Digital signal processing* (DSP) chips are used to manipulate the digital audio signals. For this reason most controllers convert all analog input signals to digital signals. The digital signals are digitally processed for bass management, level correction, and so on, and then converted back to analog signals that are sent to the power amplifiers. Some controllers have an analog bypass that avoids the conversion to digital, but its drawback is that some essential audio processing steps, such as bass management, are no longer applied (Briere and Hurley, 2009; Harley, 2015).

The sound quality of your A/V controller (or the controller built into your receiver) is mainly determined by the quality of its digital-to-analog converters (DACs). The DACs determine how accurately the digitized signals will be converted to analog signals that can be reproduced by your power amplifier and loudspeakers. High-quality DACs are available at low cost, but not all DACs sound the same (Harley, 2015; Pohlmann, 2011). The differences in sound were more pronounced a few years ago; many modern DACs sound excellent.

Pay attention to the quality of the DACs when buying an A/V controller or receiver. Product reviews, either online or in magazines, can be a good place to start zooming in on a specific product. I also recommend that you listen to the sound of the controller or receiver before you buy it. To get a good idea of how it will sound in your system, it is essential that you perform the listening test with the same amplifier and loudspeakers that you have at home. The clarity of the sound (see section 3.4.2) is an important criterion to use when evaluating the sound of a DAC. The DAC should be able to resolve all the detail in the sound without sounding analytical or aggressive. A good DAC makes it possible for you to distinguish individual sounds in time and in three-dimensional space and to recognize their unique pitch, timbre, and time structure (Harley, 2015).

Recommendation 7.35 *Buy an A/V controller or receiver that has high-quality DACs. Before buying the device listen to it to evaluate its sound quality. Use the same power amplifier and loudspeakers that you have at home.*

7.5.1 Surround Decoding and Processing

A digital audio signal that contains multichannel sound can be sent to your A/V controller (or the controller built into your receiver) in different formats. The most common ones are the formats from Dolby Laboratories and DTS, such as Dolby TrueHD, Dolby Digital, DTS-HD Master Audio, and DTS Digital Surround. An overview of the different formats was given in section 5.2. Your controller unpacks this digital signal and decodes it into several digital audio signals, one for each of the different channels: fronts, surrounds, and LFE (see section 2.1.1 for an overview of all the different surround sound channels). Almost all modern controllers automatically detect the type of incoming signal and there is no need for you to make any special menu settings.

When the controller senses a two-channel input signal, you may need to change some settings on the A/V controller. The reason is that there are essentially two different types of two-channel signals: 1) an ordinary stereo signal with a left and a right channel; and 2) a matrix-encoded surround signal, which is represented by two channels. Before the invention of digital multi-channel audio, surround sound was distributed as a two-channel matrix-encoded signal. This ensured compatibility with two-channel source components and broadcast technology. The most commonly used matrix-encoded format was *Dolby Surround*. It uses three front channels and one surround channel. From the two-channel signal, the center channel (C) can be obtained by summing the two channels, and the surround channel (S) can be obtained by taking the difference between the two channels. Of course, the separation between the four audio channels L, C, R, and S is limited: each channel also contains some information from the other three channels, though at a lower level.

Dolby Prologic decoding techniques are used to extract the four audio channels from a matrix-encoded two-channel signal (Dressler, 2000). There are two different generations of this technology: Dolby Prologic and Dolby Prologic II. Their differences are listed in table 7.6. To overcome the drawbacks of limited channel separation the surround channel in Dolby Prologic is limited to a frequency range of 100 Hz–7 kHz. By discarding the high frequencies the

Table 7.6 Dolby Prologic (PL) surround sound modes (Dressler, 2000).			
Feature	PL	PL II Movie	PL II Music
Surround bandwidth	100 Hz–7 kHz	Full-range	High-frequency shelf
Surround delay	15 ms	10 ms	None
Autobalance	On	On	Off
User controls	None	None	Optional

apparent separation between the three front channels and the surround channel increased. Furthermore, the surround channel is also slightly delayed in time. This ensures that due to the precedence effect (see section 2.3.4) dialog and other frontal sounds are actually perceived as coming from the front loudspeakers (Dressler, 2000).

Dolby Prologic II is an improved version of Dolby Prologic decoding that can be used in two different modes called *movie mode* and *music mode*. Movie mode is used to decode matrix-encoded surround sound, similar to the first generation Dolby Prologic decoders. The main difference is that it produces two surround channels that are decorrelated and also use the entire frequency range.

The music mode of Dolby Prologic II is intended to convert ordinary two-channel music into surround sound. Ordinary stereo recordings contain lots of ambient information in the form of recorded reflected sound and artificially added reverberation. In music mode the goal is to extract this ambient information and send it to the surround channels for a more spacious experience. In music mode there is no delay between the fronts and the surrounds, as the goal is to have all the sounds arrive at the same time. Music mode also slightly attenuates the high frequencies in the surround channels using a high-frequency shelf filter. This results in a more natural sound field, as the high frequencies in ambient sounds are normally partly absorbed and diffused by the room. Music mode offers three user controls that you can use to fine-tune the surround experience (Dressler, 2000):

1. *Dimension control* can be used to gradually adjust the sound field either towards the front or towards the rear.
2. *Center width control* can be used to control the width of the frontal soundstage and alter the balance between the left, center, and right front loudspeakers.
3. *Panorama mode* can be used to extend the front stereo image to include the surround speakers for a wraparound effect.

In addition to adjusting these controls, it can also be useful to fine-tune the volume level of the surrounds. With some recordings the surround effect can sound grossly exaggerated and the surround effect will sound much more natural if you turn down the relative volume of the surrounds (Toole, 2018).

The idea of generating multiple audio channels from a two-channel signal can be taken one step further. Several *surround upmixing* methods are now available that can artificially create multiple surround channels and even overhead or height channels. Some popular surround upmixing techniques from Dolby Laboratories and DTS are listed in table 7.7. They differ in the number of audio channels that they can synthesize. They are not limited to upmixing two-channel material. They can also be used to convert a 5.1 surround sound format into 7.1

Table 7.7 Surround sound upmixing techniques.

Method	Input Format	Created Surround Channels	Created Height Channels
Dolby Prologic	2.0	1	0
Dolby Prologic II	2.0	2	0
Dolby Prologic IIx	2.0 / 5.1	2/4	0
Dolby Prologic IIz	2.0 / 5.1	2/4	2
Dolby Atmos	2.0 / 5.1	2/4	2/4
DTS Neo:6	2.0	2/3	0
DTS Neo:X	2.0 / 5.1 / 7.1	2/3/4/5/6	2/4

or 9.1 channels, such that you are able to use all your surround loudspeakers regardless of the number of surround channels present in the original source material. In fact, this is one of the more useful features of the surround upmixing techniques. By using all your surround loudspeakers you are effectively creating a larger ambient bubble. Note that simply relaying the surround channels to the other unused loudspeakers doesn't work. For a natural and believable sound field it is essential that the sound from the surround loudspeakers is decorrelated such that no two surround loudspeakers reproduce the exact same audio signal (Holman, 1991; ITU-R BS.775, 2012). Without decorrelation the surround sound would be much less immersive and may sometimes even seem to be located at a specific position in the room.

Recommendation 7.36 *To use all the surround loudspeakers in your system consider using the surround upmixing modes of your A/V controller (like Dolby Prologic and DTS Neo) when watching movies with a soundtrack that has less surround channels than the number of loudspeakers in your room.*

While it is generally a good idea to always use surround upmixing when watching movies, this cannot be said for listening to two-channel music. With some two-channel recordings the upmixed surround sound may create a stunning effect, while with other recordings it may sound unbelievably bad. Your mileage may vary. In general, surround upmixing reduces the front image quality, but improves the spatial impression. It is a matter of preference. Most of the time expert listeners tend to prefer the unprocessed two-channel original (Choisel and Wickelmaier, 2007; Rumsey, 1999).

Some controllers are equipped with DSP surround modes that synthesize reverberation and add it to the surround channels. These DSP modes often carry labels such as 'stadium', 'concert hall', and 'jazz club'. It is important to distinguish these modes from the surround upmixing methods, like Dolby Prologic and DTS Neo. While the upmixing methods extract ambience that is already present in the recording, the DSP modes artificially create reverberation (Harley, 2015). Such artificial manipulation completely changes the sound such that it no longer resembles the sound that the performing artist and producers intended you to hear. Therefore, I recommend that you never use these gimmicky DSP modes.

Recommendation 7.37 *For the best reproduction of two-channel music do not use any of the DSP surround modes of your A/V controller.*

Some high-end controllers are equipped with THX signal processing (thx.com). THX processing was introduced in the early days of surround sound to improve the limited performance of Dolby Prologic decoders. The intention of THX processing was to make the home theater experience match the experience in the cinema as close as possible. Much has changed since these days and THX processing seems to be of less importance nowadays (McLaughlin, 2005; Toole, 2018). Two THX signal processing techniques that are still offered today in high-end A/V controllers are *re-equalization* and *timbre matching* (Harley, 2002; Holman, 1991). Re-equalization corrects overly bright soundtracks that were intended to be reproduced in a large movie theater. In a typical movie theater high frequencies are absorbed considerably by the large audience and other sound-absorbing materials like heavy draperies. To compensate for this loss of high frequencies, the sound engineers make the soundtrack brighter. When you play such a soundtrack in a home environment it will sound overly bright. However, you should not need to enable re-equalization on your A/V controller, because a properly produced DVD or Blu-ray disc should already be re-equalized during the final mastering stage. The other THX technique of timbre matching corrects the frequency response of the surround channels such that the sounds that arrive from the front and the back have the same timbre. Timbre matching compensates for the fact that the human ear has a different frequency response for sounds arriving from the front than for those arriving from the back of the head. As with re-equalization it could be argued that such compensation needs to be done during the mastering stage of the program material and not be left to the consumer.

7.5.2 Bass Management

Every A/V controller and receiver has a setup menu for bass management. You use this menu to determine which of the loudspeakers in your system reproduce low frequencies. As I explained in section 4.1, to achieve the best low-frequency reproduction in your room you should redirect the lowest frequencies to the subwoofers. Your main loudspeakers should not reproduce them. The subwoofers reproduce the lowest frequencies from all of the main channels and also the LFE (low-frequency effect) channel (see section 2.1). If you use multiple subwoofers, which I recommend, all subwoofers should reproduce the same mono bass signal. There is no such thing as 'stereo bass'. To achieve an optimal interaction between the subwoofers and the room (see section 8.1.3), it is essential that all subwoofers reproduce the same signal. If your controller or receiver has multiple subwoofer outputs, make sure that these outputs carry the same audio signal: look it up in your user manual and make the necessary changes in the setup menu. If you have only one subwoofer output but still want to use multiple subwoofers, use a Y-splitter RCA cable.

Usually the bass management features of your controller or receiver allow you to make the following settings:

- The size of each of the main loudspeakers.
- The handling of the LFE channel.
- The subwoofer crossover frequency.

The setting for the size of your loudspeakers determines which loudspeakers reproduce low frequencies. The loudspeakers that are set to 'large' reproduce the entire frequency range including the low frequencies. The loudspeakers that are set to 'small' will not reproduce low frequencies; the low-frequency content will be redirected to the subwoofers. Thus if you follow the recommendation from section 4.1 you need to set all your front and surround loudspeakers to 'small' even if you have physically large loudspeakers. In fact, the size setting for your loudspeaker has nothing to do with its actual size, it only determines the handling of the low-frequency content. Some controllers and receivers also allow you to determine where to send the LFE channel: 1) only to the subwoofers or 2) to the subwoofers and to the main loudspeakers. Again, the recommended setting is to only send it to the subwoofers even if your main loudspeakers are large (ITU-R BS.775, 2012; McLaughlin, 2005).

Recommendation 7.38 *Use the bass management setup menu in your A/V controller or receiver to set all the front and surround loudspeakers to 'small', set the subwoofer to 'on', and direct the LFE channel only to the subwoofers and not to the other loudspeakers.*

The subwoofer's crossover frequency determines how the low-frequency range is divided between the subwoofers and the main loudspeakers. The low bass is directed to the subwoofers, while the higher bass frequencies (up to 300 Hz) are still being reproduced by the main loudspeakers. The subwoofer's crossover frequency should be chosen between 80 Hz and 120 Hz depending on the low-frequency capabilities of your main loudspeakers and the position of the subwoofers in your room.

If you choose a low subwoofer crossover frequency your main loudspeakers need to be able to reproduce more bass than when you choose a higher crossover frequency. A good rule of thumb is that your main loudspeaker's −3 dB cutoff point should be lower than or equal to half the subwoofer crossover frequency. For example, with a crossover frequency of 100 Hz, your main loudspeakers should go down to at least 50 Hz (see also the part on frequency response in section 7.1.4). The recommended crossover frequency for THX certified systems is 80 Hz (Harley, 2015; Holman, 2008). Never choose a crossover frequency below 80 Hz.

If you choose a high subwoofer crossover frequency your main loudspeakers can be small, because they need to reproduce only a small part of the low-frequency range. However, the downside is that a higher subwoofer crossover frequency increases the chances that you can localize the subwoofers and hear them as separate sound sources. This is obviously not what you want. Good sound requires that the locations of your subwoofers are not audibly apparent. Whether the subwoofers can be localized very much depends on their distance from your main loudspeakers. If they are placed near your main front loudspeakers you are less likely to run into trouble with a higher crossover frequency than when they are placed farther away. If you place the subwoofers very close to your main loudspeakers, you could get away with a crossover frequency as high as 160 Hz (EBU-Tech 3276-s1, 2004), but to stay out of trouble it is recommended not going beyond 120 Hz (Tsakiris and Orinos, 2004). To summarize, with a higher crossover frequency you will be able to use smaller main loudspeakers, but you will be less flexible in choosing the locations of your subwoofers (EBU-Tech 3276, 1998).

It is important that your subwoofers and main loudspeakers sound well-integrated. If your main loudspeakers are not very capable of reproducing low frequencies and you set the subwoofer crossover frequency too low, part of the low-frequency range is not being reproduced by the subwoofer nor by the main loudspeakers. This may result in an undesirable dip in the overall frequency response (Harley, 2015). An example is shown in figure 7.30.

Recommendation 7.39 *Set the subwoofer crossover frequency of your A/V controller or receiver between 80 and 120 Hz, such that the subwoofers and main loudspeakers sound well-integrated and the subwoofers cannot be audibly localized as separate sound sources.*

Note that the subwoofer crossover frequency doesn't influence low frequencies in the LFE channel. Regardless of the crossover setting, the LFE channel can contain low frequencies up to 120 Hz (ITU-R BS.775, 2012).

Most active subwoofers also have a knob to set the subwoofer's crossover frequency, because they have their own built-in low-pass filter. Since you are using your A/V controller or receiver to do the filtering, you want to disable this built-in low-pass filter. Some subwoofers have a switch to do this. If yours doesn't, the next best thing is to set the frequency of the low-pass filter as high as possible. This will minimize the interaction between the low-pass filter of the subwoofer and the crossover filter of the A/V controller or receiver (DellaSala, 2010; McLaughlin, 2005).

Recommendation 7.40 *Disable the low-pass filter built into your subwoofer, or set it to the highest cutoff frequency.*

7.5.3 Delay Settings

If your main loudspeakers are placed at different distances from your prime listening position you need to delay some of the audio channels to make sure the sound from all loudspeakers

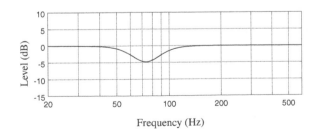

Figure 7.30
Example of a dip in the overall frequency response, which occurs if you set the subwoofer crossover frequency so low that your main loudspeakers are having trouble taking over the low-frequency reproduction.

arrives at the same time instant (Holman, 2008). Your A/V controller or receiver has an adjust-able electronic delay for each audio channel that you can use to accomplish this. If your center loudspeaker is not placed on an arc with the two other front loudspeakers, it will be at a dif-ferent distance from the prime listening position and you need to slightly delay its sound (see section 4.2.3). You also need to configure delays if your surround loudspeakers are closer or farther away from the prime listening seat than the front loudspeakers are (see section 4.2.4).

Most A/V controllers and receivers determine the appropriate delays for the different chan-nels based on the distances between the loudspeakers and the prime listening position. You simply measure those distances and input the numbers into the delay setup menu. It is as simple as that. Some A/V controllers and receivers (especially older ones) require you to enter the delay times instead of the measured distances. In this case, you also need to measure the distances between the loudspeakers and the prime listening position, and calculate the delays yourself. Increase the delay for loudspeakers located closer to the listening position and decrease the delay for those located farther from the listening position. Since sound trav-els at a speed of 344 m/s, you obtain the delay by dividing the path difference (in meters) by 344. For example if your main loudspeakers are placed at a distance of 4 m (13 ft) and your surround loudspeakers at 2 m (6.6 ft), the path difference is $4-2=2$ m and the delay for the surround loudspeakers becomes $2/344 \approx 6$ ms.

Recommendation 7.41 *Measure the distances between each main loudspeaker and the prime listening position and use these distances to set the appropriate delays in the delay setup menu of your A/V controller or receiver.*

The sound from your main loudspeakers should be in sync with the video displayed. A per-ceptible time delay between the sound and the moving images can be annoying. Sometimes you need to adjust the delay of all the main loudspeakers to achieve a good enough synchro-nization. You can use test patterns on a video test disc, such as DVE HD Basics (Advanced Patterns—A/V Timing) or HD Benchmark (Audio Calibration—A/V Sync) to check for any sync problems and correct them (see appendix A).

Recommendation 7.42 *Use a video test pattern to check audio and video synchronization. If necessary fine-tune the delay of the audio signal.*

You also need to take into account the distance between the subwoofers and the listening position. If a subwoofer is positioned at a different distance from the prime listening position than the front loudspeakers the sound from these loudspeakers doesn't arrive at the same time instant and this results in a phase shift. For example, a path difference of 1 m (3.3 ft) results in a time delay of $1/344 \approx 3$ ms and a phase shift of 105° at 100 Hz (see box 7.11). When the path difference is slightly increased and the phase shift reaches 180° the sound wave from the subwoofer is partly canceling the sound wave from the front loudspeakers, because these waves are opposing each other (see figure 2.20). As a result you will hear less bass. Therefore, it is important that around the subwoofer crossover frequency the sound from the

---

BOX 7.11 PHASE SHIFT

The phase shift $\Delta\phi$ (in degrees) that results from a time delay Δt (in seconds) equals

$$\Delta\phi = 360° \cdot f \cdot \Delta t$$

where f is the frequency in Hz.

subwoofers arrives at the listening position in phase with the sound from the main loudspeakers. To achieve this you need to correctly set the delay between the subwoofers and the front loudspeakers. There are two controls that influence this delay: 1) the loudspeaker distance or delay settings in the setup menu of your A/V controller or receiver, and 2) the phase control on the subwoofer. Some subwoofers have a phase control knob that lets you control the phase continuously between 0° and 180°, while others only have a phase switch that switches between 0° (in phase) and 180° (out of phase).

The total delay of the sound emerging from the subwoofer depends on the combined setting of the subwoofer's phase control and the delay/distance settings in the A/V controller or receiver. It is best to set these controls by listening to a test signal. Setting them purely based on measured distances is not recommended, because the electronic circuits in the subwoofer (filters and amplifiers) can also introduce a phase shift around the crossover frequency. Your goal is to set the controls such that the sound from the subwoofer and the sound from the front loudspeakers is maximally in phase, or in other words reinforces each other instead of weakening each other. You need a test signal with a pure sine that has the same frequency as the subwoofer crossover frequency, or even better a test signal with narrow-band noise centered around the subwoofer crossover frequency. You can obtain such signals from for example the Audio Check website (see appendix A). If your subwoofer has a phase switch, play the test signal and listen to the amount of bass for each position of the switch. Set the switch in the position that provides the largest amount of bass. Next, vary the subwoofer delay/distance in the A/V controller or receiver and again keep the setting that provides the largest amount of bass. If your subwoofer has a phase knob instead of a switch, set the subwoofer delay/distance in the A/V controller or receiver to the same value as those for the front loudspeakers. Then, play the test signal and listen to the amount of bass while varying the phase knob on the subwoofer and keep the setting that provides the largest amount of bass.

Recommendation 7.43 *Set the phase controls of your subwoofer and the delay/distance controls in your A/V controller or receiver such that the sound from the subwoofers and the sound from the front loudspeakers arrive in phase at the prime listening position.*

Instead of setting the subwoofer delay and phase by ear, you could also use the acoustic measurements that I describe in section 8.1.3 for more precise control.

7.5.4 Level Settings

Optimal reproduction of surround sound requires that your main loudspeakers are matched in sound level. If one of the loudspeakers plays louder than the others, the sound field gets geometrically distorted and the louder loudspeaker will draw attention to itself. There are two reasons for level mismatches between the loudspeakers. First, a level difference occurs if not all the main loudspeakers are identical models. For example if your front loudspeakers are larger and have more drivers than your surround loudspeakers they likely have a higher sensitivity and will play louder with the same amount of amplifier power (see section 7.2.1). Second, a level difference occurs with identical loudspeakers when they are placed at different distances from the prime listening position. Recall that the sound pressure level falls off with the square of the distance (see section 2.3.3).

Your A/V controller or receiver has an adjustable level for each of the audio channels in its setup menu. You use this feature to match the levels of the front and surround loudspeakers. Your goal is to make them sound equally loud at the prime listening position when they play a similar test tone. Your A/V controller or receiver will send this test tone to each loudspeaker in succession so that you can compare their different loudness levels. The test signal consists of *pink noise*. This noise contains all frequencies and sounds like a hiss. Pink noise contains less energy at high frequencies than at low frequencies (it falls off at 3 dB per octave). This property makes it less likely to damage your loudspeakers and it also makes it sound less aggressive. An even better test signal consists of filtered pink noise that only covers the range from 500 Hz to 2 kHz (Holman, 2008). The use of pink noise avoids problems with varying levels at low frequencies due to standing waves. It also avoids problems at high frequencies where the sound is only emitted in tight beams and thus the sound level can vary greatly by moving only a few centimeters to the side. To calibrate the levels of the different loudspeakers you can either use the internal test tone from your A/V controller or receiver, or you can play test tones from a test disc (see appendix A).

You can try to set the relative levels by just listening to the test signals, but that is difficult and doesn't result in very accurate level settings. When you sit facing forward in the prime listening position you will notice that the test signal will sound different when it moves around the room. Its timbre will distinctively change due to the directional characteristics of your ears. This is normal; it is the way your ears are supposed to work (see also section 3.4.4). A better way to judge the relative volumes is to turn your head and body to face each of the loudspeakers as they play the test signal (Toole, 2018). But even then, timbral differences will be apparent. Two loudspeakers of the same model never seem to sound the same when reproducing pink noise. Pink noise tends to reveal all the minor differences; it is a very sensitive indicator for timbre changes. Luckily, these minor differences between two identical models diminish when listening to music (Toole and Olive, 1988). So there is no need to worry about it.

A more accurate way to calibrate the relative levels of your loudspeakers involves the use of an SPL meter. You can use either a dedicated SPL meter or an app on your smartphone. Place the meter at you prime listening position, approximately at the height of your ears while seated. Point the meter forward, aim it slightly above your center loudspeaker. It is best to mount the meter on a tripod to keep it at a fixed position. A tripod also allows you to crouch down such that you are not in between the meter and any loudspeaker obstructing the sound. Do not move the meter during the test, and keep it pointed forward; do not aim it at each loudspeaker. Set the SPL meter to slow response and C-weighting. As the test tone cycles to each of the loudspeakers, adjust their relative levels in the setup menu of your A/V controller or receiver (Harley, 2015; McLaughlin, 2005).

It is best to calibrate the relative levels at an SPL that corresponds to your normal listening conditions. Recall from box 3.2 that the recommended listening level is 78 dB SPL (EBU-Tech 3276, 1998; ITU-R BS.1116, 2015). You should try to match the relative levels of the loudspeakers within 1 dB (EBU-Tech 3276, 1998; Toole, 2018) or even better within 0.5 dB (AES, 2008; EBU-Tech 3276-s1, 2004).

> **Recommendation 7.44** *Use an SPL meter (slow response, C-weighting) to calibrate the sound level of each main loudspeaker to 78 dB SPL. Adjust the relative levels of the loudspeakers in the setup menu of your A/V controller or receiver such that the loudspeakers are matched within 1 dB.*

Instead of a manual calibration with an SPL meter, you could also rely on the automatic calibration mode of your A/V controller or receiver. Instead of an SPL meter you would use the dedicated measurement microphone that came with your A/V controller or receiver. Place it in the same way as outlined above for the SPL meter, and make sure you are not obstructing the sound from reaching the microphone during the calibration procedure.

After calibration it is a good idea to check the relative levels by ear by listening to some music and watching some movies. Sometimes you may need to turn the center loudspeaker or surround loudspeakers a notch down to get the best overall experience. Make only small adjustments by ear. The levels found using your SPL measurement or automatic calibration should be about right. So never deviate too far from them.

A center loudspeaker that is too loud will impair the frontal stereo image. The sound will bunch up around the position of the center loudspeaker, and the width of the frontal sound-stage becomes smaller. On the other hand, a center loudspeaker that is too soft will consider-ably reduce dialog intelligibility when watching movies. You need to carefully balance these two requirements. To maintain overall sound quality, it is recommended that the level of the center loudspeaker does not exceed the level of the left and right loudspeakers by more than 3 dB (Toole, 2018).

> **Recommendation 7.45** *After calibration of the relative sound levels, play some music and watch some movies to check if the level of the center loudspeaker is not too high. The frontal soundstage should appear wide without the sound bunching up around the position of the center loudspeaker.*

Surround loudspeakers that are a bit too loud will attract unwanted attention to the sounds from the rear. The front-to-back balance will be distorted. You will hear too much ambience and reverberation and not enough direct sound from the front loudspeakers. This destroys the sense of immersion that the surrounds should create. Be careful not to reduce the level of the surrounds too much, because it will also destroy the sense of immersion and the sound field will collapse towards the front (Harley, 2015).

> **Recommendation 7.46** *After calibration of the relative sound levels, play some music and watch some movies to check if the level of the surround loudspeakers is not too high. The surrounds should blend seamlessly with the rest of the sound and never draw attention to themselves when playing immersive sounds.*

Setting the right level for the subwoofers is more difficult than setting the level for the main loudspeakers. Room resonances may cause large SPL variations at different frequencies and at different positions. For this reason the SPL meter may be hard to read. It may be neces-sary to obtain an estimate of the average SPL as the sound meter is moved around the prime listening position. Measuring at multiple positions that are about 0.5 m away from the prime position should be sufficient (EBU-Tech 3276-s1, 2004). Make sure that the noisy test signal that you use is not limited to frequencies above 500 Hz (filtered pink noise) and does contain frequencies below 100 Hz. A signal limited to frequencies above 500 Hz is only suitable for setting the levels of the main loudspeakers.

If you have multiple subwoofers, it is important to match their levels as closely as possible. All your subwoofers should contribute equally to the low-frequency sound. Try to match them within 1 or 2 dB. Larger level differences may result in poor integration of the sound with the main loudspeakers. It may also result in increased distortion during loud passages as the loudest playing subwoofer may be overdriven. First, use the gain control of the subwoofers themselves to match their relative levels; next, use the subwoofer level control to match the overall sound level of the subwoofers to the main loudspeakers (DellaSala, 2010).

Always check the overall subwoofer level by ear. The bass should not dominate the sound but serve as the foundation for music and effects. The subwoofer should blend seamlessly with the rest of the sound and never draw attention to itself. Some people tend to set the subwoofer level too high. The result is a heaviness that surrounds everything that you play. Such a sound is fatiguing to listen to and it will rob music and sound effects of their impact and surprise (Harley, 2015).

Recommendation 7.47 *Use an SPL meter to roughly set the subwoofer level, then play some music and watch some movies to fine-tune it. The subwoofer should blend seamlessly with the rest of the sound and never draw attention to itself.*

The best way to set subwoofer levels is to use the acoustic measurements that I describe in more detail in section 8.1.3. Such measurements also help you to identify and correct problems with room resonances.

7.5.5 Dynamic Range Control

To optimally reproduce music and movies you need to play them at the same SPL as they are played in the studio control room or the cinema. The standard for the cinema is 85 dB on average with peaks of 105 dB (DCI, 2008; SMPTE RP 200, 2012). Music is played at similar levels during mixing and mastering in the studio (Katz, 2015). Most people find an average SPL of 85 dB way too loud at home. Most people prefer an average SPL of about 75–78 dB (see box 3.2). In some situations even this may be too loud. For example, when you want to watch a movie late at night while other members of your household are trying to sleep in an adjacent room.

When you set the playback volume substantially lower than the reference playback level two things will happen. First, low-level sounds will no longer be audible as their SPL drops below the threshold of audibility (see figure 2.24). The loss of low-level sound will dramatically impact the presence and envelopment of the sound, because these sound-quality aspects heavily depend on the reproduction of low-level reflections (see section 3.4.5). At a low playback level, the ambience in the surround channels may completely disappear and you no longer feel immersed in the sound field. Second, a playback volume that is substantially lower than the reference playback level will distort the balance between low, mid, and high frequencies. At lower playback levels the human ear is less sensitive to high and low frequencies, as shown in figure 2.24. Lowering the playback volume has the most impact on the low frequencies. At low frequencies the equal-loudness curves in figure 2.24 converge. This means that the perceived loudness of low frequencies decreases faster than the mid and high frequencies do. Thus, at a lower playback volume you will hear less bass. Figure 7.31 shows the low-frequency boost needed to compensate for the difference in frequency balance between a playback level of 85 dB and a level of 75 dB.

Playing back music and movies at reference level is clearly the best option. If you reduce the playback level to an average of 75–78 dB, you will still have sufficient dynamic range left for the low-level sounds to be audible. But, if you lower the volume further, you should consider using a feature of your A/V controller or receiver that is called *dynamic range control* (DRC). When enabled, DRC reduces the dynamic range to help maintain an average loudness. It reduces the peaks and boosts the low-level sounds (McLaughlin, 2005; Taylor et al., 2006). DRC is intended for situations in which you are forced to considerably reduce your playback volume. You should only use it for those specific situations. It is best to leave it off when you are reproducing sound at a normal playback level (75–78 dB), as DRC will reduce the dynamic impact and make the sound appear to be flat and uninvolving.

Figure 7.31
Low-frequency boost
needed to compensate for
the difference between the
equal-loudness contours at
85 and 75 phon.

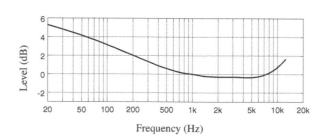

Recommendation 7.48 *Only enable the dynamic range control of your A/V controller or receiver when your average playback volume is much lower than 78 dB; keep it disabled in all other situations.*

Some modern A/V controllers and receivers are equipped with an automatic calibration mode that uses a separate microphone to measure the acoustic output of your system. These controllers and receivers are often also able to automatically compensate for the changes in frequency balance at different playback levels. They retain a proper balance of the low, mid, and high frequencies regardless of the playback level, bringing the reproduced sound closer to the sound of the studio control room and the cinema. Examples of technologies that make this possible are (Fleischmann, 2017): Audyssey Dynamic EQ (audyssey.com), Dolby Volume (dolby.com), and THX Loudness Plus (thx.com). They are the modern-day incarnations of the 'loudness' control found on some older amplifiers. This control also boosts the low frequencies to compensate for their loss at lower playback volumes, but it is a much cruder control as it doesn't take into account the actual playback level; it always applies the same compensation (Everest and Pohlmann, 2014; Toole, 2018).

7.5.6 Room Correction and Equalization

Many modern A/V controllers and receivers are equipped with an automatic *room correction* or *room equalization* feature that uses digital signal processing to correct the combined frequency response of the loudspeaker and the room. Some examples of automatic room correction systems are: Anthem ARC (anthemav.com), Audyssey MultEQ (audyssey.com), Dirac Live Room Correction Suite (dirac.com), and Trinnov Audio Optimizer (trinnov.com). Similar to the automatic calibration method, you place a measurement microphone at the prime listening position to measure the combined response of the loudspeakers and the room. The room correction system will emit some test tones, measure and analyze the response, and automatically apply some corrections to the audio signals. The idea is that these corrections counteract the errors that are present in the in-room frequency response of the loudspeaker. It holds the promise of correcting loudspeaker-response anomalies and at the same time solving room acoustical problems (Harley, 2015; Pohlmann, 2011). If this sounds too good to be true, then you might be right. Room correction systems do have their limitations.

Automatic room correction systems can be very effective at removing unwanted colorations at low frequencies, but they do not always improve the sound at mid and high frequencies. Sometimes the corrections can drastically change the sound at mid and high frequencies, making your loudspeakers sound unnatural and much worse. A room correction system is no substitute for good loudspeakers, nor can it compensate for bad loudspeaker placement and severe acoustical problems. Room correction does have its merits, but only as a final step to fine-tune the in-room response of already good loudspeakers that have been placed properly in an acoustically decent room (Harley, 2015; Newell and Holland, 2007; Toole, 2018).

Room correction systems almost always improve the quality of low-frequency reproduction. Recall that at low frequencies, below about 300 Hz, the sound that you hear is dominated by the acoustics of the room and not so much by the quality of your loudspeakers (see section 2.3.4). Room resonances and the adjacent-boundary effect dominate the combined frequency response of the loudspeaker and the room. Both can be partly corrected using a room correction system (Genelec, 2009; Toole, 2006). The peaks in the frequency response can be tamed down, but the dips, especially the deep ones, cannot be corrected. Let me explain why: a dip that results from the adjacent-boundary effect or a room resonance occurs because a sound wave is reflected off a wall and its reflection arrives with opposite phase. The original wave and its reflection cancel each other out. The only way to change this cancellation is by moving the loudspeaker or absorbing the reflection. Optimal low-frequency reproduction can only be achieved by a thoughtful combination of loudspeaker placement and electronic correction or equalization. I describe this combined approach in more detail in section 8.1.3 on subwoofer calibration.

Room correction systems do not always improve the quality of reproduction for the mid and high frequencies (above 300 Hz). Recall that at mid and high frequencies the frequency response of the loudspeaker itself dominates the perceived sound quality. A measurement of the combined frequency response of the loudspeaker and the room is not a reliable indicator for overall sound quality. The main reason is that such a measurement is blind to the direction from which sounds arrive and doesn't distinguish between the direct sound from the loudspeaker and the multiple reflections from the room. The perceived sound quality at mid and high frequencies very much depends on the combination of direct sound and early and late reflections arriving from different directions (see section 2.3.4). Therefore, using measurements of the combined frequency response of the loudspeaker and the room for automatic room correction often results in a degradation of sound quality. It is important to realize that room correction or equalization cannot change the directivity of the loudspeakers. A correction of the audio signal that is fed to the loudspeaker changes the frequency response of the loudspeaker in all directions, that is, both the on-axis and the off-axis responses are altered at the same time. Therefore, it is not possible to control the reflections in the room without also changing the direct sound from the loudspeakers (Toole, 2018).

Correction and equalization at mid and high frequencies can improve the sound if it is used to only correct the loudspeaker without taking into account the acoustics of the room. For this you would normally need to measure the anechoic response of the loudspeaker, that is, its response without any reflections from the room (see section 7.1.4). Some advanced room correction systems use a special measurement technique that yields an approximation of the anechoic response of the loudspeaker from measurements in a reflective room. This allows them to calculate and apply corrections that only take into account response anomalies of the loudspeaker itself. These systems calculate an inverse filter of the loudspeaker that they apply to the audio signal to achieve a flat on-axis anechoic frequency response. The best room correction systems combine the best of two worlds: they apply room correction at low frequencies and only loudspeaker-response correction at mid and high frequencies (Katz, 2015). Such advanced systems only work well in the hands of a skilled technician. They often require some manual tweaking. When not used correctly the inverse filtering of the loudspeakers can result in audible artifacts and distortions that audibly degrade the sound quality (Norcross et al., 2004).

Recommendation 7.49 *Use the automatic room correction feature of your A/V controller or receiver to correct the combined frequency response of your loudspeakers and the room, but limit the corrections to the frequencies below 300 Hz.*

To avoid any problems at mid and high frequencies it is often best to limit the frequency response corrections to the low frequencies below 300 Hz (EBU-Tech 3276, 1998; EBU-Tech 3276-s1, 2004). You can accomplish this by manually adjusting the settings that the automatic room-calibration system came up with. Alternatively, you can apply automatic room correction only to the subwoofers (Harley, 2015), but this has the drawback that you are missing out on correcting the higher bass frequencies (between 80 and 300 Hz). Some subwoofers are equipped with their own room correction system that you can use to optimally integrate them into the room. As I mentioned earlier, I will describe the correction of the in-room subwoofer response in more detail in the upcoming section 8.1.3.

8

ROOM ACOUSTICS

It shouldn't come as a surprise that the acoustical properties of your listening room influence the quality of the reproduced sound. Fortunately, there is no need to cover your entire room with ugly acoustical treatments. A regularly furnished domestic room provides a good starting point for high-quality sound reproduction. However, paying some attention to the acoustics of your room can take the sound from good to excellent. There is room to improve, especially in the low-frequency region. A good-sounding room lets you better enjoy the sound, because you typically hear more nuances and more importantly, the sound more closely matches the artistic intent of the artists and sound engineers.

Recall from section 2.3.4 that above about 300 Hz the sound that you hear is dominated by the quality of your loudspeakers, while below 300 Hz the room dominates. Therefore, no acoustical treatment of your room can compensate for poor-sounding loudspeakers at mid and high frequencies; having high-quality loudspeakers is essential. Conversely, at low frequencies your room easily ruins the sound of otherwise excellent loudspeakers. No matter how great your loudspeakers are the acoustics of your room will dominate the sound at low frequencies. Unfortunately, changing the acoustics of your room at low frequencies, especially below 100 Hz, is not easy. It requires either drastic changes to the geometry of the room or the installation of large acoustical devices, neither being very attractive nor practical. Not all is lost though, as good low-frequency sound can be obtained in almost any room with a clever combination of several less extreme changes. It requires the use of multiple well-positioned

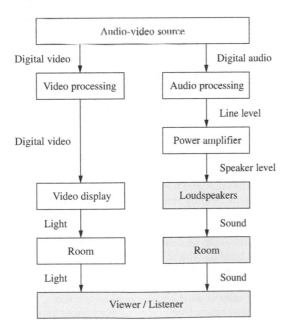

subwoofers, electronic equalization, and multiple low-frequency absorbers in the corners of the room (Toole, 2018). I will show you how to apply this combined approach in the first part of this chapter.

Above 300 Hz the acoustics of the room are dictated by reflections from hard surfaces. Reflections from the room, up to a certain degree, are a good thing. Reflections create a sense of space and enrich the timbre of musical instruments. Without any reflections the room would sound dull and uninvolving. However, too much reflected sound will make the sound hollow and impairs the clarity and definition of the reproduced sound. The amount of reflections in a room can be controlled by strategically placing sound-absorptive material and diffusers that scatter the sound in all directions. Some rooms need more fine-tuning than others. Most rooms benefit from some sound absorption on the floor and ceiling. I will discuss the different possibilities in the second part of this chapter.

8.1 Reproducing Low Frequencies in Rooms

The reproduction of low frequencies is influenced by reflections from the adjacent room boundaries and room resonances. As I described in section 4.2.5 the detrimental effect of reflections from the adjacent boundaries can be managed by carefully positioning the loudspeakers and subwoofers. Dealing with room resonances is a bit more involved.

Room resonances occur in any room. They make the distribution of low-frequency sound energy in the room very uneven. Some frequencies are strongly amplified while others are attenuated. The amount of amplification and attenuation changes with the position in the room. Therefore, the frequency response is different at different points in the room. Furthermore, resonances store energy in the room. This energy is slowly released and causes an audible extension of certain bass notes. This unwanted extension or ringing causes the bass to sound booming. It is easily detectable on short impulsive sounds such as kick drum and plucked bass. Room resonances destroy the perceived tightness of the bass. While humans are able to adapt to a certain amount of bass misbehavior, dominant resonances are difficult to ignore. It seems that we have a great desire to hear bass that is free of resonances (Toole, 2018).

A good test signal for evaluating the effect of room resonances is the Musical Articulation Test Tone (MATT) developed by Acoustic Sciences Corporation (see appendix A). It consists of a series of tone bursts with silences between them. The bursts slowly rise in frequency from 28 Hz up to 780 Hz. Every burst lasts 1/16 second and is followed by 1/16 second of silence. When you listen to this signal through headphones, you will hear a sequence of distinct bursts with each burst clearly articulated. When you reproduce this signal through your loudspeakers, the bursts at certain frequencies become smeared. You are hearing the effect of the room resonances that store and slowly release low-frequency energy (Harley, 2015).

8.1.1 Room Resonance Modes

Let's have a more detailed look at room resonances and how they affect the reproduced sound. The low-frequency sound emitted by your loudspeakers bounces around the room and is reflected by the walls, the floor, and the ceiling. The low-frequency response of the room changes as a function of time. Shortly after the start of the low-frequency sound, the room response equals the curve predicted by the adjacent-boundary effect (see section 2.3.4). The response consists of the direct sound and the reflection from the three nearest room boundaries. As time progresses the room response changes as more reflections are added and the room resonances start to build up. The total response is the sum of the adjacent-boundary effect and all the room resonances (Adams, 1989).

Room resonances occur between all the boundaries of the room, regardless of the shape of the room. In a rectangular room, three types of resonances occur: *axial*, *tangential*, and *oblique* (Everest and Pohlmann, 2014; Jones, 2015b; Walker, 1992). Axial resonance modes occur between any set of two parallel room boundaries, for example the floor and the ceiling. Tangential resonance modes occur between any set of four room boundaries, for example the four walls of the room. Oblique resonance modes occur between all six room boundaries. The axial modes make the most prominent contributions to the low-frequency sound of the room. The tangential modes are less pronounced, because they require four reflections instead of two, and at each reflection some energy is lost. Hence, tangential modes only have half the energy of the axial modes. The oblique modes, which require six reflections, have even less influence. They can safely be ignored (Jones, 2015b).

There are three sets of axial resonance modes, since there are three axes to a rectangular room (length, width, and height). As I explained in section 2.3.4, each set consists of a fundamental frequency and multiples thereof. The fundamental frequency depends on the distance between the two room boundaries that support the axial resonances. The distance between these boundaries equals half a wavelength of the fundamental frequency. The sound pressure between the two boundaries varies and is maximal at the boundaries and minimal exactly in between them, as illustrated at the top of figure 2.35. Subsequent room resonances occur at multiples of the fundamental frequency f_o (at $2f_o$, $3f_o$, $4f_o$, and so forth), and result in an alternating pattern of pressure maximums and minimums (nulls) separated a quarter wavelength apart. Figure 2.35 shows the pressure distribution along one dimension of the room. It is important to realize that room resonances influence the pressure distribution in all three dimensions of the room. It is difficult to visualize this in three-dimensional space, but for a better understanding it helps to look at a two-dimensional cross-section of the room. Two examples are shown in figure 8.1. Think of this figure as the floor plan of your room and realize that it represents just one of the slices that make up a three-dimensional rectangular space.

There are many more tangential and oblique modes. It helps to identify them using a three digit code: (n_x, n_y, n_z) where each digit corresponds to one of the dimensions of the room. The value of the digits indicates the order of the resonance. The axial modes always have two digits equal to zero and the tangential modes have one digit that equals zero. With this notation the set of axial resonance modes along the length (x) of the room is represented as (1,0,0), (2,0,0), (3,0,0), etc. The axial modes along the width (y) of the room are (0,1,0), (0,2,0), (0,3,0), and those along the height (z) are (0,0,1), (0,0,2), (0,0,3). Using this three digit code all room resonances of a rectangular room can be computed using the formula given in box 8.1.

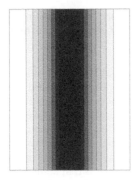
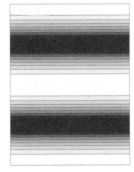

Figure 8.1
Examples of the sound pressure distribution for axial room modes. *White* indicates a pressure maximum while *black* indicates a pressure minimum. *On the left*, the distribution of the first-order axial mode between the sidewalls (along the width of the room). *On the right*, the distribution of the second-order axial mode between the front and the back wall (along the length of the room).

BOX 8.1 ROOM RESONANCES IN A RECTANGULAR ROOM

The frequencies of all resonance modes in a rectangular room can be computed from the following equation:

$$f_{n_x, n_y, n_z} = \frac{c}{2}\sqrt{\left(\frac{n_x}{\ell_x}\right)^2 + \left(\frac{n_y}{\ell_y}\right)^2 + \left(\frac{n_z}{\ell_z}\right)^2}$$

where c is the speed of sound (344 m/s), ℓ_x, ℓ_y, ℓ_z are the length, width, and height of the room, respectively, and n_x, n_y, n_z are whole numbers (0, 1, 2,...) defining the different modes.

Figure 8.2
Examples of the sound pressure distribution for tangential room modes. *White* indicates a pressure maximum while *black* indicates a pressure minimum. *On the left*, the distribution of the tangential mode (2,1,0) with a second-order resonance between the front and the back wall (along the length of the room) and a first-order resonance between the sidewalls (along the width of the room). *On the right*, the distribution of the tangential mode (2,2,0) with a second-order resonance between the front and back wall and a second-order resonance between the two sidewalls.

The tangential modes create a more complicated pressure distribution in the room. Two examples are given in figure 8.2. This figure shows the pressure distribution of the (2,1,0) and (2,2,0) tangential modes.

When you reproduce sound in a room the frequencies that correspond to the room resonances will be amplified by the room. As a result some bass notes may sound louder than others. At the positions in the room that correspond to a pressure maximum, the resonance will be clearly audible. While at the positions in the room that correspond to a pressure minimum, the low frequencies may seem to be completely absent. Thus, depending on your listening position in the room the frequency response that you experience may have several peaks and dips, and strong coloration of the reproduced sound may occur. As the resonances slowly release their energy over time, another effect that may occur is that some bass notes overhang or ring much longer than others. Pitch changes during the decay of a bass note may also occur (Wilson et al., 2003).

Whether room resonance modes are audible and appreciable as coloration depends on several factors. First, it depends on the degree of excitation of the mode, which in turn depends on the positions of the loudspeakers and listeners in the room (Everest and Pohlmann, 2014; Gilford, 1959). Maximum excitation of a room mode occurs if the loudspeaker is positioned at a pressure maximum, and minimal excitation occurs if it is positioned at a pressure minimum (null). A similar relation holds for the position of the listener. Placing a loudspeaker in the corner of a room will excite all room resonances, and is thus a bad idea. It is impossible to place the loudspeakers and listeners such that none of the room modes are excited. Loudspeakers and listeners should be placed such that the most annoying room resonances are suppressed. This raises the question, which of the modes are the most annoying?

The audibility of room resonances and the degree to which they stand out from the rest of the sound also depends on the bandwidth of the mode and its separation from the other room modes. Ideally, the room modes should be equally spaced across the entire frequency range without any large gaps between them and without any significant overlap. In rooms the size of a typical domestic living room this ideal situation never happens. In a typical room the bandwidth of a room resonance mode will be approximately 5 Hz. This means that the modes that are within 5 Hz of each other will fuse, reinforce each other and be clearly audible. Modes that are separated more than 20 Hz will also be clearly audible, because the gaps between them are too large for the adjacent modes to fuse into a smooth response (Everest and Pohlmann, 2014; Gilford, 1959; Jones, 2015b).

In conclusion, the distribution of the axial and tangential modes over the frequency range, that is their number and spacings, influences the audibility of resonances. Figure 4.31 shows an example of the distribution of the axial and tangential modes for a room with dimensions 6 × 5 × 2.5 m (20 × 16 × 8 ft). At the lowest frequencies the modes are more widely separated than at the higher frequencies. This is typical for any room. At the lowest frequencies the modes are widely spaced and thus clearly audible. Above 300 Hz the modes are so closely spaced that they fuse into one continuous sound and are never individually audible. Between these two extremes of the low-frequency range, the axial modes are spaced such that their gaps are being filled in by a large number tangential modes. Thus, the audibility of room resonance is often only a problem at the lowest frequencies of the low-frequency range. The larger the room, the lower these problematic frequencies are. Therefore, room resonances are not a problem in very large venues such as cinemas and concert halls, but they can be very annoying in a relatively small home theater or listening room (Everest and Pohlmann, 2014; Gilford, 1959).

Taking control of problematic room resonance modes is not easy. It involves a combination of different remedies, not all of them being equally successful or practical. Possible remedies are (Walker, 1992):

1. *Optimize the dimensions of a rectangular room:* Achieve an even distribution of room modes by choosing appropriate dimensional ratios. Avoid cubical rooms and rooms whose dimensions are multiples of each other's. Optimize the dimensions such that no resonances coincide or are separated by less than 5 Hz and such that resonances are not separated by more than 20 Hz.
2. *Use a nonrectangular room:* Tame room resonances by splaying walls or using an irregularly shaped room. Nonparallel room boundaries tend to skew resonance modes.
3. *Change the position of the loudspeakers and listeners:* Strategically position the loudspeakers and listeners, taking into account the pressure distribution of the resonance modes in the room. Avoid placing loudspeakers and listeners at pressure maximums and minimums.
4. *Add low-frequency absorbers to the room:* Increase the absorption at low frequencies to attenuate room resonance modes.
5. *Use electronic equalization:* Reduce the amplitude of room modes by selectively decreasing certain frequencies in the audio signal that is fed to the loudspeakers.

The first two remedies are not very practical as they require you to rebuild your listening room or do some significant remodeling. They are also not very effective. First, room resonances still exist in a room with nonparallel boundaries or any other irregular shape. Compared to a rectangular room the resonances are just more difficult to predict (Everest and Pohlmann, 2014; Winer, 2018), because they can no longer be calculated using the equation from box 8.1. Second, although room resonances for a rectangular room can be calculated quite easily, it turns out that such calculations are not a very good representation of what is really happening in the room. The low-frequency performance of the room cannot be accurately predicted based on only its dimensional ratios, because of several reasons (Toole, 2018): 1) No room is perfectly rectangular; 2) No room has infinitely rigid walls—practical walls will absorb some of the low-frequency energy, especially at windows and door openings; 3) Not all room modes will be excited by a similar amount, because the loudspeakers are generally not positioned in the corners of the room; 4) Not all room modes are equally audible, because the listeners are generally not positioned in the corners of the room.

Optimizing the shape or dimensions of the room does not help much in controlling room resonance modes. There is no uniquely good set of room dimensions. The literature is full of quasi-scientific guesses, and none of them is the absolute optimum (Everest and Pohlmann, 2014). Furthermore, it turns out that the corresponding optimized modal spacings do not correlate well with our subjective perceptions. Rooms with optimized dimensions do not always sound better (Fazenda et al., 2005; Wankling and Fazenda, 2009).

A more practical and effective approach to taking control of room resonances involves a clever combination of adding low-frequency absorbers to the room, optimizing the placement of loudspeakers and listeners, and applying some electronic equalization (Toole, 2018). The addition of low-frequency absorbers is a good way to dampen room resonances between 100 Hz and 300 Hz. It is difficult to create sufficient absorption below 100 Hz, as it requires very large acoustical devices. Reasonably sized bass absorbers are ineffective below 100 Hz (Hauser et al., 2008). Below 100 Hz it is much more practical to rely on electronic equalization and the careful placement of multiple subwoofers. I will explain this approach step-by-step in the upcoming section 8.1.3 on subwoofer calibration, but before that, let's have a more detailed look at bass absorbers.

8.1.2 Bass Absorbers

Room resonances can be tamed by increasing the absorption of low frequencies in the room. You can increase the absorption in your room by adding a couple of dedicated bass absorbers. These absorbers drain energy from the room resonances, making them less offensive. Adding low-frequency absorption is always a good idea. It lowers the pressure maximums and elevates the pressure minimums and thereby reduces the point-to-point variations in sound pressure throughout the room (Toole, 2018; Winer, 2018).

Rooms sound better when several bass absorbers are added. Not only does the bass response become flatter with less peaks and dips, also the prolonged decay times of room resonances are reduced. The time definition of the bass will be greatly improved and everything will sound much tighter and well-defined (Noxon, 1985). The more bass absorbers you add to your room, the smoother and tighter the bass will sound. It is up to you how much effort and expense you want to put in (Winer, 2018).

Most rooms benefit from broadband bass absorption, because problematic room resonances can occur throughout the entire low-frequency range from 20 Hz to 300 Hz. Most dedicated bass absorbers work over a wide range of bass frequencies falling off at some lower frequency depending on their thickness. The thicker the absorber the lower it will go. Most practical absorbers provide very little absorption below 100 Hz.

Sound absorbers come in two different types: *resistive absorbers* and *resonant absorbers*. Resistive absorbers consist of porous material like mineral wool, fiberglass, or foam. They offer resistance to the vibrations of air molecules. They must be located where particle movement exists, preferably at the location where particle velocity is at its maximum. Physics dictates that particle velocity is at its maximum when sound pressure is at its minimum. This occurs at a quarter wavelength from a room boundary. At 100 Hz the quarter-wavelength point lies at a distance of 0.86 m (2.8 ft) from the wall. Thus, a resistive absorber is most effective when placed a considerable distance from the wall. Attaching a resistive absorber directly on the wall is not effective, because at the wall the particle velocity is almost zero. Therefore, resistive absorbers are not a very practical choice for the absorption of low frequencies. They are better suited for the absorption of mid and high frequencies, as I will describe in more detail in section 8.2.2. For the absorption of low frequencies, you are much better off using the other type of bass absorbers: the resonant absorber (Cox and D'Antonio, 2016; Toole, 2018).

A resonant absorber removes sound energy from the room by moving in response to sound pressure and dissipating the sound energy mechanically. It must be located at a point in the room where sound pressure is at its maximum. This happens at the room boundaries. The most effective location for a resonant absorber is in one of the corners of the room, because the corners are high pressure points for all room resonances (Cox and D'Antonio, 2016; Toole, 2018). This makes the resonant absorber a very practical device for taming room resonance modes. There are two types of resonant absorbers that are commonly used: the *tube trap* (or *bass trap*) and the *Helmholtz resonator*. You can buy them from companies that sell acoustical treatment products or build them yourself.

A tube trap is a proprietary low-frequency absorber introduced by Acoustic Sciences Corporation (acousticsciences.com). It is a hollow tube of glass fiber given structural strength by a wire mesh frame. The ends of the tube are sealed such that the air can only move in and out through the resistive wall of glass fiber. This wall acts as a vibrating diaphragm against the spring of air inside the tube. Figure 8.3 shows the construction of a tube trap. Tube traps come in different diameters, and the larger ones with a diameter of 40 cm (16 inch) provide some good absorption even somewhat below 100 Hz (Everest and Pohlmann, 2014; Harley, 2015; Noxon, 1985).

A Helmholtz resonator is a low-frequency absorber that consists of a perforated panel in front of a closed chamber. The air trapped inside the holes of the perforated panel resonates against the spring of air trapped inside the chamber. A sheet of mineral wool is placed inside the chamber against the perforated panel to drain energy from the resonating air particles. Figure 8.4 shows the cross-section of a Helmholtz resonator. A variant of the Helmholtz resonator employs a panel that is constructed of closely spaced slats. The two types of panels are shown in figure 8.5 (Cox and D'Antonio, 2016). The slat-type absorber is easier to construct than the perforated one.

Figure 8.3
Construction of a tube trap. The tube trap is a hollow cylinder made of porous sound-absorbing material. A wire mesh is wrapped around this cylinder to give it structural strength. This wire mesh is covered with a fabric to give it a nice finished look. Both ends of the tube are sealed airtight with a rigid disk (not shown in the figure).

Figure 8.4
Cross-section of a
Helmholtz resonator.

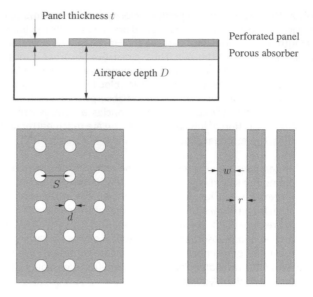

Figure 8.5
Two types of panels
employed in Helmholtz
resonators. *On the left,*
a perforated panel with
several holes with diameter
d spaced by a center-to-
center distance *S*. *On the
right,* a panel made out
of closely spaced slats of
width *w* spaced by slots of
width *r*.

BOX 8.2 PERFORATED HELMHOLTZ RESONATOR

The resonance frequency f_R of a perforated membrane resonator is given by (Szymanski, 2015):

$$f_R = \frac{c}{2\pi}\sqrt{\frac{\varepsilon}{t'D}}$$

where ε is the fraction of open area, D is the depth of the air space, t' is the effective panel thickness, and $c = 344$ m/s is the speed of sound.

For a perforated panel the fraction of open area is given by

$$\varepsilon = \frac{\pi}{4}\left(\frac{d}{S}\right)^2$$

where d is the diameter of the holes and S is the center-to-center distance between the holes (see figure 8.5). The effective panel thickness is given by

$$t' = t + \delta d$$

where t is the actual thickness of the panel and δ is a correction factor. This factor is often approximated as 0.85, but can be calculated for low values of $\varepsilon < 0.16$, as follows:

$$\delta = 0.8(1 - 1.4\sqrt{\varepsilon})$$

A Helmholtz resonator provides low-frequency absorption over only a narrow band of frequencies around its resonance frequency. The resonance frequency can be tuned quite accurately by changing the depth of the cavity behind the panel and varying the size of the holes or slats. Increasing the depth of the cavity lowers the resonance frequency, while increasing the holes or the space between the slats increases the resonance frequency. The design equations for the perforated Helmholtz absorber are given in box 8.2, and the ones for the slat Helmholtz absorber in box 8.3. You can use these equations to design and build your own

---

BOX 8.3 SLAT HELMHOLTZ RESONATOR

The resonance frequency f_R of a slat resonator is given by (Szymanski, 2015):

$$f_R = \frac{c}{2\pi}\sqrt{\frac{\varepsilon}{t'D}}$$

where ε is the fraction of open area, D is the depth of the air space, t' is the effective panel thickness, and $c = 344$ m/s is the speed of sound.

For a slat absorber the fraction of open area is given by

$$\varepsilon = \frac{r}{w+r}$$

where r is the slot width and w is the slat width (see figure 8.5). The effective panel thickness is given by

$$t' = t + 2\delta r$$

where t is the actual thickness of the panel and δ is a correction factor. This factor can be calculated as follows:

$$\delta = -\frac{1}{\pi}\ln\left(\sin\left(\frac{\pi\varepsilon}{2}\right)\right)$$

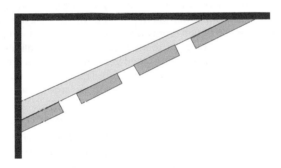

Figure 8.6
A Helmholtz resonator built in a corner of the room provides broadband low-frequency absorption because of its varying depth.

absorbers. You can build a stand-alone device with a cabinet constructed from wood or MDF (medium density fiberboard), or you can build the Helmholtz resonator directly into a wall.

Standard Helmholtz resonators are frequency selective: they only absorb low frequencies in a narrow band surrounding their resonance frequency. You can use the Helmholtz resonator to tame a particular room resonance by tuning it such that its resonance frequency equals the resonance frequency of the room mode. While this approach is useful to tame one or two obnoxious room resonances, you do not want to build a dedicated Hemlholtz resonator for every room resonance. You are better off adding some broadband low-frequency absorption to your room to deal with multiple room resonances at once. You broaden the frequency band in which the Helmholtz resonator works by using a variable depth for the air space. The most practical way to do this is to build your Helmholtz resonator in one of the corners of the room (Szymanski, 2015), as shown in figure 8.6.

Every room benefits from some broadband low-frequency absorption. It is not practical to treat each room resonance separately. Therefore, I recommend that you focus on achieving

Figure 8.7
Best places in a room to install low-frequency absorbers. The eight corners indicated by the *large dots* are prime positions for bass absorbers, because room resonances in all three dimensions are attenuated. It is common to install bass absorbers in two or more corners covering an entire wall intersection from floor to ceiling. These positions are indicated by the *solid lines*. If more bass absorption is needed the other intersections indicated by the *broken lines* are the best places to add next.

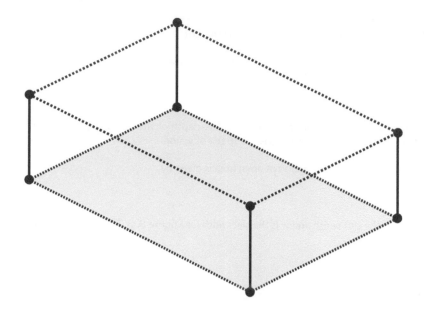

adequate broadband low-frequency absorption in your room by installing several tube traps or corner-built Helmholtz resonators. As I already explained, these absorbers are most effective when installed at the locations where the room resonances attain maximum pressure. The eight corners of the room, as indicated in figure 8.7, are the best places, because at these corners all room resonances achieve maximum pressure. The next best position is along the entire length of the intersection of two room boundaries, as indicated in figure 8.7. There are 12 of these intersections in a rectangular room: four where each wall meets another wall, another four where each wall meets the floor, and yet another four where each walls meets the ceiling (Everest and Pohlmann, 2014; Winer, 2018). These intersections and corners are often also the most practical places to install these relatively large devices.

Recommendation 8.1 *Install low-frequency absorbers at the locations in the room where the room resonances attain maximum pressure. The best positions are the eight corners of the room followed by the 12 intersections of the walls, floor, and ceiling.*

How many low-frequency absorbers do you need? For broadband absorption it is best to use either large tube traps or large corner-based Helmholtz resonators to cover an entire intersection of two room boundaries. The tube traps should have a diameter of 40 or 50 cm (16 or 20 inch). The Helmholtz resonators should be 60 cm (24 inch) wide with a design frequency of about 100 Hz for the average depth of their triangular cross-section (Everest and Pohlmann, 2014; Winer, 2018). To obtain sufficient low-frequency absorption you need at least a 1 m (3.3 ft) long tube trap or corner-based Helmholtz resonator for every 14 m³ (500 ft³) of room volume (Noxon, 1985). For example in a room with dimensions 6 × 5 × 2.5 m (20 × 16 × 8 ft) you need to cover at least 5.4 m (18 ft) of wall intersection. This means that for most domestic rooms it is sufficient to cover only two wall intersections from floor to ceiling. It is best to first treat the two intersections behind the front loudspeakers, as it can improve the stereo image (Everest and Pohlmann, 2014). If you desire more low-frequency absorption you could also cover the other two wall intersections from floor to ceiling. And, if you want to go even further you could also treat all the wall to ceiling intersections. The more bass absorption you add, the more even the low-frequency response will be.

Recommendation 8.2 *Cover an entire intersection of two room boundaries with a large tube trap or corner-based Helmholtz resonator. The tube traps should have a diameter of 40 or 50 cm (16 or 20 inch). The Helmholtz resonators should be 60 cm (24 inch) wide with a design frequency of about 100 Hz for the average depth of their triangular cross-section. You need at least a 1 m (3.3 ft) long tube trap or corner-based Helmholtz resonator for every 14 m³ (500 ft³) of room volume.*

Low-frequency sound may also be absorbed by the walls of the room and large pieces of furniture. Rigid walls of concrete or brick will absorb almost nothing, while walls made of wooden panels or gypsum board may contribute significantly to the low-frequency absorption in the room (Bradley, 1997; Toole, 2018). Nonrigid walls act as resonant absorbers and absorb sound over a narrow frequency range. This type of absorber is called a *membrane absorber*. Wall paneling, ceiling tiles, windows, doors, and the panels of large pieces of furniture all act as membrane absorbers. However, it is difficult to predict their resonance frequency as the design equations for membrane absorbers tend to be quite inaccurate (Cox and D'Antonio, 2016; Szymanski, 2015). Therefore, it is best not to rely on them and add some appreciable broadband absorption to your room using tube traps or corner-based Helmholtz absorbers.

8.1.3 Subwoofer Calibration

It is very difficult to achieve sufficient low-frequency absorption in the frequency range reproduced by the subwoofers. Below 100 Hz an adequate bass absorber would become unacceptably large. This is unfortunate, because in most rooms the room resonance modes in the subwoofer frequency range tend to be clearly audible and annoying, the reason being that the number of room resonances in this frequency range is sparse with large gaps between the modes (see for example figure 4.31). Instead of adding huge bass absorbers to your room, a different strategy for dealing with these room resonances is needed.

The best way to deal with room resonances in the subwoofer frequency range is to use a combination of proper placement of the subwoofers in the room with high-resolution electronic equalization (Fazenda et al., 2012; Toole, 2018; Welti and Devantier, 2006). Many A/V controllers and receivers and many subwoofers are equipped with a parametric equalizer that lets you alter the audio signal that is reproduced by the subwoofers. Alternatively you could use a stand-alone equalizer device that is connected between the subwoofer output of the A/V controller (or receiver) and the inputs of your subwoofers. A parametric equalizer can be used to correct peaks and dips in the frequency response. Such an equalizer consists of multiple equalization filters that can be tuned using three parameters: 1) an adjustable gain to boost dips or reduce peaks, 2) the center frequency that determines where to apply this gain, and 3) the width of the frequency band around the center frequency. Figure 8.8 illustrates these three parameters.

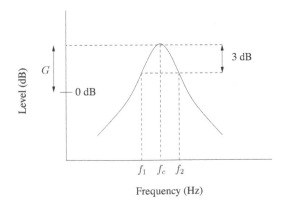

Figure 8.8
The three parameters of a filter in a parametric equalizer: the gain G, the center frequency f_c, and the bandwidth $f_2 - f_1$.

It is important to stress that it is really a combination of subwoofer placement and equalization that gets the job done. Equalization alone will improve your low-frequency response, but it has its limitations. There are three challenges in getting a better bass response, that can be better handled when using the combined approach. They are (Wilson et al., 2003):

1. The low-frequency response varies significantly with the position of the listener in the room, which means that the use of a parametric equalizer can improve the frequency response at one position in the room, but at the same time make it worse at some other position.
2. The low-frequency response is likely to have several large dips that cannot be boosted using only a parametric equalizer, because too large a boost would cause the subwoofer amplifier to run out of power and cause your subwoofer driver to reach its excursion limit.
3. Generally, some of the dips in the low-frequency response are nonminimum phase and therefore difficult to correct completely using a parametric equalizer.

Since the frequency response varies significantly from one listening location to the other, it is difficult to get consistently good bass for multiple listeners in the room. Electronic equalization changes the low-frequency response at all listening positions in exactly the same manner. Therefore, it cannot be used to change the variations between different listening positions. What you can do to reduce those variations is to use multiple subwoofers and pay attention to where you position them (Toole, 2018; Welti and Devantier, 2006).

In the combined approach of subwoofer placement and equalization, the first goal is to minimize the variation in the low-frequency response from one seat to the next in the listening area. You can reduce the seat-to-seat variation significantly by optimizing the positions of multiple subwoofers in the room. The idea is to make the responses at multiple seats as similar to each other as possible. The second goal is to use electronic equalization to make these responses as flat as possible. A practical approach is to take the average of the low-frequency responses of all the different seats and use this average to adjust the parametric equalizer. In this way, the low-frequency response can be optimized for all listeners in the room instead of only the one seated at the prime listening position (Welti, 2002; Welti and Devantier, 2006).

Since large dips in the low-frequency response are difficult or even impossible to correct using electronic equalization, they should be dealt with by carefully positioning the subwoofers. Dips in the response are usually the result of placing the subwoofer or listener too close to a pressure minimum of one of the room's resonances.

For the combined approach to work you need to be able to perform your own acoustical measurements. Without these measurements you have no idea what is happening in your room. Merely relying on general advice on subwoofer placement, such as the advice that I gave in section 4.2.6, will not provide the best results. Even in simple rectangular rooms, the room resonance patterns will deviate from the theoretical predictions given in section 8.1.1. Unequal wall constructions, door openings, windows, and large pieces of furniture will distort the patterns. The only way to gain insight into what is really happening is to make acoustical measurements at different seats. Taking such measurements with your subwoofers in different positions puts you in a strong place to make informed decisions about their optimal placement in the room (Toole, 2018).

Recommendation 8.3 *Improve the low-frequency response of your subwoofers using a combined approach in which you first select the best positions for your subwoofers and then apply parametric equalization. Use your own acoustical measurements to make informed decisions on where to put the subwoofers and how to tune the parametric equalizer.*

I do not recommend using the automatic room correction feature found in many A/V receivers, controllers, and subwoofers to optimize the in-room frequency response of your subwoofers.

Due to its automatic nature, it doesn't give you any insight into what is really happening in your room. You cannot use it to determine the best positions for your subwoofers; you just have to accept its automatically calculated electronic corrections. As I explained in section 7.5.6 the automatic room correction feature can be useful to correct frequency-response anomalies below 300 Hz. It can surely further improve the sound of your system, since it covers a larger frequency range than the range reproduced by your subwoofers. However, it is best to only apply it as a final optimization step after you have first optimized the positions of your subwoofers using the approach that I describe below.

Acoustic Measurements

Performing your own acoustic measurements might seem intimidating, but I urge you to give it a try. It is actually not that hard. It doesn't require any specialized equipment other than a measurement microphone. You could even use the built-in microphone of an SPL meter as a measurement microphone. In addition you need a computer running software that allows you to take acoustic measurements; examples are Acourate and Room Equation Wizard (see appendix B). Room Equation Wizard (REW) is a free program that is easy to use. It includes tools for generating audio test signals and measuring frequency responses. REW also includes analysis tools, and it can also automatically calculate the settings of a parametric equalizer to counter the effects of room resonance modes and match the frequency responses to a target curve. It is a versatile tool that comes with a comprehensive user manual. I highly recommend using REW to help you find out where to put your subwoofers and how to tune your parametric equalizer.

It is important that the frequency-response measurements that you use have sufficiently high resolution. Traditional measurements that are averaged over a frequency range of 1/3 of an octave are not suitable. They are blind to the rich details of room resonances. You need high-resolution measurements (accurate to at least 1/20 of an octave) to correctly identify center frequencies and bandwidths of room resonance modes (Fazenda et al., 2012; Toole, 2018; Winer, 2018). The measurement software mentioned above is able to provide such high-resolution measurements.

Recommendation 8.4 *Only use high-resolution acoustic measurements (accurate to at least 1/20 of an octave) to analyze the low-frequency response of your room and subwoofers.*

To make a measurement with the computer you need to connect the measurement microphone to the audio input of the computer's sound card. You also need to connect its audio output to an audio input on your A/V receiver or controller to be able to play the test signals that the computer generates. You should configure your A/V receiver or controller such that during the measurements only the left and right front loudspeakers and the subwoofers are active. This usually means putting the receiver or controller into two-channel stereo mode. You don't want to measure only the subwoofers, because you also want to analyze how the sound from the subwoofers integrates with the sound of the front loudspeakers. In your measurement software you should set the upper frequency limit of the measurements to 200 Hz (Winer, 2018).

Recommendation 8.5 *To analyze the reproduction of low frequencies in your room, measure the combined response of the subwoofers and the left and right front loudspeakers from 20 Hz to 200 Hz.*

To measure the low-frequency response in the room, you place the measurement microphone at the listening position that you want to measure. Recall that the response varies with position, so you need to make a measurement for each seat in your room. Point the microphone upward to the ceiling. Since small movements of the microphone can result in variations in the measurements, I recommend that you mount the microphone on a tripod. Do not stand between the subwoofers and the microphone during the measurements. Also make sure that there are no obstructions and large reflecting surfaces in the vicinity of the microphone. Refrain from talking during the measurements and make the room as quiet as possible to avoid disturbing the measurements.

Recommendation 8.6 *To measure the low-frequency response in the room mount the measurement microphone on a tripod, point it at the ceiling, and place it at one of the listening positions. Do not stand between the subwoofers and the microphone, and make the room as quiet as possible.*

Subwoofer Placement

The first step in optimizing the low-frequency response of your subwoofers is to find the best locations for them in the room. As I already explained in section 4.2.6 the position of the subwoofers and the listeners determines how the resonance modes are excited. When a subwoofer is placed close to a pressure maximum, it stimulates the corresponding room resonance, resulting in a peak in the frequency response. On the other hand, when a subwoofer is placed close to a pressure minimum, it doesn't stimulate the corresponding room resonance and a dip may occur in the frequency response. Of course, the position of the listeners also influences the frequency response that they experience. When a listener is placed close to a pressure maximum a peak occurs in the frequency response, and when a listener is placed close to a pressure minimum it results in a dip.

Finding the best place for the subwoofers and listeners involves some trial and error. You start by placing them at a position that seems to be practical for your particular room. The generic configurations shown in figure 4.45 and figure 4.46 and the guidelines presented in section 4.2.6 provide a good starting point. You then measure the frequency response at each seat in the listening area. Next, you have a look at the measurements that you obtained and identify the frequencies at which major peaks and dips occur. It is essential to try to relate these frequencies to particular room resonance modes, because it allows you to make informed changes to the positions of the subwoofers and listeners. In most rooms the axial resonances dominate the frequency response. In rooms with very stiff walls the first tangential mode may also be relevant. The other tangential and the oblique modes can be neglected (Welti and Devantier, 2006). Thus, in a rectangular room you can get a rough idea by computing the frequencies of the axial modes. The actual frequencies of the peaks and dips in the measurements might differ somewhat from the theoretical predictions due to nonrigid walls, doors, windows, and other physical differences between your room and a theoretically rectangular one. If your room has some other shape, for example an L-shape, you might still be able to relate the measurements to axial resonance modes by computing them for the different distances that exist between the walls of your room. If you have a slanted wall or sloped ceiling, you can compute the axial modes based on the average distance between the opposing room boundaries.

When you have determined which room resonances are responsible for the peaks and dips in your frequency-response measurements, you can alter the position of the subwoofers and listeners to try to improve low-frequency reproduction in your room. The goal is to obtain a good low-frequency reproduction at all the seats in the room. All the listeners in the room should be able to experience good-quality bass, not only the one seated at the prime listening location. This means that you have to make some compromises. The goal is to find positions for the subwoofers and listeners such that the frequency responses at the different listening positions are as similar as possible. In other words: your goal is to reduce the seat-to-seat variations. Your goal in this first step is not to make these frequency responses as flat as possible. Peaks that occur in all of them can be tamed using electronic equalization, which is the second step in optimizing the low-frequency response of your subwoofers. In addition to finding positions that reduce the seat-to-seat variations it is also important to try to have as few dips as possible in the frequency responses, because they are difficult to correct with electronic equalization.

Recommendation 8.7 *Experiment with different positions of the subwoofers and listeners in the room. Measure the frequency response at each listening position and try to move your subwoofers and listeners such that these frequency responses are as similar as possible and have the least number of dips.*

You can manipulate the excitation of room modes and hence change the peaks and dips in the frequency response by moving the subwoofers and listeners towards or away from pressure maximums and pressure minimums (Groh, 1974). Typically, you need to make a change of at least 0.5 m to see any appreciable effect on the low-frequency response (Fazenda et al., 2012). In most rooms, all the listeners are seated at the same height and the subwoofers are placed on the floor. In this case the influence of the floor/ceiling room modes will be the same for each listener, and thus not of any concern in this first step of optimizing the low-frequency reproduction (Welti, 2002). Moving the listeners and loudspeakers around in the horizontal plane only influences the room resonance modes between the different walls of the room. You can reduce a peak in the frequency response by moving a subwoofer closer to the corresponding pressure minimum in the room. Alternatively, you can use two subwoofers placed on opposite sides of a pressure minimum to reduce the offending room resonance by driving it destructively (Toole, 2018; Welti, 2002). Figure 8.9 shows these two possibilities. You can also move the listeners with respect to the pressure minimums in the room, but you should avoid placing them at or too close to a pressure minimum. At these positions in the room, the frequency response shows a large dip (up to −25 dB) at the corresponding resonance frequency (Welti and Devantier, 2006; Winer, 2018).

Recommendation 8.8 *Manipulate the excitation of an offending room mode by placing a subwoofer at or near a pressure minimum, or by placing two subwoofers on opposing sides of a pressure minimum.*

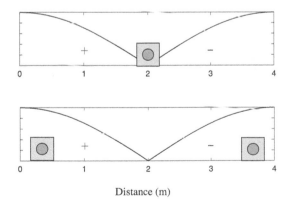

Distance (m)

Figure 8.9
Two examples of subwoofer placement with respect to the pressure distribution of room resonances. The placement is such that the first axial mode from the example in figure 4.41 is attenuated. *At the top,* **the subwoofer is placed at a pressure minimum of the room resonance.** *At the bottom,* **two subwoofers are placed at opposite sides of a pressure minimum where the sound pressure changes polarity. As a result the subwoofers drive the corresponding room resonance destructively.**

Based on Toole (2018), figure 8.13.

As I explained in section 4.2.6 the use of multiple subwoofers, preferably two or four, smooths out the bass in the room. It greatly helps to reach the goal of reducing the seat-to-seat variations in the low-frequency responses. As an added benefit it often also reduces the number of dips in the frequency responses. Let's have a look at an example. Figure 8.10 shows a rectangular room with two subwoofers and one row of three listeners. When only one of the subwoofers is active, the frequency responses at the three listening positions have a few large dips. Figure 8.11 shows the responses when only the left subwoofer is active, and figure 8.12 shows the responses when only the right subwoofer is active. By contrast, when both subwoofers are active the large dips disappear as shown in figure 8.13. Also note that with both

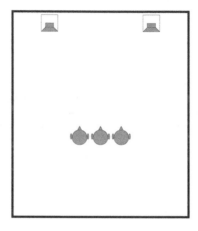

Figure 8.10
Example setup with two subwoofers and three listeners for the measurements presented in figures 8.11–8.13. The room is approximately 4.7 by 5.5 m (15 by 18 ft) and 2.4 m (8 ft) high. The listeners are 3.4 m (11 ft) from the front wall.

Figure 8.11
Subwoofer responses at three different listening positions with one subwoofer located at the front wall, 110 cm from the left sidewall.

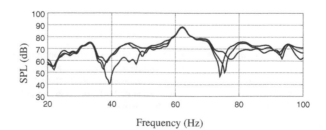

Figure 8.12
Subwoofer responses at three different listening positions with one subwoofer located at the front wall, 110 cm from the right sidewall.

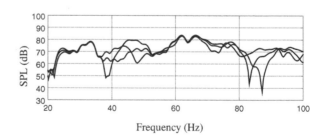

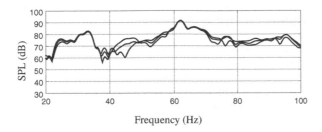

Figure 8.13
Subwoofer responses at
three different listening
positions with two
subwoofers located at the
front wall: one subwoofer
110 cm from the left
sidewall and the other 110
cm from the right sidewall.

subwoofers active the differences between the three frequency-response curves become smaller, proving that the use of multiple subwoofers reduces the seat-to-seat variations.

Subwoofer Equalization

The second step in optimizing the low-frequency response of your subwoofers is to use electronic equalization to reduce the peaks in the frequency response. Applying the same equalization to all subwoofers is called *global equalization* and is the most straightforward option. It is important to realize that global equalization changes the frequency response for all positions in the room at the same time. You cannot use global equalization to reduce the seat-to-seat variations. You can only use global equalization to reduce the peaks that are common among the frequency responses for the different seats. Once you try to counter a peak that occurs at only one seat, you create a dip in the frequency response of the other seats.

When you have multiple subwoofers, it is also possible to use a different equalization setting for each subwoofer. This gives you more flexibility, but also tends to complicate matters. Tuning the equalization for each individual subwoofer can take up a lot of time as it is an elaborate trial and error process. However, it can be used to further reduce the seat-to-seat variations. If you want to try it out, it is best to tackle it as a two-step procedure. In the first step you adjust the gain, delay, and equalization of each individual subwoofer such that the low-frequency responses at the different seats in the room are as similar as possible and contain the least number of dips. In the second step, you use global equalization to reduce the peaks in the frequency responses for the different seats. Make sure that while experimenting with different settings for different subwoofers you always measure the combined response of all subwoofers playing together. Measuring and optimizing one subwoofer at a time does not work.

The main idea of global equalization is to use the equalizer to reduce the peaks that are common among the frequency responses for the different seats. A practical way to determine those peaks is to average the different frequency responses at each measurement frequency. Measurement software, like REW (see appendix B), can calculate this average frequency response for you. REW can also automatically determine the settings for your equalizer by comparing the (averaged) measurements to a flat target curve. In order to be able to counter a number of resonance modes, your equalizer should have a sufficient number of individually configurable filter sections, preferably six or more.

Measurement software, like REW, can automatically calculate the gain, center frequency, and bandwidth for each filter of your global parametric equalizer such that the upward thrusting peaks are being countered. You use these calculated filter settings to configure your parametric equalizer. After you have configured your equalizer, it is a good idea to remeasure the frequency responses. Sometimes the automatically calculated settings need some manual fine-tuning. One or two iterations usually suffice. Some equalizers use a bandwidth parameter in octaves to set the width of a peak, while others use the Q parameter. In box 8.4 I explain the relation between the two.

BOX 8.4 BANDWIDTH AND Q

A parametric equalizer can be used to boost or reduce peaks in the frequency response. Such a filter can be characterized by three parameters: 1) the gain, either positive or negative; 2) the center frequency f_c at which it operates; and 3) the bandwidth over which it operates (see figure 8.8). The bandwidth can be specified in two different ways: in octaves N or as the quality parameter Q. The quality parameter is defined as:

$$Q = \frac{f_c}{f_2 - f_1}$$

where f_1 and f_2 are the frequencies on both sides of the center frequency f_c where the amplitude has dropped to -3 dB (see figure 8.8). The bandwidth in Hz is given by $f_2 - f_1$. This can be expressed in the number of octaves N between f_2 and f_1 as follows:

$$N = \frac{\log_{10}(f_2 / f_1)}{\log_{10} 2}$$

because one octave corresponds to the doubling of frequency, or to put it differently $f_2 = 2^N f_1$.

Figure 8.14
Subwoofer response at the prime listening position with *(thick line)* and without electronic equalization *(thin line)*.

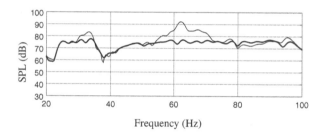

You should adjust the parametric equalizer such that the resulting frequency response is as flat as possible. However, peaks and dips should be treated differently. Narrow dips should be left alone, because they cannot be corrected by electronic equalization. These dips are the result of a pressure minimum and can only be removed by changing the pressure distribution in the room or moving the listeners to a better location in the room. Trying to correct narrow dips by electronic equalization may damage your subwoofers: the subwoofer's amplifier will run out of power and will drive the subwoofer into severe distortion. While narrow dips should be left alone, broad depressions in the frequency response may be safely boosted if the gain is limited to about 6 dB (Toole, 2018).

Recommendation 8.9 *Use a parametric equalizer to reduce the peaks in the in-room frequency response of the subwoofers. Focus on the peaks that are common among the frequency responses for the different seats. Do not attempt to correct narrow dips in the response. Broad depressions may be boosted if the gain is limited to about 6 dB.*

An example of the effect of electronic equalization on the in-room subwoofer response is shown in figure 8.14. It compares the frequency response with and without equalization for the prime listening seat in the example room of figure 8.10. Note that the equalizer removes the large bump that occurs between 55 Hz and 70 Hz. The equalizer settings have been calculated using the REW software. Figure 8.15 shows how these settings change the frequency

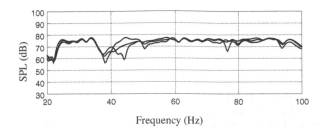

Figure 8.15
Subwoofer responses
at the three listening
positions with electronic
equalization.

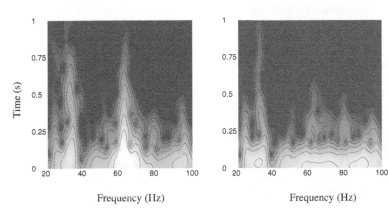

Figure 8.16
Spectrogram of the
subwoofer response at the
prime listening position
before equalization *(left)*
and after equalization
(right). The darker the plot,
the lower the SPL.

responses for the three different seats in the room. Compared to the unequalized responses in figure 8.13 the curves are much flatter now.

In the example of figure 8.15 there is still a significant dip in the response around 40 Hz, and the three responses for the different seats also differ around this frequency. It is difficult to correct this in this particular room. The dip and the response differences are due to room resonance modes that operate along the width of the room, that is from sidewall to sidewall. Since each of the three seats is located at a slightly different distance from the sidewalls, the responses at these seats differ. The dip turns out to be uncorrectable by electronic equalization as this part of the frequency response is *nonminimum phase*. Electronic equalization is only effective for parts of the response that are *minimum phase*. A response is called minimum phase if the amplitude versus frequency, and phase versus frequency are uniquely related. Correcting the amplitude of a minimum phase response also alters the phase response, and therefore improves the behavior in time. By contrast, in the case of a nonminimum phase response, amplitude correction alone cannot correct the phase response (Newell and Holland, 2007). The REW software allows you to judge whether a dip is minimum phase or nonminimum phase. It calculates the so-called *excess group delay*. The frequency regions for which the excess group delay is a flat curve are the regions that are minimum phase. If part of this curve is not flat, this part is nonminimum phase and not correctable by electronic equalization.

Recommendation 8.10 *Do not attempt to use electronic equalization to correct dips in the frequency response that are nonminimum phase.*

The peaks in the subwoofer frequency response that are the result of room resonances are always minimum phase. This means that changing the amplitude of the peak also changes the phase response. In fact, flattening the peak in the amplitude response changes the phase such that the time behavior of the room resonance is improved: it reduces the decay times of the room resonance (Toole, 2018). This is fortunate, because the reduced decay times strongly improve the perceived reproduction quality of low frequencies (Fazenda et al., 2012). Let's have a look at an example. Figure 8.16 shows how the electronic equalization applied in figure 8.14 changes the decay times of the room resonances. It shows the spectrograms of the low-frequency response before and after equalization. The spectrogram is a plot that shows how the frequency response changes in time. Figure 8.16 clearly shows that the electronic equalization has significantly reduced the decay times of the resonances at 35 Hz and 65 Hz.

The easiest way to apply electronic equalization to your subwoofers is to use the built-in parametric equalizer of your A/V receiver, A/V controller, or subwoofer. If none of these devices has an equalizer built-in, you can use a stand-alone parametric equalizer. You connect this device to the subwoofer output of your A/V receiver or controller and the line input of your subwoofer. Modern stand-alone equalizers are digital devices. They first convert the analog subwoofer signal from your A/V receiver or controller to a digital audio signal. Then they apply their correction to this digital signal and convert it back to an analog signal that can be fed to the subwoofer. These additional analog-to-digital and digital-to-analog conversions will not perceptibly degrade sound quality as the low frequencies reproduced by your subwoofers are less prone to such degradations than the mid and high frequencies (Harley, 2015). There is no need to worry about that. However, there are two other things that need your attention. First, some digital equalizers invert absolute polarity or introduce a slight time delay. Both of these will alter the phase of the sound reproduced by your subwoofers. Therefore, you should check if they are still in phase with your main loudspeakers and if not take appropriate measures (see section 7.5.3). Second, you need to carefully tune the output level of the subwoofer output of your A/V controller or receiver. This control is often found in the speaker-level menu and labeled 'subwoofer level'. To maximize the signal-to-noise ratio it is important to set it as high as possible, but not so high that you distort the input signal to the equalizer. Most equalizers have an input level indicator or a distortion indicator that can help you find the highest volume setting that will not lead to distortion. Use the volume control of the subwoofer itself (and not the subwoofer level in the menu of your A/V controller or receiver) to calibrate the output level of the subwoofers to the output level of the main loudspeakers. I explained how to match these levels in section 7.5.4.

Subwoofer Integration

After you have found the best positions for your subwoofers and reduced the peaks in the frequency response by electronic equalization, your next step is to improve the integration between the subwoofers and main loudspeakers.

First, you need to fine-tune the relative levels of the subwoofers and the main loudspeakers (see also section 7.5.4). For a seamless integration, the subwoofers should audibly extend the low-frequency range without being too loud. When the subwoofers attract too much attention it is a clear sign that they are too loud. Many people are tempted to set the subwoofer level too high, because initially lots of bass will sound impressive. However, after some time listening to it will become fatiguing and everything you play sounds heavy and bloated.

After you have flattened the low-frequency response with electronic equalization, it is usually necessary to increase the subwoofer level by a few decibels. The reduction of the major resonance peaks significantly reduces the perceived amount of bass (DellaSala, 2010). Use the acoustic measurements to find out how much. Look at the measured frequency response of one of the listening positions and check whether the average sound level stays about the same from 20 Hz to 300 Hz. The average sound level for the frequencies below the subwoofer crossover frequency should closely match the average level above the crossover frequency. If they don't match adjust the subwoofer level. The sound level measurements give you a good indication of the appropriate level settings, but it is always a good idea to verify the subwoofer level by listening to some of your favorite music. If it doesn't sound completely right, a minor adjustment of the subwoofer level should be all that is needed to correct it.

Recommendation 8.11 *Use acoustic measurements to set the subwoofer level. The average sound level for the frequencies below the subwoofer crossover frequency should closely match the average level above the crossover frequency.*

The second step in realizing proper integration between the main loudspeakers and the subwoofers is to make the frequency response in the crossover region as flat as possible. Although in theory the crossover filter in your A/V receiver or controller should ensure a smooth transition (see figure 4.39), the acoustical effects of the room can significantly change this. In the region around the crossover frequency the adjacent-boundary effect and room resonances often significantly change the response (Toole, 2018). You can only find out what is really happening by taking your own acoustical measurements. Based on such measurements you can play around with the time-delay and phase settings of the subwoofers to improve the integration with the main loudspeakers (see section 7.5.3). You could also try to improve the response by changing the subwoofer crossover frequency (see section 7.5.2) or apply some additional electronic equalization. Free-standing loudspeakers often have a dip in their frequency response around 100 Hz, which arises from the reflection of the wall behind the loudspeakers (Mäkivirta and Anet, 2001). You could try to move this dip by adjusting the distance to the nearby room boundaries as I explained in section 4.2.5. The key takeaway is to use the acoustic measurements to make some informed changes and improve the integration between the subwoofers and the main loudspeakers.

Recommendation 8.12 *Use acoustic measurements to improve the integration between the main loudspeakers and the subwoofers in the crossover region. Play around with the subwoofer's time-delay, phase, and crossover frequency.*

8.2 Reproducing High and Mid Frequencies in Rooms

At mid and high frequencies sound travels like rays. Sound rays originate from the loudspeakers and travel towards your ears, either directly or after first being reflected once or multiple times in the room. As I explained in section 2.3.4 the sound that you finally hear consists of three parts: 1) the direct sound from the loudspeakers; 2) the early reflections from the walls, the floor, and the ceiling that arrive typically within 20 ms of the direct sound; and 3) the late reflections due to sound that has been reflected multiple times within the room.

The early and late reflections alter the perception of sound in a room. Thus, you can change the quality of sound reproduction in a room by controlling these reflections. Good acoustical room design involves the proper combination and placement of reflecting surfaces, absorbers, and diffusers (Cox and D'Antonio, 2016). Absorbers remove sound energy from the room by turning it into heat. Absorbers for mid and high frequencies are made of porous materials, such as fabrics, glass fiber, and foams. Furniture made of such materials, for example upholstered chairs and sofas, acts as sound absorbers in your room. Drapes and carpets will also absorb some sound energy. Diffusers scatter a single sound ray arriving from one direction into several lower-intensity sound rays that propagate in multiple directions. Any object in the room that has an irregular reflecting surface can act as a diffuser, provided that its surface irregularities are of the same order as the wavelength of the impinging sound (Szymanski, 2015).

You can buy all sorts of acoustical products, such as acoustical tiles and specially designed diffusers to change the acoustics of your room. Fortunately, most domestic rooms do not need extensive acoustical treatment to sound great. Often a large gain in sound quality can be achieved by just adding, moving, or removing carpets, drapes, and furniture. Further improvements can be obtained installing a few acoustical tiles or diffusers at strategically chosen locations in the room. It is not necessary, nor desirable from an aesthetic point of view, to cover your entire room with acoustical materials.

Some rooms sound live, and some sound dead. It all depends on the amount and distribution of sound-absorbing material in the room. For example your bathroom will sound very live. The tiled walls and floor will not absorb any sound. The sound in your bathroom will be bright, hard, and thin. By contrast, a room that has its walls and floor completely covered in thick carpeting will sound very dead. The carpet will absorb nearly all the high and mid frequencies.

High-quality sound reproduction requires a room that sounds neither too live nor too dead. It is common to characterize the liveness or deadness of a room by the *reverberation time T_{60}*, which is the time it takes for sound to decay by 60 dB. However, for the reverberation to be meaningful all the listeners need to be within a sound field that is dominated by reverberation. This is the case in large performance spaces, such as concert halls and large cinemas, but never in smaller rooms such as the typical domestic listening room or home theater. In these smaller rooms the sound field is dominated by the direct sound and several strong early reflections. The late reflections that create reverberation are much more diminished (Everest and Pohlmann, 2014; Toole, 2006, 2018).

In a small room the impression of liveness or deadness depends on the ratio of direct to reflected sound and the timing of the early reflections. Adjusting the acoustics of small rooms involves absorbing and diffusing early reflections. Since the sound that reaches the listeners is not dominated by the reverberation of the room, the reverberation time is not a useful indicator of the quality of reproduced sound (Toole, 2018; Zhang, 2015).

8.2.1 Controlling Early Reflections

The early reflections at mid and high frequencies are the second-loudest sound sources in the room. They travel towards your ears and alter your perception of the direct sound. They also propagate further throughout the room and contribute to the sonic signature of the late reflections that will arrive at your ears. As I explained in section 2.3.4, we humans generally like reflections. The right amount of reflections creates a sense of space, enriches the timbre of the sound, and improves speech intelligibility. A room without any reflections sounds dull and lifeless. However, reflections can also have a detrimental effect on sound quality. They can reduce clarity, distort imaging, and alter the tonal balance. Whether reflections are beneficial or harmful depends on their frequency content, the angle from which they arrive, their relative intensity with respect to the direct sound, and the delay between the direct sound and the reflected sound.

There are two different schools of thought in dealing with reflections. The traditional approach is to absorb or diffuse all early reflections, such that they cannot alter the sound quality of the direct sound. The other approach is to leave the early reflections as they are, such that they enhance the spatial experience and enrich the timbre of the reproduced sound. Which approach you choose seems partly a matter of personal preference. However, another factor at play is the quality of the loudspeakers used. The quality of loudspeakers has improved enormously over the years. More than ten years ago even the best loudspeakers had trouble maintaining a consistent frequency response off-axis. As a result the early reflections from the nearest room boundaries had a frequency content that was significantly different from the direct sound. Getting good sound from these loudspeakers was only possible by careful positioning and absorbing most of the early reflections. Today's loudspeakers are much better in that their off-axis sound is very similar to the direct sound. Using these modern loudspeakers greatly reduces the detrimental effect of early reflections. Only the strongest reflections may cause some harm (Toole, 2018).

Early reflections strongly influence the spatial quality of the sound reproduction (see sections 3.4.4 and 3.4.5), in particular the sense of acoustical space and the apparent dimension and

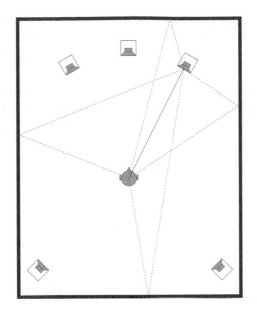

Figure 8.17
Early reflections created by the sound from the right front loudspeaker. The sidewall reflections arrive from angles that contribute to image broadening (ASW), while the reflections off the front and back wall do not.

location of sound sources within that space. One of the problems with reproducing music or movies through a loudspeaker is that many sounds in real life occupy a much larger space than the relatively small loudspeaker. The early reflections contribute to image broadening, the effect that the apparent source width (ASW) of a sound appears larger than the loudspeaker. Image broadening makes the reproduced sound bigger and more believable. It requires early reflections that arrive within 80 ms of the direct sound and have significant frequency content above 500 Hz (Toole, 2018). Image broadening is generated by sidewall reflections that arrive from an angle between 45° and 135° on either side of the listener (Hidaka et al., 1996). A large contribution comes from the reflections arriving from the front loudspeakers in the front half of the room.

Not all early reflections have a beneficial spatial effect; it strongly depends on the direction from which they arrive. In a typical loudspeaker configuration, such as the one shown in figure 8.17, the reflections from the three front loudspeakers that bounce off the sidewalls of the room arrive from the appropriate angles. They contribute to the image-broadening effect and the sense of spaciousness. By contrast, the reflections off the front and back wall are of little value spatially, because they do not arrive from the sides, they arrive from angles that fail to contribute to the image-broadening effect.

The higher its relative level and the lower the delay, the stronger a reflection and the more easily its effect can be heard. The relative level of a reflection depends on the reflective properties of the surface and on the path difference between the direct sound and the reflected sound. Most hard surfaces reflect the sound almost completely, so this leaves us with the path difference. The longer the path difference, the lower the level of the reflection. In addition to lowering the level, the path difference also introduces a time delay between the direct sound and the reflected sound. The longer the path difference, the larger the delay.

In general, for the front loudspeakers you have the option to either absorb the reflections from the sidewalls, or let them be. You should decide between these two options based on the amount of spaciousness that you prefer. It is best to absorb the reflections off the front and back walls of the room. The front wall is the most important one, because usually the front loudspeakers are quite close to this wall. Hence, the reflections will have a relatively high level and a short delay. Covering the area between the front loudspeakers on the front wall with sound-absorbing material significantly improves image localization and reduces coloration (Everest and Pohlmann, 2014). If you are situated well away from the back wall then absorption of the reflections off this wall is optional. However, if you are sitting close to the back wall then the reflections will need to be absorbed, because they have a relatively high level and a short delay. Strong reflections off the back wall will not only alter the timbre of the reproduced sound, but also significantly degrade the definition of sound images in the frontal soundstage.

Recommendation 8.13 *Absorb the reflections from the front loudspeakers that bounce off the front and back walls of the room. Absorption of the reflections that bounce off the sidewalls is optional and depends on the amount of spaciousness that you prefer.*

The surround loudspeakers also produce reflections off the walls of the room as shown in figure 8.18. The reflections from the surround loudspeakers off the back and sidewalls improve the spatial quality of the reproduced sound. Their effect is not limited to increasing spaciousness. They also improve the feeling of immersion (see section 3.4.5), because they arrive from angles that contribute to listener envelopment (LEV). The reflection off the front wall arrives from an unproductive angle. It will be reduced in level by several dBs, because of the large distance that the sound has to travel to reach the listener's ears. If you heeded the previous advice to cover the area between the front loudspeakers on the front wall with sound-absorbing material, then this reflection will be even further reduced in level. This is beneficial as this reflection does not contribute to the spatial quality of the reproduction.

The spatial effect of the reflections from the surround loudspeakers off the back and sidewalls can be further increased by scattering them in the horizontal plane. To this end special acoustical diffusers could be installed on the sidewalls and towards the sides on the back wall, as illustrated in figure 8.19. Diffusers also help to cover a larger listening area with nonlocalizable sound from the surround loudspeakers (Cox and D'Antonio, 2016; Toole, 2018).

Recommendation 8.14 *Absorb the reflections from the surround loudspeakers that bounce off the front wall of the room. The reflections off the back wall and the sidewalls should never be absorbed. These reflections can be diffused to further increase the sense of immersion in a larger listening area.*

Figure 8.18
Early reflections created by the sound from the right surround loudspeaker. Except for the reflection off the front wall, the early reflections arrive from angles that contribute to image broadening (ASW) and listener envelopment (LEV).

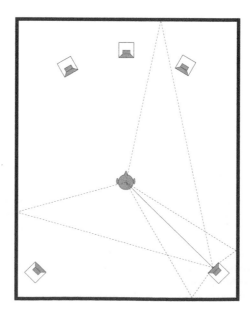

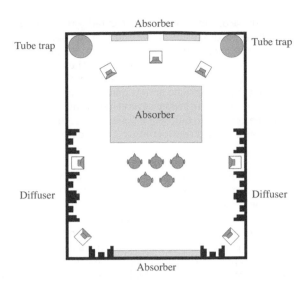

Figure 8.19
Complete acoustical
treatment of the listening
room.

Adapted from Toole (2018), figure
15.10.

Reflections from the front and surround loudspeakers will also bounce off the floor and the ceiling. Reflections from the front loudspeakers should be absorbed, while those from the surround loudspeakers can be left alone or diffused. Assuming that the high-frequency drivers of your left and right front loudspeakers are at seated ear height (where they should be as explained in section 4.2.3), the reflection points are exactly halfway between you and the loudspeakers (Winer, 2018). The center loudspeaker is usually mounted a bit lower, which means that its reflection point on the floor is a bit closer towards the loudspeaker and that its reflection point on the ceiling is a bit closer towards the listener. The gray rectangle between the front loudspeakers and the listeners in figure 8.19 indicates the area on the floor and the ceiling that should be covered with absorptive material.

Recommendation 8.15 *Absorb the reflections from the front loudspeakers that bounce off the floor and the ceiling. For the surround loudspeakers it is best to leave the reflections off the floor and the ceiling alone.*

It is common practice to absorb the floor bounce by covering the floor between the listeners and the front loudspeakers with carpet. Absorbing the reflections off the floor improves clarity and dialog intelligibility (Harley, 2015). The type of carpet can affect the sound quality; some carpet materials are more absorptive than others. I describe the different options in section 8.2.2. It is best not to cover the entire floor area in carpet as that may make the room sound too dead and dull. The same is true for the ceiling; covering only the part between the listeners and the front loudspeakers should be sufficient. A patch of acoustic ceiling tiles should do the job. An absorptive ceiling creates a sense of intimacy and a feeling of being in a small space. If the room is already small to begin with adding additional absorption to the ceiling may not be desirable. In such a case it is better to install some diffusers on the ceiling between the listeners and the front loudspeakers, such that the reflections from the front loudspeakers are scattered towards the sidewalls (Jones, 2015b; Szymanski, 2015).

A complete acoustical treatment of the listening room is summarized in figure 8.19. As I mentioned earlier, it is quite normal that in a multipurpose living room compromises have to be made. Sometimes acoustical treatments can be cleverly integrated in the design of the room, but the acoustical treatments outlined in figure 8.19 are often more at home in a dedicated listening room than in a multipurpose living room. It is often not necessary to install all these acoustical treatments in your living room to enjoy high-quality sound, but in a dedicated listening room that is sparingly furnished it can certainly improve the sound quality.

The general principle for the acoustic design of a dedicated listening room is that the front end of the room should be acoustically absorptive and that the back of the room should be acoustically reflective with plenty of diffusers. This is sometimes called the *dead-end/live-end* design principle. The only exception to this principle is that the central portion of the back wall should be absorptive to get rid of the early reflections from the three front loudspeakers. This is especially important if your listening area is relatively close to the back wall. If your room is so large that all the listeners are seated several meters away from the back wall, then the entire back wall can be covered in diffusers instead of absorbers. Such a diffusive back wall is often employed in larger studio control rooms (Cox and D'Antonio, 2016; Everest and Pohlmann, 2014).

A complete acoustical treatment of all the surfaces of the room is almost never necessary. Too much absorption will make the room sound dull and lifeless. Therefore, the guiding principle is to add only a minimum of absorptive material. Reflections only come from certain particular areas on the walls. It is sufficient to use an absorptive panel of about 60 by 60 cm (2 by 2 ft) that is centered at the reflection point. Use a mirror to easily determine the reflection points. Take a seat at the prime listening location and ask an assistant to move the mirror along the wall. The reflection point is the spot where the mirror reflects the high-frequency driver of one of the loudspeakers. If you have multiple listeners, you should repeat this procedure for every seat in the room and make sure you cover all the reflection points (Everest and Pohlmann, 2014; Harley, 2015; Rushing, 2004; Winer, 2018). I will describe different acoustical treatments and common domestic materials that can be used for absorption in section 8.2.2.

Similarly, there is no need to cover entire walls with diffusers. The optional diffusers on the back wall and the sidewalls should be installed at ear height, such that they form a horizontal band around the room that extends about 60 cm (2 ft) above and below seated ear height (Toole, 2018). I will describe suitable acoustical devices that can be used for diffusion in section 8.2.3.

8.2.2 Porous Absorbers

As I explained in the previous two sections absorbers are used to fine-tune the acoustics of your listening room. Porous absorbers are used to absorb sound energy by converting it into heat. Porous absorbers are made of any material where sound propagation occurs in a network of interconnected pores in such a way that the vibrating air molecules cause frictional losses. These absorbers are usually made of fuzzy, fibrous materials in the form of boards, foams, fabrics, carpets, or acoustical tiles. Common examples of porous materials are natural fibers (cotton and wool), glass fibers, mineral wool, and open cell foams (Cox and D'Antonio, 2016; Everest and Pohlmann, 2014; Szymanski, 2015).

Porous absorbers differ in the amount of absorption that they provide. This amount depends on the type of material, its surface area, its thickness, and the angle of incidence. The amount of absorption is called the *absorption coefficient*. It is a measure for the efficiency of a surface in absorbing sound averaged over all possible angles of incidence. It is expressed in sabins. An absorption coefficient of 1 sabin corresponds to a surface of 1 square meter of complete sound absorption: 100% of the sound is absorbed (Cox and D'Antonio, 2016; Everest and Pohlmann, 2014; Szymanski, 2015). A typical example of such a 100% absorber is an open window, because sound passing through it never returns to the room. Most materials have an absorption coefficient between 0 and 1, but values greater than 1 are also possible if diffraction around the edges makes the surface appear acoustically larger than it really is (Everest and Pohlmann, 2014). A typical example of that is a wedge-shaped foam that presents a larger surface area to the sound than its area of wall contact (Harley, 2015).

The absorption coefficient of porous absorbers varies with frequency. Porous absorbers are typically good at absorbing mid and high frequencies, but not so good at the low end of the frequency spectrum. Absorbing frequencies below 200 Hz requires special bass absorbers,

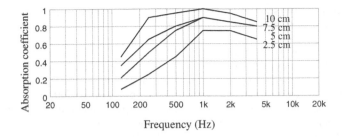

Figure 8.20
The effect of thickness on the absorption of a glass fiber panel. Increasing the thickness of the panel increases the absorption, especially at the upper-bass frequencies.

Data from Cox and D'Antonio (2016), appendix A.

such as the ones described in section 8.1.2. Absorbing low frequencies above 200 Hz is possible with porous absorbers but requires a considerable thickness (Cox and D'Antonio, 2016). Figure 8.20 shows the absorption coefficient as a function of frequency for glass fiber panels of different thickness. Significant absorption (70%) down to 300 Hz requires a panel of at least 7.5 cm (3 inch) thick.

The thickness of an absorber has the most influence on its absorption coefficient. There is little difference in absorption between glass fiber panels of different density. Natural cotton panels have absorption characteristics that are comparable to mineral fiber panels, while open cell foams are usually of a lower density than fibrous materials and tend to have slightly lower absorption coefficients (Szymanski, 2015). However, differences in absorption coefficients among these materials are small and rarely relevant as they are often simply the result of slightly different testing conditions (Toole, 2018). Rigid glass fiber boards are the most widely used porous absorbers and the product of choice for professional installers. Their mechanical strength makes them easier to install compared to pieces of flexible batting.

I recommend using porous absorber panels of at least 7.5 to 10 cm (3 to 4 inch) thick to deal with early reflections (as described in section 8.2.1). This thickness is needed to make sure that the absorber is effective at all frequencies above the transition region of the room (see section 2.3.4), which in most rooms lies somewhere between 200 and 300 Hz. Thinner absorbers don't go down to 300 Hz and only absorb the mid and high frequencies. They distort the spectral balance of the early reflections and thereby degrade sound quality (Toole, 2018).

> **Recommendation 8.16** *Absorber panels used to control early reflections should be at least 7.5 to 10 cm (3 to 4 inch) thick.*

Another way to increase the effectiveness of an absorber panel at low frequencies is to space it a few centimeters from the wall instead of mounting it directly on the wall. A 10 cm (4 inch) thick panel is about as effective around 300 Hz as a 5 cm (2 inch) thick panel that is spaced 10 cm (4 inch) from the wall. This simple trick can be used to save some money on acoustical materials (Everest and Pohlmann, 2014).

Acoustical absorber panels are often covered with fabric. One reason is that fibers of mineral wool and glass fiber panels can be respiratory and skin irritants. Stretching fabric over the panel prevents the fibers from escaping into the air. Though in practice the fibers are not likely to escape unless the material is disturbed. Another reason for covering the panel with fabric is purely aesthetic, because most acoustical treatments do not look very attractive. It is important that the fabric is acoustically transparent. You can easily test this by holding the fabric to your mouth and try to blow air through it. If you cannot blow air through it, the fabric is not suitable; it will block the sound and reflect it instead of allowing it to pass into the underlying fibrous absorber (Szymanski, 2015; Winer, 2018).

Absorptive panels are very practical to install on the walls and the ceiling, but less so on the floor. To absorb the early reflections off the floor it is way more practical to use carpet or a suitably sized rug. Carpet is an effective acoustical absorber, but the type of carpet that you use can make a huge difference in sound quality. Carpet is most effective at absorbing mid and especially high frequencies. The amount of absorption depends on the lengths of the fibers used. Clipped-pile, not loop-pile, of sufficient height yields the best results. Figure 8.21 shows that increasing the pile height increases the absorption at mid and high frequencies. Thus, again the thicker the carpet the better. Another factor that influences the amount of absorption is the type of backing that is used. If the backing is open celled, such as felt hair and foam, then the sound can further penetrate into the carpet, increasing its effective thickness for absorption. Impenetrable backings, such as rubber are to be avoided (Cox and D'Antonio, 2016; Szymanski, 2015).

Carpets have little absorption at mid and low frequencies. When they are used to absorb the early reflections off the floor they will distort the spectral balance as they are mostly absorbing high frequencies. Carpet can be turned into a more broadband absorber by using a thick felt or hair underlay to increase the absorption in the midrange. Figure 8.22 shows the effect of different underlays. It is best to use a heavy and porous underlay, such as a thick felt cushion (Cox and D'Antonio, 2016; Toole, 2018).

Recommendation 8.17 *When you use carpet as an acoustical absorber, it is best to use a clipped-pile type with high-pile fibers and an acoustically porous backing. To increase its absorption at mid frequencies it is best to lay it on a thick and heavy acoustically porous underlay, such as a felt cushion.*

Many domestic rooms and home theaters use curtains to block out the light. Curtains are essentially also porous absorbers. So, they could be used to control the acoustics of your listening room. Similar to carpet, curtains mainly absorb high frequencies. The heavier the material the more they absorb. Heavy velour is a popular choice. Lightweight open-weave fabrics absorb almost nothing (Toole, 2018). The absorption in the mid-frequency range

Figure 8.21
The effect of pile height on the absorption of woven wool carpet. Increasing the pile height increases the absorption.

Data from Cox and D'Antonio (2016), appendix A.

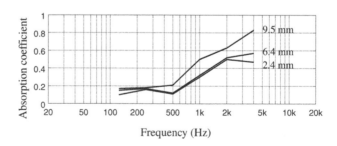

Figure 8.22
The effect of hair underlay on the absorption of loop-pile tufted carpet. Increasing the density of the underlay increases the absorption at the midrange frequencies.

Data from Cox and D'Antonio (2016), appendix A.

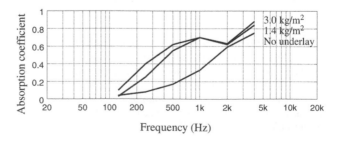

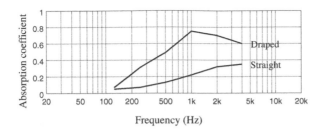

Figure 8.23
The effect of draping on the absorption of a medium velour curtain (0.475 kg/m²). Draping the curtains to half area increases the absorption.

Data from Cox and D'Antonio (2016), appendix A.

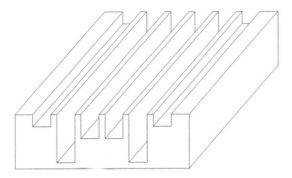

Figure 8.24
Example of a quadratic residue diffuser (QRD).

can be increased significantly by draping the curtains instead of hanging them straight. Figure 8.23 shows the difference. The deeper the drape fold, the greater the absorption. Absorption in the midrange can also be slightly increased by hanging the drapes 10 to 20 cm (4 to 8 inch) from the wall, but the effect is small compared to draping (Everest and Pohlmann, 2014; Szymanski, 2015).

8.2.3 Diffusers

Diffusers can be installed on the sidewalls and the back wall of your room to enhance the feeling of being enveloped in surround sound (see figure 8.19). This is especially useful when you have multiple listeners in the room. The diffusers scatter the sound from the surround loudspeakers in multiple directions. The ideal diffuser scatters the incident sound uniformly over a wide angular range for a broad range of frequencies. Geometrically shaped diffusers such as cylinders and triangles provide useful scattering, but only over a limited frequency range. Special surfaces with irregularities on the order of the wavelengths of the impinging sound will do a much better job (Cox and D'Antonio, 2016; D'Antonio and Cox, 1998a, 1998b; Szymanski, 2015).

Recommendation 8.18 *Consider installing diffusers on the sidewalls and the back wall if you have a large listening room with multiple seats.*

The *quadratic residue diffuser* (QRD) is a widely used special surface that diffuses sound spherically in the horizontal direction (Cox and D'Antonio, 2016; Szymanski, 2015). It consists of a sequence of wells of the same width but different depths that are separated from each other by thin dividers. An example of a QRD is shown in figure 8.24. The QRD is a member of a family of diffusers known as *reflection phase gratings* or *Schroeder diffusers*. The well depths of these diffusers are based on a special number sequence. The sound enters each of the wells, gets reflected, and propagates out of the wells. The different depths of the wells cause different delay times and the resulting interference between the delayed reflections forms a complex pattern. This pattern can be optimized by choosing the well depths according to a special mathematical number sequence. Schroeder (1975) showed that by choosing a quadratic residue sequence, the energy reflected into each direction is the same. Different quadratic residue sequences are being used in practical diffusers. Figure 8.25 shows some of them. The formula for the sequence is given in box 8.5.

Figure 8.25
Quadratic residue sequences
for $N = 5$, $N = 7$, and $N = 11$.

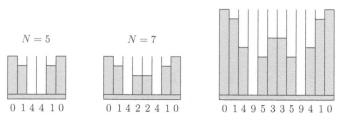

$$N = 11$$

$$N = 5$$

$$N = 7$$

0 1 4 4 1 0   0 1 4 2 2 4 1 0   0 1 4 9 5 3 3 5 9 4 1 0

BOX 8.5 QUADRATIC RESIDUE SEQUENCE

A quadratic residue sequence $s_0, s_1, s_2, \ldots, s_N$ can be computed from

$$s_n = n^2 \text{ modulo } N$$

where N is a prime number and *modulo* indicates the least nonnegative remainder. A prime number is any number other than 0 or ±1 that is not divisible without remainder by any other integer, except ±1 and the integer itself.

BOX 8.6 QRD FREQUENCY RANGE

The width of the wells of a QRD determines the highest frequency for which it is still effective (Cox and D'Antonio, 2016):

$$f_H = \frac{c}{2w}$$

where w is the width of the well in meters and $c = 344$ m/s is the speed of sound. The maximum well depth of a QRD determines its design frequency f_0 as follows:

$$f_0 = \frac{s_{max}}{N} \frac{c}{2d_{max}}$$

where d_{max} is the maximum well depth in meters, s_{max} the largest sequence number, and N the prime number on which the quadratic residue sequence is based. The design frequency f_0 determines the lowest frequency for which the QRD is still effective. QRDs are effective to about two octaves below the design frequency. Thus, the lowest frequency is given by

$$f_L \approx f_0 / 4$$

The frequency range over which a QRD works is determined by the width and maximum depth of its wells. I explain in box 8.6 how you can calculate this frequency range. The width of the wells determines the highest frequency for which the QRD still works: the wider the wells, the lower this highest frequency. Manufacturing constraints put a practical limit of 2.5 cm (1 inch) on the width of the wells (Everest and Pohlmann, 2014). This corresponds to a highest frequency of approximately 7 kHz. Above this frequency the theory on which the QRD design is based is no longer applicable, but in practice the QRD still diffuses sound above this frequency (Cox and D'Antonio, 2016).

The main goal of diffusers is to enhance the feeling of being enveloped in surround sound. Listener envelopment (LEV) is governed by late reflections that arrive more than 80 ms later than the direct sound, come at you from the side and the back, and have significant frequency content between 100 Hz and 1000 Hz (see section 3.4.5). Therefore, the diffusers should be effective between 100 Hz and 1000 Hz (Toole, 2018). For a QRD with $N = 7$, this means that the wells should be no wider than 17 cm (6.5 inch) and that the maximum well depth should be 25 cm (10 inch) or more.

Recommendation 8.19 *A diffuser that is used to diffuse the reflections from the surround loudspeakers should operate over a frequency range from 100 Hz to 1000 Hz. This means that a QRD with $N = 7$ used for this purpose should have wells that are no wider than 17 cm (6.5 inch) and have a maximum well depth of 25 cm (10 inch) or more.*

A diffuser works optimally in the far field where the different reflections from the wells combine into a uniformly dispersed sound field. Therefore, listeners should be positioned as far away from a diffuser as possible. This can be a challenge in a small listening room or home theater. Ideally the minimum distance between a listener and a diffuser with a lowest frequency of 100 Hz should be 3 m (10 ft). Consequently, the use of diffusers is only recommended in larger listening rooms that have multiple rows of listeners.

Recommendation 8.20 *Keep a minimum distance of 3 m (10 ft) between a diffuser and any of the listeners in the room.*

Different manufacturers sell diffusers in different sizes and designs, but you can also build them yourself. Make sure that your own design follows the sequences given in figure 8.25 and that it works in the frequency range between 100 Hz and 1000 Hz. To facilitate the design of your own QRD you can use the QRDude software listed in appendix B.

9

SUMMARY

In the previous chapters I have given you practical advice for optimally setting up an audio-video system in your living room, dedicated home theater, or any other small room. Each chapter focused on a particular part of the audio and video reproduction chain. In this final chapter I have grouped all the advice from the previous chapters in a different way that resembles practical tasks, such as buying equipment, designing the room, placing loudspeakers, installing equipment, setting up the system, and playing content. Based on the task at hand you will find a comprehensive overview of all the important points. Bear in mind that you do not need to rigorously follow all the advice. In real life practical constraints may often interfere with the advice given and compromises have to be made, especially when a room is used for multiple purposes, aesthetic considerations are taken into account, and budget is limited. The detailed explanations of the underlying principles in the previous chapters should help you to understand the reasons behind the advice. Understanding these reasons is key if you need to make compromises, find alternatives, and separate the rigid requirements from the more flexible ones. I am convinced you will be able to follow at least some of my advice and as a result enjoy an improved reproduction quality. And maybe this motivates you to get rid of some compromises and to implement more of the advice I have given. If you do, chances are that you will enjoy your favorite music and movies even more. Experiment and have fun!

9.1 Buying Equipment

Buying a Flat-Panel Display

- Your display should be capable of displaying 3840×2160 pixels (UHD, 4K), or at least 1920×1080 pixels (HD, 1080p) (Recommendation 6.1 on page 169).
- Choose your screen size and viewing distance using figure 4.12 such that you achieve a viewing angle of at least 21° with an HD flat-panel display and an angle of at least 31° with a UHD flat-panel display (Recommendation 4.10 on page 80).
- Your display should at least be capable of displaying all the colors in the Rec. 709 color gamut with 8-bit precision, and for UHD and HDR all the colors within the DCI-P3 gamut with 10-bit precision (Recommendation 6.4 on page 171).
- Your flat-panel display should be capable of achieving a peak luminance of 100 cd/m² for SDR content and 1000 cd/m² for HDR content (Recommendation 6.2 on page 170).
- To achieve a high-contrast ratio, the blacks on your display should have the lowest possible luminance. Aim for 0.05 cd/m² or lower (Recommendation 6.3 on page 170).
- To achieve high contrast without any blooming on a flat-panel display choose an OLED model (Recommendation 6.6 on page 174).
- To achieve high contrast with minimal blooming on a flat-panel display choose a direct-lit full-array LCD that features local dimming with many independently controlled zones (Recommendation 6.5 on page 173).
- Make sure your display supports both HDMI 2.0a and HDCP 2.2 if you want to be able to enjoy UHD content (Recommendation 5.25 on page 161).

Buying a Projector

- Your projector should be capable of displaying 3840 × 2160 pixels (UHD, 4K), or at least 1920 × 1080 pixels (HD, 1080p) (Recommendation 6.1 on page 169).
- For a more immersive viewing experience choose your screen size and viewing distance using figure 4.12 such that you achieve a viewing angle of at least 31° with an HD projector, and at least 40° with a UHD projector (Recommendation 4.10 on page 80).
- Your projector should at least be capable of displaying all the colors in the Rec. 709 color gamut with 8-bit precision, and for UHD and HDR all the colors within the DCI-P3 gamut with 10-bit precision (Recommendation 6.4 on page 171).
- Make sure that your projector provides enough lumen output for the size of your projection screen. The peak luminance on the screen should be at least 38–58 cd/m^2 (Recommendation 6.12 on page 180).
- Do not rely on the specifications of lumen output provided by the projector's manufacturer. The lumen output after calibration to a white point of 6500 K is often much lower (20–50%) (Recommendation 6.13 on page 181).
- To achieve a high-contrast ratio, the blacks on your display should have the lowest possible luminance. Aim for 0.05 cd/m^2 or lower (Recommendation 6.3 on page 170).
- Choose an LCoS projector over an LCD projector to get better black levels and increased contrast (Recommendation 6.8 on page 177).
- Choose a DLP projector over an LCoS or LCD projector to get better black levels and increased contrast (Recommendation 6.9 on page 178).
- To avoid the rainbow effect with a DLP projector, choose a single-panel model that has either a BrilliantColor color wheel or uses three separate LED light sources for the primary colors. Or alternatively, use a DLP projector that has three separate panels, one for each primary color (Recommendation 6.10 on page 178).
- Make sure your projector supports both HDMI 2.0a and HDCP 2.2 if you want to be able to enjoy UHD content (Recommendation 5.25 on page 161).
- Choose a projector that does not produce more than 30 dBA fan noise (Recommendation 6.11 on page 179).

Buying a Media/Blu-Ray/DVD player

- When you buy a media player (computer, smart TV, A/V receiver), make sure it can play all the different types of media files that you require. Check compatibility for both the file formats and the data formats (Recommendation 5.16 on page 147).
- When you buy a disc player, make sure it can play the different types of optical discs that you require. Check compatibility for the physical formats, application formats, and media file formats (Recommendation 5.23 on page 158).

Buying an A/V Controller

- Buy an A/V controller or receiver that has high-quality DACs. Before buying the device listen to it to evaluate its sound quality. Use the same power amplifier and loudspeakers that you have at home (Recommendation 7.35 on page 269).

Buying an Amplifier

- Before you buy an amplifier, always listen to how well this amplifier drives your loudspeakers (in the store or at home). It should sound neutral without any sense of strain or harshness, even at high volumes (Recommendation 7.20 on page 248).
- When you buy an amplifier, make sure it complies with the technical specifications listed in table 7.3 (Recommendation 7.21 on page 248).
- The amount of amplifier power that you need depends on the sensitivity of your loudspeakers and the listening distance. Use figure 7.16 to determine the amount of power

that you need to be able to produce a peak SPL of 96 dB at the listening position (Recommendation 7.18 on page 242).

- When comparing power specifications of multichannel amplifiers make sure that the power is listed as follows (Recommendation 7.19 on page 245):

  - Watts per channel (not the total sum).
  - With no more than 0.1% of THD (not 1% or 10%).
  - Over a bandwidth of 20 Hz to 20 kHz (not at 1 kHz, not from 40 Hz to 20 kHz).
  - Continuous power (not peak power).
  - With all channels driven (not just one or two channels).
  - Into a nominal load of 8 Ω (not 6 Ω or 4 Ω).

Buying Loudspeakers

- Before you buy loudspeakers, always listen to them to evaluate their sound quality. Select loudspeakers that sound as neutral as possible (Recommendation 7.2 on page 224).
- When evaluating and comparing loudspeakers at the dealer, make sure that during the listening tests (Recommendation 7.3 on page 225):

  - The loudspeakers under evaluation are placed at the same positions in the room.
  - You sit at the same position in the room for each evaluation.
  - You sit at the same distance from the loudspeakers as you do at home.
  - You sit in a chair that has the same height as the one you use at home.
  - The loudspeakers are driven by the same amplifier that you have at home.

- Use recordings with a broad and dense frequency content when performing listening tests to evaluate the sound quality of loudspeakers. In addition, use speech and vocals, solo piano, and violin. Listen for upper-bass, midrange, and high-frequency colorations, and other reproduction flaws (Recommendation 7.4 on page 226).
- Your main loudspeakers should have a flat and smooth on-axis frequency response between 40 Hz and 20 kHz. They should also have smooth off-axis frequency responses, with the 30°–45° response closely following the on-axis response and not deviating more than 4 dB below 10 kHz (Recommendation 7.5 on page 229).
- Your main loudspeakers should have low nonlinear distortion levels that do not exceed the values given in table 7.2 (Recommendation 7.6 on page 232).
- Your main loudspeakers should each be able to produce a peak SPL of at least 96 dB at the listening position (Recommendation 7.7 on page 233).
- Prefer a small front loudspeaker over a larger one if they both have the same retail price. The smaller one will likely have a better sound quality (Recommendation 7.8 on page 234).
- Choose front loudspeakers having a low-/mid-frequency driver that has a diameter of at least 10 cm (4 inch) (Recommendation 7.9 on page 234).
- Choose a center loudspeaker that has a driver arrangement that is horizontally symmetrical. Vertical driver arrangements (shown in figure 7.12) provide the best results. Alternatively, a horizontal design with a vertically stacked high-frequency and midrange driver (shown in figure 7.13b) can be used (Recommendation 7.11 on page 236).
- Prefer a center loudspeaker that is identical to the other two front loudspeakers. Alternatively, make sure that the center loudspeaker is a timbre-matched model from the same manufacturer and model line (Recommendation 7.12 on page 237).
- Use conventional direct-radiating loudspeakers as your surround loudspeakers. Do not use multidirectional loudspeakers such as dipoles (Recommendation 7.13 on page 238).
- In the ideal situation all your surround loudspeakers are identical and they are also identical to the front loudspeakers. Since this requirement is often too stringent, at least aim for the following (Recommendation 7.14 on page 239):

  - All ear-level surround loudspeakers should be identical.
  - All overhead surround loudspeakers should be identical.
  - Ear-level and overhead surround loudspeakers should be timbre-matched.
  - Surround loudspeakers and the front loudspeakers should be timbre-matched.

Buying Subwoofers

- Your subwoofers should be able to reproduce low frequencies down to 20 Hz when placed in your listening room. This usually means that the anechoic frequency response should extend to at least 30 Hz (Recommendation 7.22 on page 254).
- Prefer subwoofers that have a closed-box enclosure over those with a bass-reflex enclosure (Recommendation 7.23 on page 255).
- Your subwoofers should together be able to produce a peak SPL of at least 108 dB (Recommendation 7.24 on page 256).
- Choose subwoofers with large drivers. In total you need at least four 9-inch drivers or two 13-inch drivers (Recommendation 7.25 on page 257).
- Make sure that the amplifiers in your subwoofers are at least as powerful as indicated in figure 7.27 (Recommendation 7.26 on page 258).

9.2 Designing the Room

Controlling Ambient Light

- Install window treatments such that you can darken the room (Recommendation 4.35 on page 109).
- If you use a flat-panel display, light the wall behind it such that the intensity of the light on this wall is 10% to 15% of the maximum light output of the display. For watching HDR content a maximum of only 5 cd/m^2 is recommended (Recommendation 4.36 on page 110).
- If you use a projection screen make the room as dark as possible to achieve maximum contrast. Also pick neutral, matte, and subdued colors for your room's surfaces, furnishing, and decorations. Consider painting the walls and ceiling a dark color to minimize the amount of light that is reflected back to the screen (Recommendation 4.37 on page 111).
- Control the background luminance and choose a neutral color in an area around the video display that is 53° high by 83° wide. Also avoid having any bright lights or reflecting objects in this area (Recommendation 4.38 on page 113).

Controlling Background Noise

- Identify the likely source of background noise and take measures to tame it if its SPL exceeds 35 dBA or 50 dBC in your room (Recommendation 4.39 on page 114).
- Listen to a low-frequency sweep that ranges from 15 Hz to 200 Hz to identify and locate any objects in the room that rattle or buzz. Take measures to silence these objects: remove, move, dampen, or fixate them (Recommendation 4.40 on page 114).

Choosing Viewing/Listening Positions

- Determine the distance to your screen based on its resolution. Sit as close as possible to your screen to maximize the sense of reality, but don't sit any closer than the optimal viewing distances in table 4.1 (Recommendation 4.9 on page 79).
- Choose your screen size and viewing distance using figure 4.12 such that you achieve a viewing angle of at least 21° with an HD flat-panel display and an angle of at least 31° with a UHD flat-panel display. For a more immersive viewing experience use a large projection system and aim for a viewing angle of at least 31° with an HD projector, and at least 40° with a UHD projector (Recommendation 4.10 on page 80).
- Make sure the distance between you and your main loudspeakers is large enough for the sound of the different drivers to integrate,. About three to ten times the height of the front of the loudspeaker cabinet should be sufficient (Recommendation 4.8 on page 76).
- Never put your prime listening position against the back wall. Sit at least 1.5 m (5 ft) from the back wall (Recommendation 4.5 on page 74).
- Make sure you arrange your seating such that it always includes the prime listening location that has the same distance to the left and right front loudspeakers (Recommendation 4.6 on page 75).

- Position all seats at least 1.5 m (5 ft) away from the front, side, and back walls (Recommendation 4.7 on page 75).
- Keep all seats within 10° from the sides of the screen as illustrated in figure 4.13 (Recommendation 4.11 on page 81).
- Make sure every seat has a clear line of sight to the screen (Recommendation 4.12 on page 82).

Controlling Room Acoustics

- Install low-frequency absorbers at the locations in the room where the room resonances attain maximum pressure. The best positions are the eight corners of the room followed by the 12 intersections of the walls, floor, and ceiling (Recommendation 8.1 on page 290).
- Cover an entire intersection of two room boundaries with a large tube trap or corner-based Helmholtz resonator. The tube traps should have a diameter of 40 or 50 cm (16 or 20 inch). The Helmholtz resonators should be 60 cm (24 inch) wide with a design frequency of about 100 Hz for the average depth of their triangular cross-section. You need at least a 1 m (3.3 ft) long tube trap or corner-based Helmholtz resonator for every 14 m³ (500 ft³) of room volume (Recommendation 8.2 on page 291).
- Absorb the reflections from the front loudspeakers that bounce off the front and back walls of the room. Absorption of the reflections that bounce off the sidewalls is optional and depends on the amount of spaciousness that you prefer (Recommendation 8.13 on page 304).
- Absorb the reflections from the surround loudspeakers that bounce off the front wall of the room. The reflections off the back wall and the sidewalls should never be absorbed. These reflections can be diffused to further increase the sense of immersion in a larger listening area (Recommendation 8.14 on page 304).
- Absorb the reflections from the front loudspeakers that bounce off the floor and the ceiling. For the surround loudspeakers it is best to leave the reflections off the floor and the ceiling alone (Recommendation 8.15 on page 305).
- Absorber panels used to control early reflections should be at least 7.5 to 10 cm (3 to 4 inch) thick (Recommendation 8.16 on page 307).
- When you use carpet as an acoustical absorber, it is best to use a clipped-pile type with high-pile fibers and an acoustically porous backing. To increase its absorption at mid frequencies it is best to lay it on a thick and heavy acoustically porous underlay, such as a felt cushion (Recommendation 8.17 on page 308).
- Consider installing diffusers on the sidewalls and the back wall if you have a large listening room with multiple seats (Recommendation 8.18 on page 309).
- Keep a minimum distance of 3 m (10 ft) between a diffuser and any of the listeners in the room (Recommendation 8.20 on page 311).
- A diffuser that is used to diffuse the reflections from the surround loudspeakers should operate over a frequency range from 100 Hz to 1000 Hz. This means that a QRD with $N = 7$ used for this purpose should have wells that are no wider than 17 cm (6.5 inch) and have a maximum well depth of 25 cm (10 inch) or more (Recommendation 8.19 on page 311).

9.3 Placing Loudspeakers

- Do not place loudspeakers horizontally, unless they are properly designed for it (Recommendation 7.10 on page 236).
- Remove the grilles from your loudspeakers for an improved reproduction of high frequencies (Recommendation 7.1 on page 224).
- Mount small loudspeakers on a rigid, nonresonant, and mechanically stable loudspeaker stand. The stand can be mass loaded to make it less prone to vibrations (Recommendation 7.15 on page 239).
- Use hardened steel spikes with narrow points to rigidly couple the loudspeaker stand to the floor (Recommendation 7.16 on page 240).

- Use four small lumps of compressed mastic, Blu-Tack, or a similar sticky gumlike material to couple the loudspeaker to the stand (Recommendation 7.17 on page 240).

Placing Front Loudspeakers

- Ideally all your main loudspeakers should be placed symmetrical with respect to the left and right room boundaries. Beyond that, symmetry for the front loudspeakers is more important than symmetry for the surround loudspeakers (Recommendation 4.3 on page 71).
- If your room has nonparallel surfaces (sidewalls or ceiling) position your front loudspeakers at the narrow end of the room (Recommendation 4.4 on page 73).
- Never compromise on the symmetry of the front loudspeaker arrangement. The prime listening position should form a triangle with the front left and right loudspeakers. Place the center loudspeaker and the video display centered between the left and right loudspeakers (Recommendation 4.2 on page 70).
- Put the center loudspeaker at the same distance from the prime listening position as the left and right front loudspeakers. The three front loudspeakers should form a gentle arc with the center loudspeaker exactly centered on the arc between the left and right loudspeakers. Alternatively, place the center loudspeaker between this position on the arc and the line connecting the left and right loudspeakers and use an appropriate time delay and level setting in your A/V controller or receiver (Recommendation 4.15 on page 83).
- Place the left and right front loudspeakers at the same height, such that the high-frequency driver is at ear height when you sit in the prime listening location: about 1.2 m or 4 ft from the floor (Recommendation 4.16 on page 84).
- Place the center loudspeaker beneath the video screen such that its height is as close as possible to the heights of the left and right front loudspeakers. The height difference between the high-frequency drivers should be less than 10 cm for each meter of distance between the center loudspeaker and the listener (1.26 inch for each foot) (Recommendation 4.17 on page 84).
- Place the left and right front loudspeakers 2 to 3 meters apart (6.6 to 10 ft), or up to 5 m apart (16 ft) in larger rooms (Recommendation 4.18 on page 85).
- Place the left and right front loudspeakers within 15° from the sides of the screen (Recommendation 4.19 on page 85).
- Place the left and right front loudspeakers at 30° from the center loudspeaker. If they end up too far away from the sides of the video screen, you can slightly reduce this angle up to a minimum of 22° (Recommendation 4.20 on page 86).
- Fine-tune the distances to the walls for the front loudspeakers to obtain the smoothest possible frequency response. Use figure 4.36 or 4.37 as a guide and make the distances to the nearby walls as different as possible (Recommendation 4.32 on page 99).
- Aim the front left and right loudspeakers at the audience and position the audience such that everyone is covered by the 30° beams from each of the three front loudspeakers (Recommendation 4.21 on page 87).
- Ensure that the front left and right loudspeakers have the same amount of toe-in (Recommendation 4.22 on page 88).
- Use a listening test to fine-tune the amount of toe-in of the left and right front loudspeakers. Listen with a disabled center loudspeaker to check for two-channel imaging and listen with an active center loudspeaker to check for three-channel imaging (Recommendation 4.23 on page 89).
- Avoid putting reflective objects in the direct vicinity of your front loudspeakers, especially in the area between the left and right loudspeakers (Recommendation 4.24 on page 90).

Placing Surround Loudspeakers

- Ideally all your main loudspeakers should be placed symmetrical with respect to the left and right room boundaries. Beyond that, symmetry for the front loudspeakers is more important than symmetry for the surround loudspeakers (Recommendation 4.3 on page 71).

- Put the surround loudspeakers at the angular positions recommended in table 4.3, such that they are placed symmetrically on the left and right sides of the audience (Recommendation 4.25 on page 92).
- Place all the surround loudspeakers at the same height, such that the high-frequency driver is about 1.2 m (4 ft) from the floor. Alternatively place them a bit higher to facilitate covering the entire audience, but never more than 1.5–1.8 m (5–6 ft) from the floor (Recommendation 4.27 on page 93).
- Aim the surround loudspeakers at the listening positions such that all audience members are within the 30° beams emerging from the surround loudspeakers. Use a listening test to fine-tune the toe-in of the surround loudspeakers to ensure that all audience members experience an enveloping surround sound (Recommendation 4.26 on page 92).
- Fine-tune the distances to the walls for the surround loudspeakers to obtain the smoothest possible frequency response. Use figure 4.36 or 4.37 as a guide and make the distances to the nearby walls as different as possible (Recommendation 4.32 on page 99).

Placing Overhead Surround Loudspeakers

- Place the overhead surround loudspeakers on the left and right sides of the audience at the same distance from the sidewalls as the front loudspeakers (Recommendation 4.28 on page 93).
- Mount the overhead loudspeakers in a 5.1.2 and 7.1.2 system at an angle of 80° as shown in figure 4.28, or alternatively within the 65°–100° range. Install the overhead loudspeakers in a 5.1.4 and 7.1.4 system towards the front and the back of the listener at an angle of 45° as shown in figure 4.29, or alternatively within the 30°–55° range (Recommendation 4.29 on page 93).
- Mount the overhead surround loudspeakers at least 1.2 m (4 ft) above the ears of the highest seated listener (Recommendation 4.30 on page 95).
- Dolby Atmos enabled loudspeakers can be used as an alternative to overhead loudspeakers in the configurations listed in table 4.4. Ensure that the area on the ceiling that lies between the listener and these loudspeakers is reflective and that the height of the ceiling is between 2.4 and 4 m (between 8 and 13 ft) (Recommendation 4.31 on page 96).

Placing Subwoofers

- Keep the frequencies below 80 Hz out of your main loudspeakers and use one or more well-positioned subwoofers to reproduce these low frequencies (Recommendation 4.1 on page 67).
- Use at least two subwoofers placed at two different positions in the room to get the smoothest bass response. Use the configurations from figure 4.45 as a starting point. If you have multiple rows of seats, consider using four subwoofers distributed around the room to minimize the variations from row to row. Use the configurations from figure 4.46 as a starting point (Recommendation 4.34 on page 108).
- If you only use a single subwoofer place it on the floor between the left and right loudspeakers close to the front wall and away from the sidewalls. Do not place it further than 60 cm (2 ft) away from the front wall, and do not place it at the pressure maximums and minimums of the room resonances that occur between the two sidewalls (Recommendation 4.33 on page 105).

9.4 Installing Equipment

Installing a Video Display

- Place the video display centered between the left and right loudspeakers (Recommendation 4.2 on page 70).
- Put the video screen flush with the front of the center loudspeaker, or put it behind the center loudspeaker, but never in front of it (Recommendation 4.14 on page 82).

- Place the video screen such that viewers don't have to look up more than 30° to see the top of the screen (Recommendation 4.13 on page 82).

Installing a Front-Projection System

- For maximum image brightness, do not zoom the projector's lens too far in, but instead place the projector closer to the screen. Also mount the projector on the ceiling and not on the floor (Recommendation 4.41 on page 116).
- Keep the front of the projector parallel to the projection screen. Use the lens shift controls to move the image up and down. Do not tilt the projector and do not use the digital 'keystone correction' feature from the projector's setup menu (Recommendation 4.42 on page 117).
- Focus the lens of the projector to achieve maximum sharpness. Use a test pattern with a lot of small details to check the result (Recommendation 4.43 on page 118).
- If you have a three-panel projector, check if the panels are correctly aligned using a test pattern that consists of a grid of white lines on a black background. Use the projector's setup menu to adjust the panels if necessary (Recommendation 6.7 on page 176).
- Use a projection screen that remains perfectly flat. Use either a fixed flat screen or a tab-tensioned retractable screen (Recommendation 6.14 on page 181).

Installing Cables

- Choose loudspeaker cables that have closely spaced conductors (Recommendation 7.28 on page 261).
- Keep the electrical resistance of your loudspeaker cables below 0.1 Ω by choosing thicker cables for longer lengths such that you do not exceed the maximum recommended lengths listed in tables 7.4 and 7.5. Keep the inductance of the cables low by avoiding cables that have a cross-sectional area that exceeds 4 mm² or that have an AWG number lower than 12 (Recommendation 7.29 on page 264).
- Keep your loudspeaker cables as short as possible, but also keep the cables for the left, center, and right front loudspeakers the same length and diameter. Similarly, keep the cables for each pair of left and right surround loudspeakers the same length and diameter (Recommendation 7.30 on page 264).
- Use loudspeaker cables that are terminated with air-sealed connectors that are gold-plated (Recommendation 7.31 on page 264).
- Use interconnect cables with heavy-braided copper shields, and keep them as short as possible (Recommendation 7.33 on page 267).
- Use interconnect cables with air-sealed RCA or XLR connectors that are gold-plated (Recommendation 7.34 on page 268).
- Organize your cables in the following way to keep potential electromagnetic interference to a minimum (Recommendation 7.32 on page 267):

  - Separate and bundle your cabling according to their type: loudspeaker cables, analog interconnects, digital interconnects, and power cables. Cable bundles should only contain cables of a single type.
  - Keep analog interconnects away from digital interconnects and power cables. Separate them as much as possible, but at least 25% of their parallel distance.
  - Keep loudspeaker cables away from digital interconnects and power cables. Separate them as much as possible, but at least 2.5% of their parallel distance.
  - Always keep a minimum distance of at least 2.5 cm (1 inch) between the different bundles.
  - If possible, cross cables from different bundles only at right angles.
  - Keep all cables as short as possible. Wind any excess cables into a figure-8 pattern.

- For reliable transmission of HD video over HDMI, use an HDMI cable that supports at least 10.2 Gbit/s (high speed). For UHD video, use an HDMI cable that supports at least 18 Gbit/s (premium high speed) (Recommendation 5.26 on page 164).

9.5 Setting Up the System

Setting Up a Video Display

- Only use a full-screen mode to display letterbox material. Do not attempt to get rid of the black bars using another display mode, because it will result in a cropped or geometrically distorted picture (Recommendation 5.1 on page 122).
- To avoid unnatural distortions of the picture, always use the pillarbox format to display material with a 4:3 aspect ratio on a 16:9 screen (Recommendation 5.2 on page 122).
- Configure your display such that it shows the entire video image with a one-to-one pixel mapping and without any overscan (Recommendation 6.33 on page 202).
- Turn off all automatic contrast enhancements on your display (Recommendation 6.38 on page 214).
- If possible, turn off automatic backlight adjustment and automatic iris adjustment on your display (Recommendation 6.39 on page 214).
- Turn on local dimming on LCD displays. Adjust the amount of local dimming such that you achieve the darkest blacks without seeing any noticeable halos or blooming around smaller bright parts in a dark scene (Recommendation 6.40 on page 215).
- Turn off all automatic color enhancements on your display (Recommendation 6.41 on page 215).
- Only use noise reduction on your display when you need it on low-quality material. Don't leave it on by default (Recommendation 6.42 on page 215).
- Use a test pattern that has black lines on a gray background to adjust the sharpness control of your display. Set the sharpness control as high as possible without introducing white halos around dark lines (Recommendation 6.34 on page 204).
- If your display has an item in the setup menu for the deinterlacing mode or progressive mode, set it such that film mode with cadence detection is engaged (Recommendation 6.36 on page 211).
- Try out the different motion processing modes of your display, and choose the one that suits your personal taste with respect to motion blur and the soap opera effect (Recommendation 6.37 on page 214).

Calibrating a Video Display

- Turn off all automatic picture-adjustment features of your display, such as ambient light sensor, dynamic contrast, contrast enhancement, color enhancement, live color, auto black level, and shadow detail (Recommendation 6.18 on page 185).
- Perform your display calibration under the same lighting conditions that you generally use for watching the display (Recommendation 6.17 on page 184).
- Allow your display to stabilize before you start calibration. Your display needs to warm up for approximately 10 minutes (Recommendation 6.19 on page 185).
- Display controls influence each other. Therefore, go back and forth between the different calibration steps until no further adjustments are needed (Recommendation 6.16 on page 184).
- Choose a color temperature preset that sets the display to 6500 K (Recommendation 6.21 on page 185).
- Use a video test pattern to calibrate the black level of your display. Set the black level of the display as low as possible without losing detail in the dark parts (above reference black) (Recommendation 6.22 on page 188).
- Use a video test pattern to calibrate the white level of your display. Set the white level of the display as high as possible without losing detail in the bright parts and without introducing color shifts (Recommendation 6.23 on page 189).
- After setting the black and white level of your display, check if a grayscale ramp is displayed as a smooth gradient. If it is not, try to make the ramp smoother by reducing the white level (Recommendation 6.24 on page 190).
- Adjust the maximum luminance output of your display to 100–250 cd/m² for a flat-panel display and to 38–58 cd/m² for a projection system. Directly change the intensity of the

light and try to avoid using the white level control for this purpose (Recommendation 6.25 on page 190).

- Use a video test pattern and a blue filter to calibrate the color and tint controls of your display (Recommendation 6.26 on page 192).
- Calibrate your display for SDR video such that gamma is approximately equal to 2.4 and stays between 2.3 and 2.5 for the entire luminance range (Recommendation 6.27 on page 197).
- Calibrate your display for HDR10 and Dolby Vision video such that the grayscale tracks the SMPTE ST 2084 transfer function up to the maximum luminance capability of the display (Recommendation 6.29 on page 198).
- Calibrate your display for HLG video such that the grayscale tracks the HLG transfer function (Recommendation 6.30 on page 198).
- Calibrate your display such that the white point has chromaticity coordinates $x = 0.3127$ and $y = 0.3290$ (6500 K) and such that this white point is approximately constant across the entire luminance range with a reproduction error ΔE_{uv}^* smaller than 4 (Recommendation 6.28 on page 197).
- If your display has an adjustable CMS, use it to calibrate your display such that the primary and secondary colors have a reproduction error ΔE_{uv}^* smaller than 7 (Recommendation 6.31 on page 200).
- Make sure that the picture controls are set to the optimal values found during the calibration process for each individual input of the display (Recommendation 6.15 on page 184).
- Calibrate your projector at regular intervals, for example after every 200 hours of use (Recommendation 6.20 on page 185).

Setting Up a Blu-Ray/DVD Player

- Only use a full-screen mode to display letterbox material. Do not attempt to get rid of the black bars using another display mode, because it will result in a cropped or geometrically distorted picture (Recommendation 5.1 on page 122).
- To avoid unnatural distortions of the picture, always use the pillarbox format to display material with a 4:3 aspect ratio on a 16:9 screen (Recommendation 5.2 on page 122).
- Turn off all automatic contrast enhancements on your player (Recommendation 6.38 on page 214).
- Turn off all automatic color enhancements on your player (Recommendation 6.41 on page 215).
- Only use noise reduction on your player when you need it on low-quality material. Don't leave it on by default (Recommendation 6.42 on page 215).
- If your player has an item in the setup menu for the deinterlacing mode or progressive mode, set it such that film mode with cadence detection is engaged (Recommendation 6.36 on page 211).
- Configure your player to process and output digital video with a bit depth of 30 bits (10 bits per channel) or 36 bits (12 bits per channel) (Recommendation 6.32 on page 201).
- Use the HD Benchmark test disc to determine the output color space of your Blu-ray disc player (4:2:0, 4:2:2, 4:4:4, or $R'\ G'\ B'$) that results in the best-quality chroma upsampling (Recommendation 6.35 on page 207).
- Disable the secondary audio stream in the setup menu of your Blu-ray disc player to achieve the best possible sound quality (Recommendation 5.24 on page 159).

Setting Up Surround Sound

- Listen to audio test signals to verify that each loudspeaker is reproducing the appropriate audio channel and that each loudspeaker is connected with the correct polarity (Recommendation 7.27 on page 261).
- Measure the distances between each main loudspeaker and the prime listening position and use these distances to set the appropriate delays in the delay setup menu of your A/V controller or receiver (Recommendation 7.41 on page 274).

- Use a video test pattern to check audio and video synchronization. If necessary fine-tune the delay of the audio signal (Recommendation 7.42 on page 274).
- Use an SPL meter (slow response, C-weighting) to calibrate the sound level of each main loudspeaker to 78 dB SPL. Adjust the relative levels of the loudspeakers in the setup menu of your A/V controller or receiver such that the loudspeakers are matched within 1 dB (Recommendation 7.44 on page 276).
- After calibration of the relative sound levels, play some music and watch some movies to check if the level of the center loudspeaker is not too high. The frontal soundstage should appear wide without the sound bunching up around the position of the center loudspeaker (Recommendation 7.45 on page 277).
- After calibration of the relative sound levels, play some music and watch some movies to check if the level of the surround loudspeakers is not too high. The surrounds should blend seamlessly with the rest of the sound and never draw attention to themselves when playing immersive sounds (Recommendation 7.46 on page 277).
- To use all the surround loudspeakers in your system consider using the surround upmixing modes of your A/V controller (like Dolby Prologic and DTS Neo) when watching movies with a soundtrack that has less surround channels than the number of loudspeakers in your room (Recommendation 7.36 on page 271).
- For the best reproduction of two-channel music do not use any of the DSP surround modes of your A/V controller (Recommendation 7.37 on page 271).
- Only enable the dynamic range control of your A/V controller or receiver when your average playback volume is much lower than 78 dB; keep it disabled in all other situations (Recommendation 7.48 on page 279).
- Use the automatic room correction feature of your A/V controller or receiver to correct the combined frequency response of your loudspeakers and the room, but limit the corrections to the frequencies below 300 Hz (Recommendation 7.49 on page 280).

Setting Up Subwoofers

- Use the bass management setup menu in your A/V controller or receiver to set all the front and surround loudspeakers to 'small', set the subwoofer to 'on', and direct the LFE channel only to the subwoofers and not to the other loudspeakers (Recommendation 7.38 on page 272).
- Set the subwoofer crossover frequency of your A/V controller or receiver between 80 and 120 Hz, such that the subwoofers and main loudspeakers sound well-integrated and the subwoofers cannot be audibly localized as separate sound sources (Recommendation 7.39 on page 273).
- Disable the low-pass filter built into your subwoofer, or set it to the highest cutoff frequency (Recommendation 7.40 on page 273).
- Set the phase controls of your subwoofer and the delay/distance controls in your A/V controller or receiver such that the sound from the subwoofers and the sound from the front loudspeakers arrive in phase at the prime listening position (Recommendation 7.43 on page 275).
- Use an SPL meter to roughly set the subwoofer level, then play some music and watch some movies to fine-tune it. The subwoofer should blend seamlessly with the rest of the sound and never draw attention to itself (Recommendation 7.47 on page 278).

Calibrating Subwoofers

- Improve the low-frequency response of your subwoofers using a combined approach in which you first select the best positions for your subwoofers and then apply parametric equalization. Use your own acoustical measurements to make informed decisions on where to put the subwoofers and how to tune the parametric equalizer (Recommendation 8.3 on page 292).
- Only use high-resolution acoustic measurements (accurate to at least 1/20 of an octave) to analyze the low-frequency response of your room and subwoofers (Recommendation 8.4 on page 293).

- To analyze the reproduction of low frequencies in your room, measure the combined response of the subwoofers and the left and right front loudspeakers from 20 Hz to 200 Hz (Recommendation 8.5 on page 293).
- To measure the low-frequency response in the room mount the measurement microphone on a tripod, point it at the ceiling, and place it at one of the listening positions. Do not stand between the subwoofers and the microphone, and make the room as quiet as possible (Recommendation 8.6 on page 294).
- Experiment with different positions of the subwoofers and listeners in the room. Measure the frequency response at each listening position and try to move your subwoofers and listeners such that these frequency responses are as similar as possible and have the least number of dips (Recommendation 8.7 on page 295).
- Manipulate the excitation of an offending room mode by placing a subwoofer at or near a pressure minimum, or by placing two subwoofers on opposing sides of a pressure minimum (Recommendation 8.8 on page 295).
- Use a parametric equalizer to reduce the peaks in the in-room frequency response of the subwoofers. Focus on the peaks that are common among the frequency responses for the different seats. Do not attempt to correct narrow dips in the response. Broad depressions may be boosted if the gain is limited to about 6 dB (Recommendation 8.9 on page 298).
- Do not attempt to use electronic equalization to correct dips in the frequency response that are nonminimum phase (Recommendation 8.10 on page 299).
- Use acoustic measurements to set the subwoofer level. The average sound level for the frequencies below the subwoofer crossover frequency should closely match the average level above the crossover frequency (Recommendation 8.11 on page 301).
- Use acoustic measurements to improve the integration between the main loudspeakers and the subwoofers in the crossover region. Play around with the subwoofer's time-delay, phase, and crossover frequency (Recommendation 8.12 on page 301).

Setting Up Computer Audio

- Do not use the built-in sound card of your computer to convert digital audio to analog audio. For the best audio quality send the digital audio from the computer to an A/V controller, A/V receiver, or dedicated external DAC, and let them handle the conversion (Recommendation 5.18 on page 151).
- Preferably use the asynchronous transfer mode with a USB audio interface. If this mode is not supported, use adaptive mode. Never use synchronous mode (Recommendation 5.27 on page 166).
- To get the best sound quality from your computer, take the following steps (Recommendation 5.19 on page 152):

    - If possible, give your music player exclusive access to the sound driver.
    - Set the digital volume control of your music player to its maximum level.
    - Set the master volume of the operating system's sound mixer to its maximum level.
    - Set the volume of your music player in the mixer to its maximum level and turn down or mute all the other sources.
    - Disable all operating system sounds.
    - Disable all sound enhancements in the sound driver, operating system, and music player.

- Always set the audio bit depth of your computer to the maximum value that your DACs can handle (usually 24 bits) (Recommendation 5.20 on page 153).
- Set the audio sampling rate of your computer either to the native sampling rate of the files that you play (using software that automatically switches the sampling rate of the sound driver), or set it to the maximum sampling rate that your DACs can handle (96 kHz or 192 kHz). Listen and compare the two methods and choose the one that sounds best (Recommendation 5.21 on page 155).

9.6 Playing Content

Playing Video Files and Streams

- Prefer high-resolution video sources over low-resolution ones. That is, if possible use HD or UHD video instead of SD video, choose HDTV broadcasts over standard television broadcasts, and Blu-ray discs over DVDs (Recommendation 5.3 on page 124).
- When you are given a choice between high bit rate and low bit rate digital video files (that are encoded with the same video codec), always choose the high bit rate ones (Recommendation 5.4 on page 128).
- Make sure your Internet connection speed does not limit the quality of the streaming video that you receive. A minimum speed of 15 Mbit/s would be a good starting point for HD and UHD video (Recommendation 5.14 on page 144).

Playing Audio Files and Streams

- For the best sound quality use either uncompressed or losslessly compressed digital audio. When given a choice, always prefer lossless audio compression over lossy compression (Recommendation 5.9 on page 137).
- For the best audio quality use lossless LPCM files like AIFF, ALAC, or FLAC with at least 16 bits and a sampling rate of at least 44.1 kHz (Recommendation 5.17 on page 148).
- When you are given a choice between HRA with 88.2 kHz or 96 kHz sampling and standard audio with 44.1 kHz or 48 kHz sampling, consider choosing the HRA version. It might sound better, if it has been mastered with greater care and has a larger dynamic range (Recommendation 5.5 on page 132).
- There is absolutely no compelling reason to use a sampling frequency higher than 96 kHz for digital audio (Recommendation 5.6 on page 132).
- When you are given a choice between HRA with 24 bits and standard audio with 16 bits, consider choosing the HRA version. It might sound better, if it has been mastered with greater care and has a larger dynamic range (Recommendation 5.7 on page 135).
- DSD recordings often sell for a higher price than equivalent LPCM ones. There is absolutely no compelling reason to pay extra for the DSD version if an LPCM version with at least 20 bits and 96 kHz sampling is available (Recommendation 5.8 on page 136).
- To obtain the best audio quality with music streaming services make sure the service delivers lossless CD-quality streaming: 44.1 kHz sampling with 16 bits (Recommendation 5.15 on page 146).
- When you are given a choice between high bit rate and low bit rate perceptually coded audio streams (that are encoded with the same audio codec), always choose the high bit rate ones (Recommendation 5.11 on page 140).
- Prefer MPEG-2 AAC over MPEG-2 Layer III (MP3). Use two-channel MPEG-2 AAC streams or files with a bit rate of at least 128 kbit/s, and preferably higher: 256 kbit/s or 320 kbit/s (Recommendation 5.12 on page 140).
- Convert your CDs into digital media files using DAE software that supports secure ripping and AccurateRip (Recommendation 5.22 on page 156).

Playing Audio From Blu-Ray Discs and DVDs

- Blu-ray discs usually contain multiple audio tracks. For the best audio quality select the LPCM track (sometimes referred to as 'uncompressed') or an audio track with lossless compression: either DTS-HD Master Audio or Dolby TrueHD (Recommendation 5.10 on page 137).
- DVDs usually contain multiple perceptually coded audio tracks. For the best audio quality, select the multichannel DTS Digital Surround track. If this track is not available, choose the multichannel Dolby Digital track instead (Recommendation 5.13 on page 141).

APPENDIX A

Test Tones and Patterns

- **Audio Calibration Disc & HD Music Sampler** (Blu-ray disc)

  AIX Records 82002 (2009)
  aixrecords.com
  Section 4.3.2: Low-frequency sweep.
  Section 7.4: Loudspeaker channel and polarity check.

- **Audio Check** (Files)

  www.audiocheck.net
  Section 4.2.3: Aiming front loudspeakers.
  Section 4.3.2: Low-frequency sweep.
  Section 7.4: Loudspeaker channel and polarity check.
  Section 7.5.3: Setting subwoofer delay and phase.

- **AVS HD 709—Blu-ray & MP4 Calibration** (Files)

  AVS Forum (2010)
  www.avsforum.com
  Section 4.3.1: Ambient light and contrast measurements.
  Section 6.1.3: Projector panel alignment.
  Section 6.2: Video display calibration.
  Section 6.3.1: Setting aspect ratio and overscan.
  Section 6.3.2: Setting sharpness.

- **Digital Video Essentials (DVE) HD Basics** (Blu-ray disc)

  Joe Kane Productions (2008)
  www.videoessentials.com
  Section 4.2.3: Aiming front loudspeakers.
  Section 4.3.1: Ambient light and contrast measurements.
  Section 4.3.2: Low-frequency sweep.
  Section 6.1.3: Projector panel alignment.
  Section 6.2: Video display calibration.
  Section 6.3.1: Setting aspect ratio and overscan.
  Section 6.3.2: Setting sharpness.
  Section 7.4: Loudspeaker channel and polarity check.
  Section 7.5.3: Setting audio and video synchronization.
  Section 7.5.4: Matching relative loudspeakers levels.

- **DVS UHD/HDR-10 Test Pattern Suite** (Blu-ray disc/files)

  Diversified Video Solutions (2017)
  diversifiedvideosolutions.com

Section 4.3.1: Ambient light and contrast measurements.
Section 6.2: Video display calibration.
Section 6.3.1: Setting aspect ratio and overscan.
Section 6.3.2: Setting sharpness.

- **DVS UHD/HLG-HDR Test Pattern Suite** (Files)

  Diversified Video Solutions (2018)
  diversifiedvideosolutions.com
  Section 4.3.1: Ambient light and contrast measurements.
  Section 6.2: Video display calibration.
  Section 6.3.1: Setting aspect ratio and overscan.
  Section 6.3.2: Setting sharpness.

- **HD Benchmark 2nd Edition** (Blu-ray disc)

  Stacey Spears and Don Munsil (2013)
  www.spearsandmunsil.com
  Section 4.3.1: Ambient light and contrast measurements.
  Section 6.1.3: Projector panel alignment.
  Section 6.2: Video display calibration.
  Section 6.3.1: Setting aspect ratio and overscan.
  Section 6.3.2: Setting sharpness.
  Section 6.3.3: Color space selection.
  Section 6.3.5: Setting frame rate interpolation.
  Section 7.4: Loudspeaker channel and polarity check.
  Section 7.5.3: Setting audio and video synchronization.
  Section 7.5.4: Matching relative loudspeakers levels.

- **Music and Audio Guide** (Audio files/Blu-ray disc)

  Mark Waldrep (2017)
  musicandaudioguide.com
  Section 4.3.2: Low-frequency sweep.
  Section 7.4: Loudspeaker channel and polarity check.
  Section 7.5.4: Matching relative loudspeakers levels.

- **Musical Articulation Test Tones (MATT)** (Files)

  Acoustic Sciences Corporation
  www.acousticsciences.com/matt
  Section 8.1: Listening test for room resonances.

APPENDIX B

Software

- **Acourate**

  Windows
  www.audiovero.de
  Section 8.1.3: Acoustic measurements for subwoofer calibration.

- **Amarra**

  macOS
  sonicstudio.com/amarra
  Section 5.3.4: Bit perfect media player for iTunes.

- **Audacity**

  Linux/macOS/Windows
  www.audacityteam.org
  Section 5.3.3: Analyze time and frequency content of digital audio files.

- **Audirvana Plus**

  macOS
  audirvana.com
  Section 5.3.4: Bit perfect media player software.

- **BitPerfect**

  macOS
  bitperfectsound.blogspot.com
  Section 5.3.4: Bit perfect media player for iTunes.

- **Bitter**

  macOS/Windows
  www.stillwellaudio.com/plugins/bitter
  Section 5.3.3: Analyze bit depth of digital audio files.

- **CalMAN Home Video**

  Windows
  www.spectracal.com
  Section 6.2: Color and grayscale measurements for video display calibration.

- **ChromaPure**

  Windows
  www.chromapure.com
  Section 6.2: Color and grayscale measurements for video display calibration.

- **Computer Aided Room Acoustics (CARA)**

  Windows
  www.cara.de
  Section 4.2.5: Computer simulation of loudspeaker-room interaction.

- **dBpoweramp**

  macOS/Windows
  www.dbpoweramp.com
  Section 5.3.5: Secure CD ripping with AccurateRip support.

- **Exact Audio Copy (EAC)**

  Windows
  www.exactaudiocopy.org
  Section 5.3.5: Secure CD ripping with AccurateRip support.

- **Foobar2000**

  Windows
  foobar2000.org
  Section 5.3.4: Bit perfect media player software.

- **HCFR**

  macOS/Windows
  sourceforge.net/projects/hcfr
  www.homecinema-fr.com/colorimetre-hcfr/hcfr-colormeter
  Section 6.2: Color and grayscale measurements for video display calibration.

- **iTunes**

  macOS/Windows
  apple.com/nl/itunes
  Section 5.3.4: Media player software.

- **JRiver MediaCenter**

  Linux/macOS/Windows
  jriver.com
  Section 5.3.4: Bit perfect media player software.

- **LightSpace**

  Windows
  www.lightillusion.com
  Section 6.2: Color and grayscale measurements for video display calibration.

- **Lossless Audio Checker**

  Linux/macOS/Windows
  losslessaudiochecker.com
  Section 5.3.3: Detect upsampling and bit depth usage of digital audio files.

- **MediaMonkey**

  Windows
  mediamonkey.com
  Section 5.3.4: Bit perfect media player software.

- **MusicScope**

  macOS/Windows
  www.xivero.com
  Section 5.3.3: Analyze time and frequency content of digital audio files.
  Section 5.3.3: Analyze the dynamic range of digital audio files.

- **Pure Music**

  macOS
  channld.com/puremusic
  Section 5.3.4: Bit perfect media player for iTunes.

- **QRDude**

  Windows
  www.subwoofer-builder.com/qrdude.htm
  Section 8.2.3: Design of 1D and 2D quadratic residue diffusers (QRDs).

- **Rip**

  macOS
  sbooth.org/Rip
  Section 5.3.5: Secure CD ripping with AccurateRip support.

- **Room EQ Wizard (REW)**

  Linux/macOS/Windows
  www.roomeqwizard.com
  Section 8.1.3: Acoustic measurements for subwoofer calibration.

- **Roon**

  Linux/macOS/Windows
  roonlabs.com
  Section 5.3.3: Analyze the dynamic range of digital audio files.
  Section 5.3.4: Bit perfect media player software.

- **Signalyst HQPlayer**

  Linux/macOS/Windows
  signalyst.com
  Section 5.3.4: Bit perfect media player software.

- **TT Dynamic Range Meter**

  macOS/Windows
  dr.loudness-war.info
  pleasurizemusic.com
  Section 5.3.3: Analyze the dynamic range of digital audio files.

- **X Lossless Decoder (XLD)**

  macOS
  tmkk.undo.jp/xld/index_e.html
  Section 5.3.5: Secure CD ripping with AccurateRip support.

BIBLIOGRAPHY

Adams, Glyn (1989). Time dependence of loudspeaker power output in small rooms. *Journal of the Audio Engineering Society* 37(4), 203–209. Also in AES (1996b), 415–421.

AES (1980). *An Anthology of Articles on Loudspeakers Volume 1* (Second edition). New York: Audio Engineering Society.

AES (1984). *An Anthology of Articles on Loudspeakers Volume 2.* New York: Audio Engineering Society.

AES (1996a). *An Anthology of Articles on Loudspeakers Volume 3: Systems and Crossover Networks.* New York: Audio Engineering Society.

AES (1996b). *An Anthology of Articles on Loudspeakers Volume 4: Transducers, Measurements and Evaluation.* New York: Audio Engineering Society.

AES (2001). Multichannel surround sound systems and operations. AESTD1001.1.01–10. *Technical Committee on Multichannel and Binaural Audio Technology.* Audio Engineering Society. Also in AES (2006), 236–262.

AES (2006). *An Anthology of Articles on Spatial Sound Techniques Part 2: Multichannel Audio Technologies.* New York: Audio Engineering Society.

AES (2008). AES recommended practice for professional audio: Subjective evaluation of loudspeakers. AES20–1996 (s2008). Audio Engineering Society.

AES (2012). Technology trends in audio engineering: A report by the AES technical council. *Journal of the Audio Engineering Society* 60(1/2), 90–107.

Allen, Ioan (2000). Screen size: The impact on picture and sound. White Paper. Dolby Laboratories, Inc. www.dolby.com.

Allison, Roy F. (1974). The influence of room boundaries on loudspeaker power output. *Journal of the Audio Engineering Society* 22(5), 314–320. Also in AES (1980), 353–359.

Allison, Roy F. and Edgar Villchur (1982). On the magnitude and audibility of FM distortion in loudspeakers. *Journal of the Audio Engineering Society* 30(10), 694–700. Also in AES (1984), 350–356.

Anderson, Joseph and Barbara Anderson (1993). The myth of persistence of vision revisited. *Journal of Film and Video* 45(1), 3–12.

Archimago's Musings (2015). Measurements: Apple Mac OS X ('Yosemite') software audio upsampling. www.archimago.blogspot.com/2015/11.

Ashihara, Kaoru, Shogo Kiryu, Nobuo Koizumi, Akira Nishimura, Juro Ohga, Masaki Sawaguchi and Sho-kichiro Yoshikawa (2005). Detection threshold for distortions due to jitter on digital audio. *Acoustical Science and Technology* 26(1), 50–54.

ASTM (2008). Standard specification for standard nominal diameters and cross-sectional areas of AWG sizes of solid round wires used as electrical conductors. ASTM B258–02. American Society for Testing and Materials. www.astm.org.

Atkinson, John (1992). The sound of surprise (The loudspeaker/stand interface). Online Publication. Stereophile. www.stereophile.com.

Ballagh, Keith O. (1983). Optimum loudspeaker placement near reflecting planes. *Journal of the Audio Engineering Society* 31(12), 931–935. Also in AES (1980), 458–462.

Ballou, Glen (2015). Fundamentals and units of measurement. In Glen Ballou (Ed.), *Handbook for Sound Engineers* (Fifth edition), Chapter 50, pp. 1627–1670. Oxford: Focal Press.

Ballou, Glen and Joe Ciaudelli (2015). Microphones. In Glen Ballou (Ed.), *Handbook for Sound Engineers* (Fifth edition), Chapter 20, pp. 597–702. Oxford: Focal Press.

Barbour, James L. (2005). Subjective consumer evaluation of multichannel audio codecs. In *Audio Engineering Society Preprints 119th Convention*, New York, USA (October). Preprint 6558.

Barnes, John R. (2004). *Robust Electronic Design Reference Book*. Boston: Kluwer Academic Publishers.

Baskind, Alexis and Jean-Dominique Polack (2000). Sound power radiated by sources in diffuse field. In *Audio Engineering Society Preprints 108th Convention*, Paris, France (February). Preprint 5146.

Bayer, Bryce E. (1976). Color imaging array. Patent 3971065. www.patentscope.wipo.int.

BDA (2011). Blu-ray disc read-only format: 2.B audio visual application format specifications for BD-ROM version 2.5. White Paper. Blu-ray Disc Association. www.blu-raydisc.com.

BDA (2015a). Blu-ray disc format: General. White Paper. Blu-ray Disc Association. www.blu-raydisc.com.

BDA (2015b). Blu-ray disc read-only format (ultra hd blu-ray): Audio visual application format specifications for BD-ROM version 3.0. White Paper. Blu-ray Disc Association. www.blu-raydisc.com.

Bech, Søren (1994). Perception of timbre of reproduced sound in small rooms: Influence of room and loudspeaker position. *Journal of the Audio Engineering Society* 42(12), 999–1007.

Bech, Søren (1998). The influence of stereophonic width on the perceived quality of an audiovisual presentation using a multichannel sound system. *Journal of the Audio Engineering Society* 46(4), 314–322.

Benade, Arthur H. (1985). From instrument to ear in a room: Direct or via recording. *Journal of the Audio Engineering Society* 33(4), 218–233.

Benjamin, Eric and Benjamin Gannon (1998). Theoretical and audible effects of jitter on digital audio quality. In *Audio Engineering Society Preprints 105th Convention*, San Francisco, USA (September). Preprint 4826.

Blech, Dominic and Min-Chi Yang (2004). DVD-Audio versus SACD: Perceptual discrimination of digital audio coding formats. In *Audio Engineering Society Preprints 116th Convention*, Berlin, Germany (May). Preprint 6086.

Blomberg, Les and Noland Lewis (2008). What's the ear for? How to protect it. In Glen Ballou (Ed.), *Handbook for Sound Engineers* (Fourth edition), Chapter 47, pp. 1631–1643. Oxford: Focal Press.

Bodrogi, Peter and Tran Quoc Khanh (2012). *Illumination, Color and Imaging: Evaluation and Optimization of Visual Displays*. Weinheim: Wiley-VCH.

Borer, Tim and Andrew Cotton (2015). A 'display independent' high dynamic range television system. White Paper WHP 309. BBC Research & Development. www.bbc.co.uk/rd.

Bradley, John S. (1997). Sound absorption of gypsum board cavity walls. *Journal of the Audio Engineering Society* 45(4), 253–259.

Brawn, Alan C. (2015). The fundamentals of display technologies. In Glen Ballou (Ed.), *Handbook for Sound Engineers* (Fifth edition), Chapter 48, pp. 1585–1599. Oxford: Focal Press.

Brennesholtz, Matthew S. and Edward H. Stupp (2008). *Projection Displays* (Second edition). Chichester: John Wiley and Sons.

Briere, Danny and Pat Hurley (2009). *Home Theater for Dummies* (Third edition). Hoboken: John Wiley and Sons.

Brown, Pat (2015). Fundamentals of audio and acoustics. In Glen Ballou (Ed.), *Handbook for Sound Engineers* (Fifth edition), Chapter 5, pp. 95–114. Oxford: Focal Press.

Bücklein, Roland (1981). The audibility of frequency response irregularities. *Journal of the Audio Engineering Society* 29(3), 126–131. This is a translation from German. The original manuscript 'Hörbarkeit von Unregelmäßigkeiten in Frequenzgängen bei akustischer Ubertragung' appeared in *Frequenz* 16, 1962, pp. 103–108.

Cermak, Gregory, Margaret Pinson and Stephen Wolf (2011). The relationship among video quality, screen resolution, and bit rate. *IEEE Transactions on Broadcasting* 57(2), 258–262.

Chéenne, Dominique J. (2015). Acoustical modeling and auralization. In Glen Ballou (Ed.), *Handbook for Sound Engineers* (Fifth edition), Chapter 13, pp. 305–330. Oxford: Focal Press.

Choi, Byeong-Doo, Jong-Woo Han, Chang-Su Kim and Sung-Jea Ko (2007). Motion-compensated frame interpolation using bilateral motion estimation and adaptive overlapped block motion compensation. *IEEE Transactions on Circuits and Systems for Video Technology* 17(4), 407–416.

Choisel, Sylvain and Florian Wickelmaier (2007). Evaluation of multichannel reproduced sound: Scaling auditory attributes underlying listener preference. *Journal of the Acoustical Society of America* 121(1), 388–400.

CIE S 014–1 (2006). Colorimetry-Part 1: CIE standard colorimetric observers. International Commission on Illumination. Also published as ISO 11664–1.

CIE S 014–2 (2006). Colorimetry-Part 2: CIE standard illuminants for colorimetry. International Commission on Illumination. Also published as ISO 11664–2.

CinemaSource (2001a). Home theater seating: Designing audience seating in home theater rooms. Technical Bulletin. The CinemaSource Press. www.cinemasource.com.

CinemaSource (2001b). Understanding aspect ratios. Technical Bulletin. The CinemaSource Press. www.cinemasource.com.

CinemaSource (2002). Installing and using in-wall speakers. Technical Bulletin. The CinemaSource Press. www.cinemasource.com.

Colloms, Martin (2006). Do we need an ultrasonic bandwidth for higher fidelity sound reproduction? *Proceedings of the Institute of Acoustics* 28(8), 84–91.

Colloms, Martin (2018). *High Performance Loudspeakers* (Seventh edition). Chichester: John Wiley and Sons.

Cordell, Bob (2011). *Designing Audio Power Amplifiers*. New York: McGraw-Hill.

Cox, Trevor J. and Peter D'Antonio (2016). *Acoustic Absorbers and Diffusers: Theory, Design, and Application* (Third edition). New York: CRC Press.

Craven, Peter Graham, Malcolm Law and John Robert Stuart (2013). Doubly compatible lossless audio bandwidth extension. Patent WO/2013/186561. www.patentscope.wipo.int.

D'Antonio, Peter and Trevor Cox (1998a). Two decades of sound diffusor design and development part 1: Applications and design. *Journal of the Audio Engineering Society* 46(11), 955–975.

D'Antonio, Peter and Trevor Cox (1998b). Two decades of sound diffusor design and development part 2: Prediction, measurement, and characterization. *Journal of the Audio Engineering Society* 46(12), 1075–1091.

Davis, Fred E. (1991). Effects of cable, loudspeaker, and amplifier interactions. *Journal of the Audio Engineering Society* 39(6), 461–468.

Davis, Mark F. (2003). History of spatial coding. *Journal of the Audio Engineering Society* 51(6), 554–569. Also in AES (2006), 60–74.

Dawson, John (2011). Audio transport over HDMI. In *Proceedings of the 24th AES Conference: The Ins and Outs of Audio*, Surrey, United Kingdom (June).

DCI (2008). Digital cinema system specification version 1.2. Technical Report. Digital Cinema Initiatives, LLC. www.dcimovies.com.

DEG (2014). High resolution audio definition announcement. Brochure. The Digital Entertainment Group. www.degonline.com.

de Greef, Pierre and Hendriek Groot Hulze (2007). Adaptive dimming and boosting backlight for LCD-TV systems. *SID Symposium Digest of Technical Papers* 38(1), 1332–1335.

DellaSala, Gene (2010). Home theater multiple subwoofer set-up & calibration guide. Online Publication. Audioholics Online A/V Magazine. www.audioholics.com.

Derra, Guenther, Holger Moench, Ernst Fischer, Hermann Giese, Ulrich Hechtfischer, Gero Heusler, Achim Koerber, Ulrich Niemann, Folke-Charlotte Noertemann, Pavel Pekarski, Jens Pollmann-Retsch, Arnd Ritz and Ulrich Weichmann (2005). UHP lamp systems for projection applications. *Journal of Physics D: Applied Physics* 38(17), 2995–3010.

Deruty, Emmanuel and Damien Tardieu (2014). About dynamic processing in mainstream music. *Journal of the Audio Engineering Society* 62(1/2), 42–55.

Dressler, Roger (2000). Dolby surround pro logic II decoder principles of operation. White Paper. Dolby Laboratories, Inc. www.dolby.com.

Drewery, J. O. and Richard A. Salmon (2004). Tests of visual acuity to determine the resolution required of a television transmission system. White Paper WHP 092. BBC Research & Development. www.bbc.co.uk/rd.

Duncan, Ben (1997). *High Performance Audio Power Amplifiers for Music Performance and Reproduction* (Revised edition). Oxford: Newnes.

Dunn, Julian (2000). Jitter theory. Technote TN-23. Audio Precision. www.ap.com.

EBU R 128 (2014). Loudness normalisation and permitted maximum level of audio signals. European Broadcasting Union.

EBU-Tech 3276 (1998). Listening conditions for the assessment of sound programme material: Mono-phonic and two-channel stereophonic. European Broadcasting Union.

EBU-Tech 3276-s1 (2004). Listening conditions for the assessment of sound programme material: Multi-channel sound. European Broadcasting Union.

EBU-Tech 3286-s1 (2000). Assessment methods for the subjective evaluation of the quality of sound pro-gramme material: Multichannel. European Broadcasting Union.

EBU-Tech 3320 (2017). User requirements for video monitors in television production. European Broad-casting Union.

EBU-Tech 3324 (2007). EBU evaluations of multichannel audio codecs. European Broadcasting Union.

EBU-Tech 3342 (2016). Loudness range: A measure to supplement EBU R 128 loudness normalisation. European Broadcasting Union.

EDCF (2005). Digital cinema: The EDCF guide for early adopters. European Digital Cinema Forum. www.edcf.net.

EPA (1978). Protective noise levels: Condensed version of EPA levels document. EPA 550/9–79–100. United States Environmental Protection Agency. www.epa.gov.

Everest, F. Alton and Ken C. Pohlmann (2014). *Master Handbook of Acoustics* (Sixth edition). New York: McGraw-Hill.

Fause, Kenneth R. (1995). Fundamentals of grounding, shielding and interconnection. *Journal of the Audio Engineering Society* 43(6), 498–516.

Fazenda, Bruno, Mark R. Avis and William J. Davies (2005). Perception of modal distribution metrics in critical listening spaces: Dependence on room aspect ratios. *Journal of the Audio Engineering Society* 53(12), 1128–1141.

Fazenda, Bruno, Matthew Wankling, Jonathan Hargreaves, Lucy Elmer and Jonathan Hirst (2012). Subjec-tive preference of modal control methods in listening rooms. *Journal of the Audio Engineering Society* 60(5), 338–349.

Feng, Xiao-Fan (2006). LCD motion-blur analysis, perception, and reduction using synchronized backlight flashing. In *Proceedings SPIE 6057, Human Vision and Electronic Imaging XI*, San Jose, USA (February), pp. 1–14.

Fielder, Louis D. and Eric M. Benjamin (1988). Subwoofer performance for accurate reproduction of music. *Journal of the Audio Engineering Society* 36(6), 443–455. Also in AES (1996b), 401–414.

Fisekovic, Nebojsa, Tore Nauta, Hugo Cornelissen and Jacob Bruinink (2001). Improved motion-picture quality of AM-LCDs using scanning backlight. In *Proceedings of the 8th International Display Work-shops*, Nagoya, Japan (October), pp. 1637–1640.

Fleischmann, Mark (2017). *Practical Home Theater: A Guide to Video and Audio Systems* (2018 edition). New York: Quiet River Press.

Gabrielsson, Alf and Björn Lindström (1985). Perceived sound quality of high-fidelity loudspeakers. *Jour-nal of the Audio Engineering Society* 33(1/2), 33–52. Also in AES (1996b), 307–327.

Gander, Mark R. (1981). Moving-coil loudspeaker topology as an indicator of linear excursion capability. *Journal of the Audio Engineering Society* 29(1/2), 10–26. Also in AES (1984), 204–219.

Gander, Mark R. (1986). Dynamic linearity and power compression in moving-coil loudspeakers. *Journal of the Audio Engineering Society* 34(9), 627–646. Also in AES (1996b), 24–43.

Geddes, Earl R. (1989). An introduction to band-pass loudspeaker systems. *Journal of the Audio Engineer-ing Society* 37(5), 308–342. Also in AES (1996a), 153–188.

Genelec (2009). Monitor placement in small rooms. Genelec Oy. www.genelec.com.

Gerzon, Michael A, Peter G. Craven, J. Robert Stuart, Malcolm J. Law and Rhonda J. Wilson (2004). The MLP lossless compression system for PCM audio. *Journal of the Audio Engineering Society* 52(3), 243–260.

Gilford, C. L. S. (1959). The acoustic design of talk studios and listening rooms. *Proceedings of the IEE B* 106(27), 245–258.

Goodwin, Teresa (2012). Are your high resolution recordings really high resolution? *Positive Feedback* (60). www.positive-feedback.com.

Greiner, Richard A. (1980). Amplifier-loudspeaker interfacing. *Journal of the Audio Engineering Society* 28(5), 310–315. Also in AES (1984), 135–140.

Greiner, Richard A. and Michael Schoessaw (1983). Electronic equalization of closed-box loudspeakers. *Journal of the Audio Engineering Society* 31(3), 125–134.

Groh, Allen R. (1974). High-fidelity sound system equalization by analysis of standing waves. *Journal of the Audio Engineering Society* 22(10), 795–799.

Haas, Helmut (1972). The influence of a single echo on the audibility of speech. *Journal of the Audio Engineering Society* 20(2), 145–159. This is a translation from German. The original manuscript 'Über den Einfluss des Einfachechos auf die Horsamkeit von Sprache' appeared in *Acustica* 1(2), 1951.

Hamasaki, Kimio, Toshiyuki Nishiguchi, Kazuho Ono and Akio Ando (2004). Perceptual discrimination of very high frequency components in musical sound recorded with a newly developed wide frequency range microphone. In *Audio Engineering Society Preprints 117th Convention*, San Francisco, USA (October). Preprint 6298.

Harley, Robert (2002). *Home Theater for Everyone: A Practical Guide to Today's Home Entertainment Systems* (Second edition). Albuquerque: Acapella Publishing.

Harley, Robert (2015). *The Complete Guide to High-End Audio* (Fifth edition). Carlsbad: Acapella Publishing.

Hatada, Toyohiko, Haruo Sakata and Hideo Kusaka (1980). Psychophysical analysis of the 'Sensation of reality' induced by a visual wide-field display. *SMPTE Journal* 89(8), 560–569.

Hauser, Gabriel, Dirk Noy and John Storyk (2008). Commercial low frequency absorbers: A comparative study. In *Audio Engineering Society Preprints 124th Convention*, Amsterdam, the Netherlands (May). Preprint 7431.

Hidaka, Takayuki, Leo L. Beranek and Toshiyuki Okano (1996). Some considerations of interaural cross correlation and lateral fraction as measures of spaciousness in concert halls. In Yoichi Ando and Dennis Noson (Eds.), *Music and Concert Hall Acoustics: Conference Proceedings from MCHA 1995*, Chapter 32, pp. 315–326. London: Academic Press.

Hiyama, Koichiro, Setsu Komiyama and Kimio Hamasaki (2002). The minimum number of loudspeakers and its arrangement for reproducing the spatial impression of diffuse sound field. In *Audio Engineering Society Preprints 113th Convention*, Los Angeles, USA (October). Preprint 5674. Also in AES (2006), 313–323.

Holman, Tomlinson (1991). New factors in sound for cinema and television. *Journal of the Audio Engineering Society* 39(7/8), 529–539.

Holman, Tomlinson (2008). *Surround Sound: Up and Running* (Second edition). Oxford: Focal Press.

Hong, Sunkwang, B. Berkeley and Sang Soo Kim (2005). Motion image enhancement of LCDs. In *Proceedings of the IEEE International Conference on Image Processing*, Genoa, Italy (September), pp. 17–20.

Hornbeck, Larry J. (1983). 128×128 deformable mirror device. *IEEE Transactions on Electron Devices* ED-30(5), 539–545.

IEC 60228 (2004). Conductors of insulated cables. International Electrotechnical Commission.

IEC 60958 (2016). Digital audio interface. International Electrotechnical Commission.

IEC 61672 (2013). Electroacoustics: Sound level meters. International Electrotechnical Commission.

ISO 226 (2003). Normal equal-loudness-level contours. International Organization for Standardization.

ISO/IEC 10918 (1994). Digital compression and coding of continuous-tone still images. International Organization for Standardization.

ISO/IEC 11172 (1993). Coding of moving pictures and associated audio for digital storage media at up to about 1.5 Mbit/s-Part 3: Audio. International Organization for Standardization.

ISO/IEC 13818-2 (2013). Generic coding of moving pictures and associated audio information-Part 2: Video. International Organization for Standardization.

ISO/IEC 13818-3 (1998). Generic coding of moving pictures and associated audio information-Part 3: Audio. International Organization for Standardization.

ISO/IEC 14496-10 (2014). Coding of audio-visual objects-Part 10: Advanced video coding (AVC). International Organization for Standardization.

ISO/IEC 14496-2 (2004). Coding of audio-visual objects-Part 2: Visual. International Organization for Standardization.

ISO/IEC 14496-3 (2009). Coding of audio-visual objects-Part 3: Audio. International Organization for Standardization.

ISO/IEC 23008-2 (2017). High efficiency coding and media delivery in heterogeneous environments-Part 2: High efficiency video coding. International Organization for Standardization.

ITU (2009a). Large screen digital imagery. Report ITU-R BT.2053. International Telecommunications Union.

ITU (2009b). User requirements for a flat panel display (FPD) as a master monitor in an HDTV programme production environment. Report ITU-R BT.2129. International Telecommunications Union.

ITU (2017). High dynamic range television for production and international programme exchange. Report ITU-R BT.2390. International Telecommunications Union.

ITU-R BS.775 (2012). Multichannel stereophonic sound system with and without accompanying picture. International Telecommunications Union.

ITU-R BS.1116 (2015). Methods for the subjective assessment of small impairments in audio systems including multichannel sound systems. International Telecommunications Union.

ITU-R BS.1284 (2003). General methods for the subjective assessment of sound quality. International Telecommunications Union.

ITU-R BS.1770 (2015). Algorithms to measure audio programme loudness and true-peak audio level. International Telecommunications Union.

ITU-R BT.601 (2011). Studio encoding parameters of digital television for standard 4:3 and wide-screen 16:9 aspect ratios. International Telecommunications Union.

ITU-R BT.709 (2015). Parameter values for the HDTV standards for production and international programme exchange. International Telecommunications Union.

ITU-R BT.710 (1998). Subjective assessment methods for image quality in high-definition television. International Telecommunications Union.

ITU-R BT.1543 (2015). 1280 × 720, 16:9 progressively-captured image format for production and international programme exchange in the 60 Hz environment. International Telecommunications Union.

ITU-R BT.1769 (2008). Parameter values for an expanded hierarchy of LSDI image formats for production and international programme exchange. International Telecommunications Union.

ITU-R BT.1886 (2011). Reference electro-optical transfer function for flat panel displays used in HDTV studio production. International Telecommunications Union.

ITU-R BT.2020 (2015). Parameter values for ultra-high definition television systems for production and international programme exchange. International Telecommunications Union.

ITU-R BT.2100 (2017). Image parameter values for high dynamic range television for use in production and international programme exchange. International Telecommunications Union.

ITU-R SM.1446 (2000). Definition and measurement of intermodulation products in transmitter using frequency, phase, or complex modulation techniques. International Telecommunications Union.

ITU-T H.262 (2012). Generic coding of moving pictures and associated audio information: Video. International Telecommunications Union.

ITU-T H.263 (2005). Video coding for low bit rate communication. International Telecommunications Union.

ITU-T H.264 (2017). Advanced video coding for generic audiovisual services. International Telecommunications Union.

ITU-T H.265 (2018). High efficiency video coding. International Telecommunications Union.

Iverson, James K. (1973). The theory of loudspeaker cabinet resonances. *Journal of the Audio Engineering Society* 21(3), 177–180. Also in AES (1980), 312–315.

Jack, Keith (2007). *Video Demystified: A Handbook for the Digital Engineer* (Fifth edition). Eagle Rock: Newnes.

Jones, Doug (2015a). Acoustical noise control. In Glen Ballou (Ed.), *Handbook for Sound Engineers* (Fifth edition), Chapter 7, pp. 137–166. Oxford: Focal Press.

Jones, Doug (2015b). Small room acoustics. In Glen Ballou (Ed.), *Handbook for Sound Engineers* (Fifth edition), Chapter 6, pp. 117–136. Oxford: Focal Press.

Jones, Douglas R., William L. Martens and Gary S. Kendall (1986). LEDR: A subjective approach to evaluating and modifying control rooms. In *Audio Engineering Society Preprints 81st Convention*, Los Angeles, USA (November). Preprint 2369.

Kang, Suk-Ju, Kyoung-Rok Cho and Young Hwan Kim (2007). Motion compensated frame rate up-conversion using extended bilateral motion estimation. *IEEE Transactions on Consumer Electronics* 53(4), 1759–1767.

Katz, Bob (2015). *Mastering Audio: The Art and the Science* (Third edition). Burlington: Focal Press.

Kessler, Ken and Andrew Watson (2011). *KEF 50 Years of Innovation in Sound*. Hong Kong: GP Acoustics.

Komiyama, Setsu (1989). Subjective evaluation of angular displacement between picture and sound directions for HDTV sound systems. *Journal of the Audio Engineering Society* 37(4), 210–214.

Lacroix, Julien, Yann Prime, Alexandre Remy and Olivier Derrien (2015). Lossless audio checker: A software for the detection of upscaling, upsampling and transcoding in lossless musical tracks. In *Audio Engineering Society Preprints 139th Convention*, New York, USA (October). Preprint 9416.

Lampen, Steve and Glen Ballou (2015). Transmission techniques: Wire and cable. In Glen Ballou (Ed.), *Handbook for Sound Engineers* (Fifth edition), Chapter 18, pp. 491–554. Oxford: Focal Press.

Lee, Jiun-Haw, David N. Liu and Shin-Tson Wu (2008). *Introduction to Flat Panel Displays*. Chichester: John Wiley and Sons.

Lesso, Paul (2006). A high performance S/PDIF receiver. In *Audio Engineering Society Preprints 121st Convention*, San Francisco, USA (October). Preprint 6948.

Letowski, Tomasz (1989). Sound quality assessment: Concepts and criteria. In *Audio Engineering Society Preprints 87th Convention*, New York, USA (October). Preprint 2825.

Levitin, Daniel (2006). *This Is Your Brain on Music: Understanding a Human Obsession*. London: Atlantic Books.

Lipshitz, Stanley P. and John Vanderkooy (1981). The great debate: Subjective evaluation. *Journal of the Audio Engineering Society* 29(7/8), 482–491.

Lipshitz, Stanley P. and John Vanderkooy (2001). Why 1-bit sigma-delta conversion is unsuitable for high-quality applications. In *Audio Engineering Society Preprints 110th Convention*, Amsterdam, the Netherlands (May). Preprint 5395.

Mäkivirta, Aki V. and Christophe Anet (2001). A survey study of in-situ stereo and multi-channel monitoring conditions. In *Audio Engineering Society Preprints 111th Convention*, New York, USA (November). Preprint 5496.

Martens, William L., Jonas Braasch and Wieslaw Woszczyk (2004). Identification and discrimination of listener envelopment precepts associated with multiple low-frequency signals in multichannel sound reproduction. In *Audio Engineering Society Preprints 117th Convention*, San Francisco, USA (October). Preprint 6229. Also in AES (2006), 340–352.

Masters, Ian G. (1999). The magic of film sound. *Audio* (May), 30–36.

McLaughlin, Brett (2005). *Home Theater Hacks*. Sebastopol: O'Reilly Media.

Meridian (2012). Meridian loudspeakers: The DSP path. White Paper. Meridian Audio Limited. www.meridian-audio.com.

Meyer, E. Brad and David R. Moran (2007). Audibility of a CD-standard A/D/A loop inserted into high-resolution audio playback. *Journal of the Audio Engineering Society* 55(9), 775–779.

Milner, Greg (2009). *Perfecting Sound Forever: The Story of Recorded Music*. London: Granta Publications.

Moore, Brian C. J. (2013). *An Introduction to the Psychology of Hearing* (Sixth edition). Leiden: Brill.

Moulton, David (1990). The creation of musical sounds for playback through loudspeakers. In *Audio Engineering Society 8th International Conference: The Sound of Audio*, Washington, DC, USA (May).

Munsil, Don and Brian Florian (2000). DVD benchmark-Part 5: Progressive scan DVD. Online Publication. Secrets of Home Theater and High Fidelity. www.hometheaterhifi.com.

Newell, Philip and Keith Holland (2007). *Loudspeakers: For Music Recording and Reproduction*. Oxford: Focal Press.

Niquette, Patty (2006). Hearing protection for musicians. *The Hearing Review* 13(3), 52–58.

Nishimura, Akira and Nobuo Koizumi (2004). Measurement and analysis of sampling jitter in digital audio products. In *Proceedings of the 18th International Congresses on Acoustics*, Kyoto, Japan (April), pp. 2547–2550.

Nishimura, Akira and Nobuo Koizumi (2010). Measurement of sampling jitter in analog-to-digital and digital-to-analog converters using analytic signals. *Acoustical Science and Technology* 31(2), 172–180.

Norcross, Scott G., Gilbert A. Soulodre and Michel C. Lavoie (2004). Subjective investigations of inverse filtering. *Journal of the Audio Engineering Society* 52(10), 1003–1028.

Noxon, Arthur (1985). Listening room: Corner loaded bass trap. In *Audio Engineering Society Preprints 79th Convention*, New York, USA (October).

Noxon, Arthur (1994). Home theater acoustics volume 1–5. *Home Theater Magazine*. www.tubetrap.com.

Olive, Sean E. (2004a). A multiple regression model for predicting loudspeaker preference using objective measurements: Part I-Listening test results. In *Audio Engineering Society Preprints 116th Convention*, Berlin, Germany (May). Preprint 6113.

Olive, Sean E. (2004b). A multiple regression model for predicting loudspeaker preference using objective measurements: Part II-Development of the model. In *Audio Engineering Society Preprints 117th Convention*, San Francisco, USA (October). Preprint 6190.

Olive, Sean E., Peter L. Schuck, Sharon L. Sally and Marc E. Bonneville (1994). The effects of loudspeaker placement on listener preference ratings. *Journal of the Audio Engineering Society* 42(9), 651–669.

Otala, Matti and Pertti Huttunen (1987). Peak current requirement of commercial loudspeaker systems. *Journal of the Audio Engineering Society* 35(6), 455–462. Also in AES (1996a), 116–123.

Pinson, Margaret H., William Ingram and Arthur Webster (2011). Audiovisual quality components. *IEEE Signal Processing Magazine* 28(6), 60–67.

Piper, Jim (2008). *Get the Picture? The Movie Lover's Guide to Watching Films* (Second edition). New York: Allworth Press.

Pohlmann, Ken C. (2011). *Principles of Digital Audio* (Sixth edition). New York: McGraw-Hill.

Poynton, Charles (2012). *Digital Video and HD: Algorithms and Interfaces* (Second edition). Waltham: Morgan Kaufmann.

Pras, Amandine and Catherine Guastavino (2010). Sampling rate discrimination: 44.1 kHz vs. 88.2 kHz. In *Audio Engineering Society Preprints 128th Convention*, London, UK (May). Preprint 8101.

Purves, Dale and R. Beau Lotto (2011). *Why We See What We Do Redux: A Wholly Empirical Theory of Vision* (Second edition). Sunderland: Sinauer Associates.

Queen, Daniel (1979). The effect of loudspeaker radiation patterns on stereo imaging and clarity. *Journal of the Audio Engineering Society* 27(5), 368–379. Also in AES (1984), 69–80.

Ratliff, P. A. (1974). Properties of hearing related to quadraphonic reproduction. Report 1974–38. BBC Research & Development. www.bbc.co.uk/rd. Also in AES (2006), 353–375.

Reiss, Joshua D. (2016). A meta-analysis of high resolution audio perceptual evaluation. *Journal of the Audio Engineering Society* 64(6), 364–379.

Rives, Richard (2005). Loudspeaker placement guide. Online Publication. Audioholics Online A/V Magazine. www.audioholics.com.

Roberts, A. (2002). The film look: It's not just jerky motion . . . White Paper WHP 053. BBC Research & Development. www.bbc.co.uk/rd.

Rumsey, Francis (1999). Controlled subjective assessments of two-to-five-channel surround sound processing algorithms. *Journal of the Audio Engineering Society* 47(7/8), 563–582.

Rumsey, Francis (2002). Spatial quality evaluation for reproduced sound: Terminology, meaning, and a scene-based paradigm. *Journal of the Audio Engineering Society* 50(9), 651–666.

Rumsey, Francis, Slawomir Zieliński, Rafael Kassier and Søren Bech (2005). On the relative importance of spatial and timbral fidelities in judgments of degraded multichannel audio quality. *Journal of the Acoustical Society of America* 118(2), 968–977.

Rushing, Krissy (2004). *Home Theater Design: Planning and Decorating Media-Savvy Interiors*. Gloucester: Quarry Books.

Russell, Roger (2012). Speaker wire: A history. www.roger-russell.com/wire/wire.htm.

Salmon, Richard, Mike Armstrong and Stephen Jolly (2011). Higher frame rates for more immersive video and television. White Paper WHP 209. BBC Research & Development. www.bbc.co.uk/rd.

Schroeder, Manfred R. (1975). Diffuse sound reflection by maximum-length sequences. *Journal of the Acoustical Society of America* 57(1), 149–150.

Schroeder, Manfred R. (1996). The 'Schroeder frequency' revisited. *Journal of the Acoustical Society of America* 99(5), 3240–3241.

Schulte, Tom and Joel Barsotti (2016). HDR demystified: Emerging UHDTV systems. White Paper. SpectraCal. www.calman.spectracal.com.

Self, Douglas (2009). *Audio Power Amplifier Design Handbook* (Fifth edition). Oxford: Focal Press.

Shaw, Edgar A. G. (1974). Transformation of sound pressure level from the free field to the eardrum in the horizontal plane. *Journal of the Acoustical Society of America* 56(6), 1848–1861.

Siau, John (2015). Audio myth: DSD provides a direct stream from A/D to D/A. Application Note. Benchmark Media. www.benchmarkmedia.com.

Small, Richard H. (1972a). Closed-box loudspeaker systems part I: Analysis. *Journal of the Audio Engineering Society* 20(10), 798–808. Also in AES (1980), 285–295.

Small, Richard H. (1972b). Direct-radiator loudspeaker system analysis. *Journal of the Audio Engineering Society* 20(5), 383–395. Also in AES (1980), 271–283.

Small, Richard H. (1973a). Closed-box loudspeaker systems part II: Synthesis. *Journal of the Audio Engineering Society* 21(1), 11–18. Also in AES (1980), 296–303.

Small, Richard H. (1973b). Vented-box loudspeaker systems part I: Small-signal analysis. *Journal of the Audio Engineering Society* 21(5), 363–372. Also in AES (1980), 316–325.

Small, Richard H. (1973c). Vented-box loudspeaker systems part II: Large-signal analysis. *Journal of the Audio Engineering Society* 21(6), 438–444. Also in AES (1980), 326–332.

Small, Richard H. (1973d). Vented-box loudspeaker systems part III: Synthesis. *Journal of the Audio Engineering Society* 21(7), 549–554. Also in AES (1980), 333–338.

Small, Richard H. (1973e). Vented-box loudspeaker systems part IV: Appendices. *Journal of the Audio Engineering Society* 21(8), 635–639. Also in AES (1980), 339–343.

Small, Richard H. (1974a). Passive-radiator loudspeaker systems part I: Analysis. *Journal of the Audio Engineering Society* 22(8), 592–601. Also in AES (1996a), 19–28.

Small, Richard H. (1974b). Passive-radiator loudspeaker systems part II: Synthesis. *Journal of the Audio Engineering Society* 22(9), 683–689. Also in AES (1996a), 29–35.

SMPTE RP 120 (2005). Measurement of intermodulation distortion in motion-picture audio systems. Society of Motion Picture and Television Engineers.

SMPTE RP 200 (2012). Relative and absolute sound pressure levels for motion picture multichannel sound systems applicable for analog photographic film audio, digital photographic film audio and D-Cinema. Society of Motion Picture and Television Engineers.

SMPTE RP 431–2 (2011). D-Cinema quality: Reference projector and environment. Society of Motion Picture and Television Engineers.

SMPTE ST 421 (2013). VC-1 compressed video bitstream format and decoding process. Society of Motion Picture and Television Engineers.

SMPTE ST 2084 (2014). High dynamic range electro-optical transfer function of mastering reference displays. Society of Motion Picture and Television Engineers.

Soulodre, Gilbert A., Theodore Grusec, Michel Lavoie and Louis Thibault (1998). Subjective evaluation of state-of-the-art two-channel audio codecs. *Journal of the Audio Engineering Society* 46(3), 164–177.

Spears, Stacey and Don Munsil (2003). DVD benchmark: Special report: Chroma upsampling error. Online Publication. Secrets of Home Theater and High Fidelity. www.hometheaterhifi.com.

Spears, Stacey and Don Munsil (2013). HD benchmark 2nd edition articles. Online Publication. Spears & Munsil. www.spearsandmunsil.com.

Ståhl, Karl E. (1981). Synthesis of loudspeaker mechanical parameters by electrical means: A new method for controlling low-frequency loudspeaker behavior. *Journal of the Audio Engineering Society* 29(9), 587–596. Also in AES (1980), 241–250.

Steinke, Gerhard (2004). Surround sound: Relations of listening and viewing configuration. In *Audio Engineering Society Preprints 116th Convention*, Berlin, Germany (May). Preprint 6019.

Stuart, J. Robert and Peter G. Craven (2014). A hierarchical approach to archiving and distribution. In *Audio Engineering Society Preprints 137th Convention*, Los Angeles, USA (October). Preprint 9178.

Sugawara, Masayuki, Kenichiro Masaoka, Masaki Emoto, Yasutaka Matsuo and Yuji Nojiri (2007). Research on human factors in ultra-high-definition television to determine its specifications. NHK Science and Technology Research Laboratories. www.nhk.or.jp/digital/en.

Suzuki, Yoiti and Hisashi Takeshima (2004). Equal-loudness-level contours for pure tones. *Journal of the Acoustical Society of America* 116(2), 918–933.

Szymanski, Jeff (2015). Acoustical treatment for indoor areas. In Glen Ballou (Ed.), *Handbook for Sound Engineers* (Fifth edition), Chapter 8, pp. 167–198. Oxford: Focal Press.

Tappan, Peter W. (1962). Loudspeaker enclosure walls. *Journal of the Audio Engineering Society* 10(3), 224–231. Also in AES (1980), 88–95.

Taylor, Jim, Charles G. Crawford, Christen M. Armbrust and Michael Zink (2009). *Blu-Ray Disc Demystified*. New York: McGraw-Hill.

Taylor, Jim, Mark R. Johnson and Charles G. Crawford (2006). *DVD Demystified* (Third edition). New York: McGraw-Hill.

Texas Instruments (2013). Top 10 reasons to choose DLP projector technology. Brochure. Texas Instruments. www.ti.com.

Theile, Günther (1991). HDTV sound systems: How many channels? In *Proceedings of the Audio Engineering Society 9th International Conference: Television Sound Today and Tomorrow*, Detroit, USA (February), pp. 217–232. Also in AES (2006), 202–217.

Theile, Günther and Georg Plenge (1977). Localization of lateral phantom sources. *Journal of the Audio Engineering Society* 25(4), 196–200. Also in AES (2006), 376–380.

Thiele, A. Neville (1971a). Loudspeakers in vented boxes: Part I. *Journal of the Audio Engineering Society* 19(5), 382–392. Also in AES (1980), 181–191.

Thiele, A. Neville (1971b). Loudspeakers in vented boxes: Part II. *Journal of the Audio Engineering Society* 19(6), 471–483. Also in AES (1980), 192–204.

Tittel, Ed and Mike Chin (2006). *Build the Ultimate Home Theater PC*. Indianapolis: Wiley Publishing.

Toole, Floyd E. (1975). Damping, damping factor, and damn nonsense. *AudioScene Canada* (February), 16–17.

Toole, Floyd E. (1985). Subjective measurements of loudspeaker sound quality and listener performance. *Journal of the Audio Engineering Society* 33(1/2), 2–32. Also in AES (1996b), 276–306.

Toole, Floyd E. (1986a). Loudspeaker measurements and their relationship to listener preferences: Part 1. *Journal of the Audio Engineering Society* 34(4), 227–235. Also in AES (1996b), 336–344.

Toole, Floyd E. (1986b). Loudspeaker measurements and their relationship to listener preferences: Part 2. *Journal of the Audio Engineering Society* 34(5), 323–348. Also in AES (1996b), 357–382.

Toole, Floyd E. (1988). Principles of sound and hearing. In K. Blair Benson (Ed.), *Audio Engineering Handbook*, Chapter 1, pp. 1.1–1.71. New York: McGraw-Hill.

Toole, Floyd E. (2006). Loudspeakers and rooms for sound reproduction: A scientific review. *Journal of the Audio Engineering Society* 54(6), 451–476.

Toole, Floyd E. (2018). *Sound Reproduction: The Acoustics and Psychoacoustics of Loudspeakers and Rooms* (Third edition). New York: Routledge.

Toole, Floyd E. and Sean E. Olive (1988). The modification of timbre by resonances: Perception and measurements. *Journal of the Audio Engineering Society* 36(3), 122–142.

Tsakiris, Vassilis Bill and Chris Orinos (2004). Optimum loudspeaker system with subwoofer and digital equalization. In *Audio Engineering Society Preprints 117th Convention*, San Francisco, USA (October). Preprint 6266.

Van Daele, Bert and Wilfried Van Baelen (2011). Auro-3D Octopuss codec: Principles behind a revolutionary codec. White Paper. Auro Technologies NV. www.auro-3d.com.

VESA (2017). VESA high-performance monitor and display compliance test specification (DisplayHDR CTS). Video Electronics Standards Association.

Vickers, Earl (2010). The loudness war: Background, speculation, and recommendations. In *Audio Engineering Society Preprints 129th Convention*, San Francisco, USA (November). Preprint 8175.

Villchur, Edgar (1994). Speaker cables: Measurements vs psychoacoustic data. *Audio* (July), 35–37.

Waldrep, Mark (2017). *Music and Audio: A User Guide to Better Sound*. Los Angeles: AIX Media Group. www.musicandaudioguide.com.

Walker, R. (1992). Low-frequency room responses: Part 1-Background and qualitative considerations. BBC Research Department Report 1992/8. BBC Research & Development. www.bbc.co.uk/rd.

Wallach, Hans, Edwin B. Newman and Mark R. Rosenzweig (1973). The precedence effect in sound localization. *Journal of the Audio Engineering Society* 21(10), 817–826.

Wankling, Matthew and Bruno Fazenda (2009). Subjective validity of figures of merit for room aspect ratio design. In *Audio Engineering Society Preprints 126th Convention*, Munich, Germany (May). Preprint 7746.

Waterhouse, Richard V. (1958). Output of a sound source in a reverberation chamber and other reflecting environments. *Journal of the Acoustical Society of America* 30(1), 4–13.

Welti, Todd (2002). How many subwoofers are enough? In *Audio Engineering Society Preprints 112th Convention*, Munich, Germany (May). Preprint 5602.

Welti, Todd and Allan Devantier (2006). Low-frequency optimization using multiple subwoofers. *Journal of the Audio Engineering Society* 54(5), 347–364.

Whitlock, Bill (1996). Hum & buzz in unbalanced interconnect systems. Application Note AN-004. Jensen Transformers. www.jensen-transformers.com.

Whitlock, Bill (2015). Grounding and interfacing. In Glen Ballou (Ed.), *Handbook for Sound Engineers* (Fifth edition), Chapter 36, pp. 1261–1302. Oxford: Focal Press.

WHO (1999). Guidelines for community noise. Guideline document. World Health Organization. www. who.int.

Wilson Audio (2012). Alexandria XLF. Owner's Manual. Wilson Audio Specialities. www.wilsonaudio.com.

Wilson, Rhonda J., Michael D. Capp and J. Robert Stuart (2003). The loudspeaker-room interface: Controlling excitation of room modes. In *Audio Engineering Society 23rd International Conference: Signal Processing in Audio Recording and Reproduction*, Copenhagen, Denmark (May).

Winer, Ethan (2018). *The Audio Expert: Everything You Need to Know about Audio* (Second edition). New York: Routledge.

Zacharov, Nick (1998). Subjective appraisal of loudspeaker directivity for multichannel reproduction. *Journal of the Audio Engineering Society* 46(4), 288–303. Also in AES (2006), 440–455.

Zhang, Peter Xinya (2015). Psychocoustics. In Glen Ballou (Ed.), *Handbook for Sound Engineers* (Fifth edition), Chapter 3, pp. 55–80. Oxford: Focal Press.

INDEX

Printed and bound by CPI Group (UK) Ltd, Croydon, CR0 4YY

21/10/2024

01777099-0001